THE
SECRET
MELODY

THE

SECRET

MELODY

AND MAN CREATED THE UNIVERSE

Trinh Xuan Thuan

TRANSLATION BY STORM DUNLOP

Templeton Foundation Press
Philadelphia & London

Templeton Foundation Press
300 Conshohocken State Road, Suite 550
West Conshohocken, Pennsylvania 19428
www.templetonpress.org

Original French edition: La mélodie secrète © 1991 Librairie Arthème Fayard
English edition: © 1995 Oxford University Press, Inc.
2005 Templeton Foundation Press edition

*Templeton Foundation Press helps intellectual leaders and others learn
about science research on aspects of realities, invisible and intangible.
Spiritual realities include unlimited love, accelerating creativity, worship, and
the benefits of purpose in persons and in the cosmos.*

Library of Congress Cataloging-in-Publication Data

Thuan, Trinh Xuan.
[Mélodie secrète. English]
The secret melody : and man created the universe / Trinh Xuan Thuan ; translation
by Storm Dunlop.
p. cm.
Originally published: New York : Oxford University Press, 1995.
Includes bibliographical references and index.
ISBN-10: 1-932031-95-2 (alk. paper)
ISBN-13: 978-1-932031-95-9
1. Cosmology. 2. Solar system. 3. Astrophysics. I. Title.
QB981.T863813 2005
523.1—dc22
2005048637

Printed in the United States of America

05 06 07 08 09 10   10 9 8 7 6 5 4 3 2 1

This book is dedicated to my parents
and to all the builders of universes.

Contents

Preface to the 2005 Edition

The 10 years that have elapsed since the first publication in 1995 of *The Secret Melody* in the United States have witnessed remarkable new developments in the field of cosmology. This preface is intended to bring the reader up to date on the latest scientific discoveries on the large scale properties of the universe. This preface will put these new findings in perspective so as to give the interested person a sense of how they may impact the story told in this book. The first key thing to remember is that these new discoveries have not invalidated the Big Bang theory of the universe, told in all its glory in *The Secret Melody*. On the contrary, they have enriched it and put it on a firmer ground. The idea that the universe started its existence from an extremely small, hot, and dense state is still the best we have in town for explaining the properties of the bewilderingly diverse content of the cosmos, the fossil radiation, planets, stars, galaxies, clusters, and galaxies. Second, the idea that the universe went through an early inflationary phase, i.e. that it underwent a breathtaking expansion shortly after its creation, tripling its size every 10^{-34} second during the fraction of a second that followed its birth, is still valid.

As so often in astronomy, new discoveries have come from the development of new observational techniques and telescopes. The first key new discovery, one that took all astronomers by surprise, occurred in 1998 and it concerns the acceleration of the universe. After its very short inflationary phase, the universe is thought to have settled down into a more sedate expansion rate, one that we still observe today. But because gravity is attractive, astronomers thought that the gravitational attraction of all the material content of the universe would gradually slow down its expansion over time, i.e. it would be decelerating. To verify that assumption, astronomers needed a standard yardstick that can be used to measure the distance of far-away galaxies carried away from the Milky Way by the expansion of the universe. Using such a yardstick, they would be able to derive distances to these galaxies and compare them to distances obtained from the redshifted light of the same galaxies. This comparison would allow them to measure the expansion rate of the universe at different epochs, in recent times by studying nearby galaxies and in the distant past by studying remote galaxies. Astronomers found such a yardstick in a particular type of exploding star known as Type IA supernova. After observing dozens of type IA supernovae, they found to their great surprise, that contrary to their expectation that the expansion rate of the universe is smaller now than in the past, i.e. the universe is decelerating, the present expansion rate is actually larger than in the past, i.e. the universe is in acceleration! Observations of more supernovae (now some 150 in total) have only confirmed and strengthened that result. The conclusion is inescapable: the universe is permeated by a "dark energy" that exerts a repulsive antigravity force that has overcome the attractive force of gravity, making the universe's galaxies flying away from each other faster and faster as time passes. The antigravity force introduced by Einstein in 1915 to produce a static universe—and which he dismissed in 1929 when Hubble discovered the expansion of the universe, calling it "the greatest blunder of his life"—rears its head again! The supernova measurements imply that the dark energy accounts for about 70% of the energy in the universe, with the other 30% contributed by matter.

Because this discovery was so unexpected, there was some doubt about it at first. Questions were raised, many concerning the validity of type IA supernovae as distance yardsticks. The notion of an accelerating universe propelled by a vast amount of dark energy got a big boost with the observations of NASA's Wilkinson Microwave Anisotropy Probe (WMAP) in 2003. WMAP is the space mission which succeeded the COBE mission in 1992, the groundbreaking discoveries of which are detailed in *The Secret Melody*. Like COBE, WMAP's object of study is the afterglow of the Big Bang, the fossil radiation left over from the period when the universe was only some 300,000 years old (WMAP has now refined this number to be 379,000 years). COBE first discovered tiny temperature fluctuations in this fossil radiation, of the order of 30 millionths of a degree, caused by variations in the density of matter. These tiny density fluctuations act as seeds of cosmic structure. Thanks to the relentless work of gravity, they will grow into the majestic galaxies that make up the cosmic tapestry we see today.

By studying the fossil radiation in unprecedented detail, WMAP captured a portrait of the infant universe which permits to describe it with unsurpassed accuracy. One of the biggest surprises was the revelation that the first stars were born only 200 million years after the Big Bang, much earlier than was thought before. The age of the universe was pinned down to be at 13.7 billion years. But foremost, the WMAP observations supported the Big Bang theory as well as an inflationary phase, and were in accord with the results of the Type IA supernova observations, that about 70% of the mass-energy content of the universe is contributed by a mysterious dark energy, with the remaining 30% contributed by matter. The exact nature of the dark energy still baffles physicists. Dark energy can be as mundane as the energy of the "quantum vacuum," from which the universe sprang, or something so exotic that it has thus far eluded the minds of the most creative scientists.

As detailed in *The Secret Melody*, much of the matter in the universe is dark. The luminous matter of stars and galaxies contribute only about 0.5% of the total matter content of the universe and the remaining 29.5% is dark—not emitting any type of radiation. Out of the 29.5% of dark matter, only about 4% is made of ordinary matter like us (namely protons and neutrons, collectively called baryons), and the remaining 25.5% consists of nonordinary (nonbaryonic) matter, the nature of which is still totally unknown. The universe we see is truly the tip of the iceberg. Neutrinos, if they have mass, have been suggested to constitute at least part of this nonordinary matter. In 1998, the mass of the neutrino was measured to be about one millionth that of the electron, far too small to account for all the dark nonbaryonic mass. The total mass contributed by neutrinos is comparable to the mass contributed by luminous stars and galaxies. The discovery by WMAP that the first stars formed early also support the conclusion that fast-moving neutrinos (the so-called "hot dark matter") do not play a major role in the evolution of structures in the universe. The fast-moving neutrons would have prevented the early clumping of gas in the universe, delaying the emergence of these first stars. The leading candidates for the dark nonordinary matter are thought to be slowly moving elementary particles (dubbed "cold dark matter"), with the most promising candidates bearing the exotic names of "axion" and "neutralino."

The total amount of dark matter and energy in the universe plays a fundamental role in determining the geometry of space. If the density of matter and energy is less than the critical density, then space is open and is negatively curved like the surface of a saddle. If the density is exactly equal to the critical density, then space is flat like a sheet of paper. If the density is greater than critical density, then space is closed and positively curved like the surface of a sphere. WMAP's observations imply that the universe is flat. The flatness of the universe gives strong support to the idea that it went through an inflationary phase a tiny fraction of a second after its birth. The universe, by growing so dramatically, became flat, just as a small region on the surface of a balloon flattens when it is inflated. If the amount of dark energy remains constant with time, then the universe will expand forever. However, because the nature of the dark energy remains one of the great mysteries in all of science, we cannot be sure. If the amount of dark energy changes with time, or if other unknown and unexpected things happen in the universe, this conclusion could change.

For a cosmologist, these are heady times. We are working with a new representation of the universe that incorporates the Big Bang model, as well as an early inflationary period and the present stage of accelerated expansion. Nature keeps sending us new notes of music, challenging us to unravel its Secret Melody.

Charlottesville, Va.   T.X.T.
April 2005

Preface to the Original Edition

This book is intended for the inquisitive mind, curious about the surrounding world and interested in the latest advances in the study of the cosmos, without necessarily possessing the technical background of the specialist. The book recounts the evolution of man's view of the universe throughout the ages, with a particular emphasis on the present world view, the Big Bang universe. The book also touches on questions which go beyond the purely scientific realm, but which arise inevitably in any discussion of the origin of the universe: Are we here by chance or does our presence imply a Designer?

The first two chapters (1 and 2) tell the evolution of cosmological thought, from the magic universe, at the dawn of humanity, to the Big Bang universe in the twentieth century. They describe how the universe has considerably increased in size, from a mere solar system with the Earth at its center, to a vast immensity stretching across some 15 light-years in distance, and where the Earth has been relegated to a small corner of the Milky Way, itself lost among hundreds of billions of galaxies. Chapters 3 and 4 introduce the actors of the drama in the Big Bang universe: the space-time couple, the four fundamental forces, the elementary particles, and the galaxies. They describe how these actors must obey very strict rules dictated by general relativity and quantum mechanics.

Perhaps the most important achievement of modern cosmology is the realization that the universe has a history, a past, a present, and a future. The next two chapters tell the story of the Big Bang as it is presently known.

Chapter 5 describes how the infinitely small has spawned the infinitely large, how the whole universe, with its hundreds of billions of galaxies, has sprung from a subatomic void. It tells how galaxies have woven an immense cosmic tapestry, and how, thanks to the creative alchemy of stars and the formation of planets, life and conscience have emerged, after a long climb toward complexity. Chapter 6 concerns itself with the future of the universe. This future is not known with certainty because it depends on the total amount of matter in the universe, which cannot be easily measured, 90 to 98% of its mass being invisible. Will the universe expand forever to become a glacial and dark immensity or will it expand to a maximum size before collapsing on itself and ending in a state with infinite temperature and density?

Discussing the creation of the universe leads unavoidably to the question of a Creator. Chapters 7 and 8 show how modern science has demolished all the classic arguments concerning the existence of God, but that it has redeemed itself by making us aware that the very fact we are here is extraordinary: the universe has been very precisely fined-tuned to allow our existence: change the physical laws ever so slightly, and we shall no longer be here to talk about them! Is this astonishingly precise fine-tuning the result of chance or the manifestation of the will of a Supreme Being?

There are no absolute truths in science. The last chapter (9) reviews the rival theories of the Big Bang theory. It shows why these competing theories have not been endorsed by the majority of the cosmologists because they do not possess the predicting power, the beauty, simplicity, and elegance of the Big Bang theory. Chapter 9 also postulates that the Big Bang universe will not be the last one and that man will keep on creating more universes which will get closer to the real Universe without ever reaching it. The melody formed by the musical notes that Nature sends us will remain forever secret.

In writing the text, I have made every effort to use a clear and simple language, devoid of any scientific and technical jargon. To explain difficult concepts, I often have had recourse to examples from everyday life. But it is impossible to tell the story of the universe with rigor without introducing some basic scientific concepts. To help the reader, I have assembled in a glossary at the end of the book a list of words which are specific to astrophysics (they are signaled by * in the text), and for which I have given succinct definitions. I have also provided tables, drawings, and astronomical photographs to illustrate the important points in the text. Finally, for readers more familiar with the scientific language who desire to delve further into the subject, I have also added a few mathematical appendices. These appendices are not necessary for understanding the text.

Acknowledgments

I am deeply to several persons: Bruno and Laurence Bardèche, Hélène Boullet, Marie-Françoise Foudrat, Michel Jacasson, and Michel Cassé, who have read and commented on various parts of the preliminary version of the manuscript. My gratitude goes also to Françoise Warin who took charge of the photographic work and Gérand Berton who was responsible for the drawings.

The book was written while I was on sabbatical leave from the University of Virginia, as a Sesquicentennial Associate of the University of Virginia Center for Advanced Studies and as a Fulbright scholar in residence at the Institut d'Astrophysique in Paris. I thank its director, Jean Audouze, for welcoming me there. I also thank Catherine Cesarsky and Thierry Montmerle for their hospitality at the Service d'Astrophysique at the Centre d'Etudes de Saclay. I express my appreciation to Eric Vigne, who suggested I write the book and gave advice and encouragement during its writing, and to Jeffrey Robbins for undertaking to bring the book to an English-speaking public

Charlottesville, Va.   T.X.T.
February 1995

THE
SECRET
MELODY

|| 1 ||

Past Universes

To carry out my astronomical observations, I frequently visit Kitt Peak National Observatory, which is in a Papago Indian reservation at an altitude of about 2100 meters, on one of the highest peaks of a range of sacred mountains in Arizona. During the course of a night's observation, while the giant telescope (Fig. 1) is capturing the tiny particles of light known as photons* that have been emitted by some distant galaxy*, and bringing them to a focus on the electronic detector, I often step outside the dome that shelters the telescope and look at the sky. My main concern is to check that the sky is still perfectly clear and that there is no band of cloud lurking on the horizon, threatening to interrupt my observations and thus wreck the few precious nights that I have been allocated. But I also like to enjoy the sheer pleasure that comes from contemplating the splendor and immensity of the night sky.

Unfortunately, with modern observational techniques, astronomers no longer have direct contact with the sky. The romantic image of a scientist sitting in the dark, eye glued to the eyepiece of a telescope, and suffering in the cause of science, no longer holds true. (From my own personal experience, I can assure you that there is nothing glamorous about having to sit in one place for hours on end throughout a long winter night, guiding a telescope while fighting both the cold and an intense desire for sleep.) Nowadays, I can work in a well-lighted, warm room, controlling everything electronically with push buttons on a control panel and sophisticated computers. Once the telescope has been correctly positioned, the image of the galaxy I want to study appears, magnified thousands of times, on a television screen

3

Fig. 1. *The 4-meter telescope at Kitt Peak.* The photograph shows the dome, which is as high as a ten-story building, and which protects the 4-meter-diameter telescope. Kitt Peak is a mountain in a Papago Indian reservation in Arizona. This telescope is able to detect stars 100 million times fainter than the faintest visible to the naked eye (photograph: Kitt Peak National Observatory).

in front of me. The sky comes to me through a maze of electronic circuits. The greater accuracy that results, and the higher efficiency that goes with greater comfort, more than compensates for the loss of direct contact with the sky.

On this moonless night far from the blinding lights of towns and cities, the sky is a wonderful sight, full of thousands of brilliant points of light. Two, low in the West, are slightly brighter than the others, and appear to twinkle less. They are the planets* Mars and Jupiter, Earth's nearest neighbors farther away from the Sun: Mars, where man-made machines have already landed and where the search for life has proved to be unsuccessful, and Jupiter, the giant of the Solar System, with 11 times the size and 318 times the mass of our Earth. The light reaching my eyes from Mars and Jupiter is not their own, but the reflected light of the Sun. Because of the

Earth's rotation, that Sun, around which all the planets revolve, appears to rise in the East and set below the horizon in the West, giving rise to day and night. The Earth's rotation will also soon make Mars and Jupiter set below the horizon, and I shall only be able to see them again tomorrow night. The light from the two planets did not take very long to reach my eyes: just 12 minutes for Mars, and 42 minutes for Jupiter. The Solar System is no more than a tiny speck of sand on the shore of the vast ocean that forms the universe.

My attention turns to the other points of light. They are stars, just like the Sun, producing their own light and energy from intense nuclear reactions taking place in their inner cores. They are part of our Galaxy, which the ancients called the Milky Way, because of the whitish band of light running across the sky, and which I can see crossing the constellation of Orion. The soft light comes from billions[1] of stars scattered throughout the plane of our Galaxy. Our own Sun is just a perfectly ordinary star, one of the 100 billion that go to make up the Milky Way; like all the other stars it is orbiting the center of the Galaxy, carrying us with it, and making a complete revolution every 250 million years.

As I admire the beautiful constellation of Orion, my thoughts turn to the birth of stars. Orion is an immense stellar nursery, where large interstellar clouds collapse under their own gravity, giving birth to new stars.

The starry sky above me, which appears so serene and unchanging, is very far from tranquil in reality. Not only is everything moving—the planets around the Sun, and the Sun around the center of the Galaxy—but the stars*, just like people, are born, live out their lives, and die. Their changes, however, take place on a cosmic time scale, reckoned in millions or billions of years, and are imperceptible in terms of our own lifetimes. The light from the stars that I can see with the naked eye has taken, at most, no more than a few tens of years to reach me. Yet our own Milky Way galaxy is far larger, with a diameter of about 90,000 light years*, and the light emitted by the stars close to its edge is far too faint for me to see with the unaided eye. Yet, despite its size, the Galaxy is no more than a child's tiny sandcastle on the shores of the cosmic ocean.

I turn my attention to the constellation of Andromeda and strain my eyes to see a faint, hazy spot that differs from the pinpoints of the surrounding stars. This is the Andromeda Galaxy*, and the light from its 100 billion stars that I can see now has taken 2.3 million years to cross the intervening space. This galaxy is very similar to our own. The Milky Way and Andromeda galaxies are so massive that they dominate our own, small, "local group" of galaxies.

As I stand absorbing the sight of the sky full of brilliant stars, I am, nevertheless, surrounded by almost total darkness. Only with difficulty can

---

1. The American value of "billion," i.e., one thousand million (10^9 in scientific notation) is used throughout this book, rather than the European terms "milliard" (one thousand million, 10^9) or the larger "billion" (one million million, 10^{12}).

I make out the vague shape of the telescope dome in the darkness. Yet I realize that I am fortunate to know at long last the answer to a question that has bothered many scientists before me, and that is not nearly as simple as it appears: Why is the sky dark, despite being full of so many stars? I now know that the night is dark because the universe had a beginning, and that it has not existed for all time.

After having drunk my fill of the serene beauty of the night, I return to my observations of the remote galaxy, at a distance of more than 5 billion light years. Thanks to the light-gathering power of the enormous bucket of light that is the 4-meter-diameter mirror, and the miracles of modern electronics and computers, I can capture light that left its source more than 5 billion years ago, before the Sun and Earth even existed, and when many of the atoms in my body were still deep within the interior of a star, which will later explode and scatter them in interstellar space.

I find it quite extraordinary that the simple sight of a sky full of stars should have started me thinking along these lines, and that my brain should have immediately felt an irresistible urge to organize the fragments of information about the external world, communicated to me through my senses, into a single, coherent whole. Nature is by no means silent. Like some distant orchestra, it tantalizes us with individual notes and fragments of music. But it is not willing to hand everything to us on a plate. The melody linking the individual fragments of music is missing. The overall theme is hidden. Somehow we have to unravel the secrets of that hidden melody, so that we can listen to the composition in all its glory.[2]

This desire for a unified, coherent view of the world, this need to discover the hidden melody, is an essential part of the human spirit. Faced with the world that surrounds us, we try to conquer our anxiety and fear of the apparently endless void by trying to reduce it to some form of order, and by endowing it with a familiar face. When this organization of the external world is applied to the whole universe, it is known as cosmology*. I am a cosmologist when I try to reconcile such apparently disparate fragments of information as the rising and setting of the Sun, the starry night sky, and the changing seasons of the passing year—from the blooming radiance of spring to the vivid reds and golds of autumn. By building up a system of coherent ideas to explain the external world, in accordance with our own particular society and culture, we create a universe of our own. This universe gives us a common outlook, and helps to unify our society by giving us a collective view of the origin and evolution of the world, and also by bestowing upon us a distinct identity, different from that of any other group. We are what we believe. Naturally, the universe that we create is not necessarily unique, but can take many different aspects. It changes, depending on the epoch and culture involved. It has its own life and history, which frequently evolve in parallel with the life and history of the society from which it sprang.

2. F. Jacob, *La statue intérieure* (*The Statue Within*), Le Seuil, 1987, p. 305.

A universe, a world-view, is like a human being. It is born, attains its prime, goes into a decline, and disappears, to be replaced by another. This decline and disappearance are often caused by contact with another, more dynamic, society or culture; by facts and discoveries that are incompatible with the current universe; or, finally, by the emergence of new ideas that cast doubt on the currently accepted view.[3]

The Magic Universe

Various concepts of the world have followed one another throughout history, and the secret melody has changed with the passage of time. The first universe emerged perhaps some hundreds of thousands of years ago, at the same time as the appearance of language. Cavemen lived in a magic world filled with spirits. Picture for yourself one of the men who lived in the caves at Lascaux, in the southwest of the region that, much later, was to become France. After finishing his meal, he steps outside the cave for a breath of fresh air and to look at the sky. He knows that the Sun-spirit has laid down to rest, because his blinding light is no longer to be seen. Yet the Moon-spirit has awakened, because she can be seen over there, just above the horizon. The Star-spirits have also roused, because they are glittering all over the sky. The spirits of the Earth, of the trees, of the plants, and of the streams are slumbering. Everything is calm and serene. Our Lascaux caveman feels perfectly at ease in this magical world, where every object is endowed with its own, distinct spirit, because to him the world of the spirits mirrors that of humans, with the same desires, impulses, and customs. Our man knows that he can address the spirits in just the same way as he talks to other human beings. He knows that he can flatter them, praise them, and make them offerings to obtain their favor.

The Mythic Universe

The magic universe was simple, familiar, and on a human scale. But with the passage of time, it increasingly lost that simplicity and familiarity. As knowledge grew, so innocence disappeared. Mankind increasingly realized its insignificance and impotence in the face of the immensities of the universe. The latter became more and more complex, and soon acquired superhuman aspects. Such complexity could only be ruled by beings with powers far exceeding those of humans. A world of spirits similar to that of humans was no longer enough. Some 10,000 years ago, the magic universe transformed into a mythic, superhuman universe. The trees, flowers, and streams were no longer inhabited by spirits. The world lost its human scale. Henceforth, the universe was ruled by gods living far beyond the earthly realm. Now the Sun-god ruled the day, and the Moon-god, the gods of the planets, and the gods of the stars ruled the night. Cosmology consisted of mythical

---

3. E. Harrison, *Masks of the Universe*, New York, Macmillan, 1985.

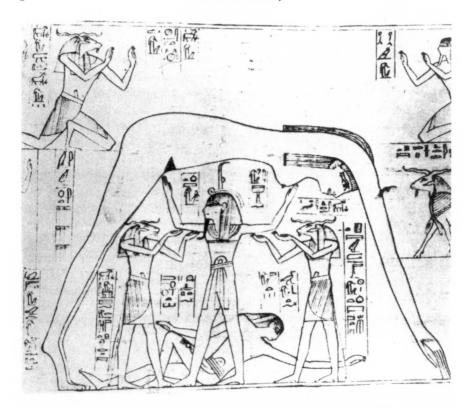

Fig. 2. *The Egyptian mythic universe.* The body of the goddess Nut, jewelled with planets and stars and supported by Shu, the god of the air, forms the vault of the sky. Each day, the Sun traveled across the back of Nut, above Geb, the god of the Earth, who appears lying at the feet of the goddess (photograph: British Museum).

tales that recounted the stories of these gods, their loves, their unions, their hatreds, and their feuds. In the mythic universe, natural phenomena, including the creation of the universe itself, were the results of the actions of these gods, driven by very human emotions such as love, hate, and passion, but endowed with superhuman powers. Religion made its appearance with the mythic universe. It was no longer possible to communicate directly with these superhuman beings, as had been the case with the world of the spirits. Communication had to be carried out through privileged intermediaries, the priests, by means of offerings of food or sacrifices. This association between cosmology and religion, between cosmologist and priest, lasted about 3000 years, until the mythic universe was displaced by the scientific universe.

There were numerous, diverse, mythic universes, which varied according to the cultures and epochs. In many of them woman's reproductive power inspired the creation myth. To the Babylonians, who lived 5000 years ago in the Tigris and Euphrates delta (modern-day Iraq), the primordial mother, Tiamat, gave birth to Anu, the sky god, as the result of her union with Apsu, the god of the oceanic abyss. The offspring of Anu and Tiamat—incestuous relationships were common in the mythical world—was Ea, the earth god. Innumerable couplings in every possible combination of partners produced 600 gods and goddesses, whose family tree would fill several pages. Despite constant quarrelling, and being continually at war, each of these gods controlled one aspect of human existence.

At almost the same time, but along the banks of the Nile, the Egyptian mythology arose. As in the Babylonian universe, water was the source of life. The primordial being, Atum, who contained within himself all existence, lived in the primordial ocean, Nun. He, in the form of Atum-Ra, created the world and the some 800 gods and goddesses of the Egyptian mythology. Atum later became Ra, the Sun-god. Geb was the Earth, a flat disk surrounded by mountains that floated on the primordial ocean, Nun. The heavenly vault was formed by the body of the beautiful goddess Nut, which was supported by Shu, the god of the air. Nut was clad in countless brilliant jewels that were the planets and the stars. A boat carried Ra, the Sun-god, on his daily journey across the body of Nut during the daytime hours, to return at night through the waters beneath the earth (Fig. 2).

The Heavenly Bureaucracy

Perhaps my favorite mythic universe is that of the Chinese. It arose around 2000 B.C. and clearly illustrates the often anthropomorphic character of many of these mythic universes, where the godly hierarchy mirrored the human one. In the Chinese universe, the gods and goddesses were part of a gigantic bureaucracy, spending their time in compiling dossiers, writing reports, and issuing directives, just like the bureaucrats of the Chinese empire. Somewhat later, around 500 B.C., the philosopher Confucius introduced the notion of the two opposing forces, the Yin and the Yang. In the Chinese view, the world was the result of the dynamical interaction of these two contrasting forces. The sky was Yang, the powerful, masculine and creative force. The earth was Yin, feminine and maternal. The sky above was in constant motion. The earth beneath was still. The Sun was Yang, bright, hot, and dry; the Moon was Yin, dark, cold, and moist. The universe went through a perpetual cycle: The Yin reached its apogee before yielding its place to the Yang. The night that succeeded day, the Moon that replaced the Sun, and the cold, dark winter that followed the hot, blazing summer: All were examples of the interaction between Yin and Yang.

The intellectual successes of the people who believed in these mythic universes were very striking. The Egyptians mastered geometry in building their pyramids. The Babylonians conquered the science of numbers to record the positions of the heavenly bodies, establish calendars, and predict eclipses of the Moon. But the Egyptian and Babylonian priests did not observe the heavens merely out of intellectual curiosity, but to learn the destinies of men: Their interests were astrological rather than astronomical. They were not concerned with using their mathematical knowledge to discover the underlying laws that governed the motions of the celestial bodies. Because the motions of astronomical objects are cyclic, it was only necessary to follow their positions on the sky over a sufficiently long period of time to be able to predict their future movements. Knowledge of the positions of the Earth relative to that of the Sun, the Moon, the stars, or the rest of the universe was not needed. Like their predecessors, who built megalithic temples at Carnac in Brittany, in western France, and at Stonehenge in southern Britain, which marked the rising and setting positions of the Sun and Moon, the Babylonian and Egyptian cosmologists were more priests than astronomers.

The Greek Miracle

Around the sixth century B.C., a most unlikely development, which has been called "the Greek miracle," and which was to last some 800 years, took place in Ionia, on the shores of Asia Minor. In the midst of the mythic universe, a handful of exceptional men managed to reverse the trend and sow the seeds of a new universe that was to sound the death knell of the old.

The Greeks introduced the scientific universe, which still prevails today. Instead of blindly accepting the gods, and being content to observe natural phenomena without trying to understand them, the Greeks had the revolutionary idea that the world could be understood in terms of its different components, and that human reason was capable of understanding the laws that governed the behavior of the different parts and of their interactions. Nature was a fit subject for reflection and speculation. The understanding of the natural laws that had been the sole prerogative of the gods in the mythic universe was now shared, in the scientific universe, with humans.

Armed with an unshakeable confidence in the power of human reason, the Greeks set to work. The structure of matter, the nature of time, biological, geological, and meteorological phenomena, nothing escaped their curious, probing attention. Leucippus and Democritus broke matter down into indivisible atoms, an idea that remains in force today. Pythagoras, by constructing his theorems, founded mathematics; Euclid built one of the most harmonious and impressive structures of human thought by developing his geometry.

From all this intellectual ferment, a new universe arose that broke away from the mythic universe. Over a period of 4 centuries, many cosmologies, successively more and more sophisticated, were elaborated. They finally reached their peak in the Ptolemaic universe, which was to reign supreme for 2000 years. The foundations of the scientific method were gradually established. Although the earliest cosmologies were derived from purely philosophical speculations, the constraints imposed by observations of the motions of the planets gradually came to assume a greater and greater importance. The early cosmologies still retained many mythological influences. To Thales (around 560 B.C.), everything was water. A flat Earth floated on a primordial ocean surmounted by the heavens, which also consisted of water. This was all very similar to the Babylonian mythic universe, where water was the primary element that vitalized the whole world. Anaximander (about 545 B.C.) rejected the idea of a single primary element. To him, the universe was the result of the interaction and intermingling of opposites: heat and cold, light and darkness—an idea that closely resembles the Chinese concept of Yin and Yang. The Earth was a squat column floating in the air within a series of rings of fire, which were the Sun, the Moon, and the planets. Already the germ of the idea of a geocentric* universe had been sown.

The Mathematical Universe

Mathematical rigor was introduced to cosmological thought by Pythagoras, in the sixth century B.C. According to him, the universe, the "Kosmos," had a geometrical harmony, governed by mathematical laws and numbers. Numbers were the basis and origin of everything, the reflection of the perfection of God. The Earth was no longer a flattened disk, but took the most "perfect" mathematical form—a sphere. Unlike later, geocentric universes where an immobile Earth was located at the center, the Pythagorean universe had an invisible, central fire, around which ten objects followed perfect circles. Their motions were in perfect musical harmony and gave rise to the "music of the spheres." The ten objects were, in increasing order from the central fire: the anti-Earth, the Earth, the Moon, the Sun, the five other known planets (Mercury, Venus, Mars, Jupiter, and Saturn), and the sphere of fixed stars. The existence of the anti-Earth, which protected the Earth from the intense heat of the central fire, was required to bring the total number of bodies to ten, the perfect number. Pythagoras would have been very unhappy to learn that there are only nine known planets in the modern Solar System. To him, the universe was governed by numbers, and could therefore be deduced by pure reasoning. Observation was of no use. The modern scientific method, which maintains that the unity of the world may be perceived by means of observations and experiments, was yet to appear.

The Geocentric Universe

Plato, in the fourth century B.C., utilized certain Pythagorean ideas to devise another universe. The Earth retained its perfect, spherical form, and heavenly bodies continued to trace out perfect circles. Their motions were regular in every way: The planets revolved around the Earth at a constant speed, with no acceleration or deceleration. The heavens inhabited by the gods must be perfect, and the perfection of the heavens required perfection of form and of motion.

The anthropomorphism of the mythic universe resurfaced, and the Earth regained its central, immobile place. The temptation to assume that the universe was geocentric is quite comprehensible. What is more natural, when we see the heavenly objects pursuing their path across the sky from east to west night after night, than to assume that the Earth is immobile at the center of the universe and that the Sun, Moon, planets, and the stars revolve around it? Plato conceived a universe where the Earth was at the center of an enormous sphere that contained the planets and the stars. The universe was finite and bounded by that sphere, which rotated once a day, thus accounting for the motion of the heavenly bodies.

Retrograde Motions

But this universe of just two spheres could not explain the strange and remarkable motions of the planets. When the planets were visible, they moved across the sky from east to west, like all the stars. But from night to night they changed their positions relative to the stars, moving inexorably from west to east. This motion was the origin of their name, because *planetos*

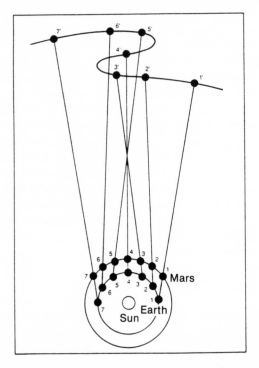

Fig. 3. *The retrograde motion of the planets.* A terrestrial observer sees Mars reverse its motion with respect to the distant stars (position 4′) when the Earth overtakes Mars in its orbit around the Sun. This retrograde motion is an illusion; it is caused by the motion of the terrestrial observer who is carried by the Earth around the Sun. An extraterrestrial, observing from a fixed position, would not see any retrograde motion. Once the Earth has overtaken Mars, the latter resumes its normal motion from west to east relative to the distant stars.

means "wandering" in Greek. The stars, however, appeared to be fixed with respect to one another. Nowadays, we know that this difference in the relative motion of the planets and the stars is actually an effect of distance. The stars are so far away that they appear stationary, whereas nearby planets move visibly.

Remarkably, from time to time, a planet appears to come to a halt and then reverse its direction of motion relative to the stars. For a period, the planet undergoes what is known as retrograde* motion, from east to west, after which it resumes its normal motion, from west to east. The modern, heliocentric universe can easily account for this behavior. It arises because the planets are being observed from a vantage point that is itself in motion. Retrograde motion occurs each time the Earth, in the course of its orbit around the Sun, overtakes a superior* planet (one that is farther away from the Sun), or is overtaken by an inferior* one (closer to the Sun). The motion is only apparent. Extraterrestrials viewing the Solar System from a spaceship would not see any retrograde motions (Fig. 3).

The Scientific Universe

Eudoxos, a young contemporary of Plato, had no means of knowing that the retrograde motions of the planets were not real. He tried to devise a universe that would faithfully reproduce the motions of the planets. In the Platonic view, it was necessary to "save the appearances," no matter what the cost. To Eudoxos, reality was not determined by pure reason alone. It had to agree with observation. Eudoxos, therefore, was the first person to devise a scientific universe. He transformed Plato's dual-sphere universe into one that consisted of multiple spheres. Between the immobile Earth at the center, and the outer sphere of stars that bounded the universe, he added a set of concentric spheres carrying the planets. He realized that any movement may be explained by combining several of the circular and regular motions that were "natural" to the planets. To account for the retrograde motion of the planets, the planetary spheres, which were themselves rotating, carried additional spheres, which rotated about their own inclined axes. In total, Eudoxos required 33 spheres to account for the observations available in his day.

The next step was taken by Aristotle (around 350 B.C.), who developed Eudoxos' universe of multiple spheres by adding both an additional physical dimension and a more spiritual one. To reproduce the more accurate observations of the planets that had then been made, the total number of spheres was increased to 55. The Moon, Mercury, Venus, the Sun, Mars, Jupiter, and Saturn were all attached to crystalline spheres centered on the Earth, which remained immobile. Each planetary sphere was linked to four or five other spheres, all of which rotated around different axes, so that their combined motion reproduced the motions of their respective planets. The universe was still bounded by the outer sphere of the fixed stars (Fig. 4).

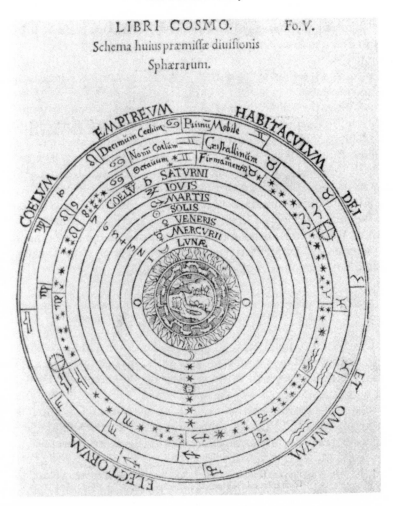

Fig. 4. *Aristotle's geocentric universe.* The Moon, Mercury, Venus, the Sun, Mars, Jupiter, Saturn and the stars are located on concentric, crystalline spheres centered on the immobile Earth (photograph: Bibliothèque Nationale, Paris).

The Aristotelian universe was divided into two regions, separated by the sphere of the Moon. The Earth and the Moon formed part of the sublunary, changeable, and imperfect world, pervaded by life, decay, and death. In this region consisting of the four elements earth, water, air, and fire, the natural motion was vertical. Everything moved in a straight line, either upward or downward. Air and fire rose toward the sky, while earth and water fell downward. Because circular motion was not permitted, the Earth was immobile, and did not rotate. On the other hand, the perfect world, that of the other planets, the Sun, and the stars, was unchanging and eternal. Consisting of a substance known as ether, its natural motion was a rotation around the Earth, thus explaining the eternal rotation of the crystalline, planetary

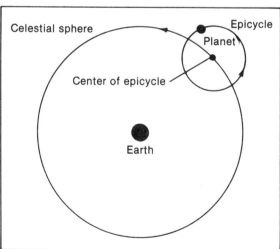

Fig. 5. *Ptolemy's universe.* The geocentric universe according to Ptolemy, whose portrait is shown here (photograph: Bibliothèque Nationale, Paris). This view was to reign supreme for more than 15 centuries. To account for the motion of the planets, Ptolemy envisaged each planet as moving on a circle, known as an epicycle, the center of which revolved itself on a celestial sphere that was centered on the Earth.

spheres. In this world-view, any changes in the sky, such as comets, which were balls of fire that appeared from time to time, could only belong to the sublunary imperfect world: They were phenomena produced by disturbances in the terrestrial atmosphere.

The Platonic and Aristotelian universes reached their apogee 2 centuries later with Ptolemy, about 140 B.C. (Fig. 5). He created a synthesis of all the knowledge acquired in the previous 4 centuries, expounding a geometrical universe, which was to be accepted without question for over 1500 years. The three main properties of previous universes were retained: The universe was geocentric. The Earth was spherical and at the very center. The natural motions of the planets were circular and regular.

The Earth Is Round

The spherical nature of the Earth had been demonstrated in the previous century by Eratosthenes (about 250 B.C.), who lived in Alexandria and who had read that at Syene, a small town in southern Egypt, the columns of the temples did not cast a shadow at midday on June 21, the first day of summer and the longest day of the year. On that day, Eratosthenes observed the columns of the temples at Alexandria at midday. He found that, unlike those in Syene, they cast considerable shadows. He concluded from this simple observation that the Earth must be curved. It could not be flat, because if it were, then any columns, anywhere on Earth, would cast similar shadows at any one time. By measuring the length of the shadows at Alexandria, and obtaining the distance between Alexandria and Syene from the number of paces between the two towns, Eratosthenes was able to calculate the circumference of the Earth, which he worked out to be 40,000 km, a value that is quite close to the modern value—a very considerable achievement.

Circles Upon Circles

Ptolemy set himself the task of accounting for certain failings of the Aristotelian universe. In particular, he wanted to provide a solution to two problems that had been revealed by careful observation. The first concerned the "abnormal" motions of the planets. These exhibited variations in the speeds at which they traced out their orbits, contrary to the doctrine of perfectly uniform speeds that was so dear to Aristotle. The second problem concerned the variation in the distance of the Moon and the planets from the Earth. This was revealed by changes in the angular size of the Moon, and in the brightness of the planets. This variation in distance could not be explained in the Aristotelian universe, because the latter stated that the planets were attached to spheres that were centered on the Earth. By definition, the distance between a planet and the Earth was always equal to the radius of the planet's sphere, and could not vary. Ptolemy had to solve these two problems, while also accounting for the occasional retrograde motion of the planets. He had the ingenious idea of detaching the planets from the celestial spheres and making them move in small circles, known as epicycles*, the centers of which were themselves located on the surface of the celestial spheres. The motion of a planet across the sky therefore consisted of a combination of two motions: the uniform revolution of the planet on the epicycle, the center of which itself rotated uniformly around a circle on the celestial sphere (Fig. 5b). The effect is similar to what a spectator at the center of a circular ring would see, when watching a cowboy galloping round on his horse, whirling a lasso with a luminous knot above his head. By introducing epicycles, Ptolemy was able to resolve all the problems posed by the Aristotelian universe. He was not only able to account for all the past motions of the planets in a quantitative, accurate, and detailed fashion, but was also able to predict their future positions. The *Almagest* ("The Greatest [Treatise]" in Arabic), where Ptolemy recorded his calculations of the positions of the planets, is undoubtedly a major work. It served as the basis of Arab astronomy for 7 centuries, and portrayed a universe that, together with Aristotle's physical world, was to remain paramount until the sixteenth century. Naturally, there were some who did not agree. In the third century B.C., Aristarchus had already rejected the geocentric view in favor of a heliocentric universe with the Sun located in the center, around which the Earth and the other planets revolved. But his opinion was quickly stifled.

The Medieval Universe

One night in the winter of 1300, a Franciscan monk leaves his warm and cozy bed to carry out the office of matins. He paces briskly across the monastery courtyard and stops outside the cloister to contemplate the sky. In the cold, still darkness of the night, he is captivated by the sight of the starry sky. Well above the horizon, he can see a point of light that is much brighter than the others. He knows that it is the planet Jupiter, carried by its crystalline sphere on its neverending circular path around the Earth, and accompanied in its journey by the angels that inhabit the higher spheres of heaven. He also knows that, between the Earth and the Moon, not visible in the sky tonight, lies Purgatory, where the spirits of the righteous are purified before they can ascend to the higher spheres and approach God. God, whose abode is the empyrean realm of Heaven, beyond the crystalline sphere of the fixed stars that he can see glittering above him. The thought of God automatically brings a prayer to his lips. He hopes that one day his own spirit will rise to the higher spheres and that he will never know the fires of Hell, deep within the bowels of the Earth. The Aristotelian universe of the crystalline spheres is vividly present in the medieval universe of the Franciscan monk. What is new is the presence of beings and concepts that are part of the Christian religion: angels, Purgatory, Hell, Heaven, and God.

Fifteen centuries have elapsed since the great synthesis of the Greek universe by Ptolemy. Many major events have come to pass. Greece had been annexed by the Roman Empire toward the end of the second century B.C., and Christianity was declared the official religion in the Empire around the year 300. The brilliance and clarity of Greek thought declined over that period. The Romans were not interested in abstract speculations. Despite brilliant technological inventions and innumerable practical innovations, such as the reform of the calendar, they contributed very little to cosmological thought. The repeated invasions of the "barbarian" hordes of the Goths and the Huns from the east during the fifth and sixth centuries dealt the final blow to the Roman Empire, which was already considerably weakened by political corruption and economic chaos. The knowledge of the Greeks disappeared from the western world.

In parallel with the decline and fall of the Roman Empire came the rise of the Islamic empire, which extended from Spain to India. The torch of civilization and science passed into the hands of the caliphs of Baghdad, who, during the period between A.D. 750 and 1000, built observatories and translated into Arabic the great works of the Greeks, such as the *Almagest*. From about A.D. 1000, Spain became the primary intellectual center of the Islamic world and, through her, Christian Europe rediscovered Greek thought. The translation of the great works of the Greeks from Arabic into Latin was undertaken, and many words of Arab origin, such as *algebra, azure, zenith,* and *zero* came to be part of the common language.

In the medieval world, learning was in the hands of the Church. All the manuscripts were to be found in the libraries of the monasteries, and only the monks had access to them. The Aristotelian concept of the universe posed some difficult and extremely serious problems for churchmen. How could the ancient Greek universe be reconciled with the Christian one? In Aristotle's universe, God did not appear in any explicit form. The planets, once set in motion, continued perpetually. There was neither beginning nor end. The role of God was far more explicit in the Christian universe: "God created the Heaven and the Earth." The universe had a beginning.

God and the Angels

It was the Dominican monk Thomas Aquinas who managed to unify the Aristotelian and Christian universes in the thirteenth century. He took the Aristotelian ideas and introduced God in an explicit fashion. The Earth remained the center of everything. It was spherical, a shape that it had regained in the ninth century, with the rejection of the flattened concept that had been accepted in the early centuries of the Christian era, through a too literal interpretation of certain passages in the Bible. Just as in the Aristotelian universe, the Moon, the Sun, the planets, and the stars all revolved around the Earth on crystalline spheres. There was an additional sphere beyond the sphere of the stars that had been envisaged by the Arabs. This was the *Primum Mobile,* the ultimate sphere, endowed by God with perpetual, uniform rotation.

God was now present. After having created the universe, he watched over the affairs of a hierarchical world, assisted by a heavenly host of angels.

The abode of God was in the empyrean, the domain of eternal fire, beyond the *Primum Mobile,* but at a finite distance, the universe itself being finite. The angels were next in the hierarchy. They inhabited the spheres of the planets and of the Sun, and were responsible for their revolution. Like heavenly mechanics, they kept the heavenly machine in working order (Fig. 6). Their divinity decreased the farther their abode was from the empyrean realm. Then came the sphere of the Moon, which was the boundary between the sublunary world and the higher spheres, passage through which was jealously guarded by the angels. Purgatory, the region where souls were purified before ascending to Heaven, and Earth, the abode of Man and mortality, were located in the sublunary world. At the bottom of the hierarchy, within the bowels of the Earth, was Hell, the region of demons and Evil, into which sinners were cast after their life on Earth. The mythic element, which had disappeared in the geometrical universes of Aristotle and Ptolemy, had now

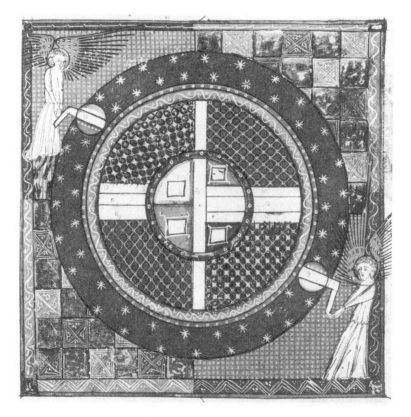

Fig. 6. *Heavenly mechanics.* St Thomas Aquinas infused a spiritual dimension into Aristotle's geocentric universe. In this universe, God watched over all creation, assisted by a host of angels. This medieval engraving shows angels, who, like mechanics, turn the handles of a device that rotates the celestial spheres of the planets (photograph: Bibliothèque Nationale, Paris).

surfaced in a religious form. The blue of the daytime sky was the ethereal light of God. Night was the time of devils and demons. The succession of day and night was a result of the unending conflict between Good and Evil. However, the scientific and rational element, introduced by the Greeks, still persisted and was to be reinforced in the coming centuries.

What If the Earth Were Moving?

Paradoxically, it was the reintroduction of religion into cosmology that was to advance the scientific element in later universes. The ecclesiastical authorities—particularly Étienne Tempin, Bishop of Paris in the thirteenth century—who scrutinized the universe proposed by Thomas Aquinas, progressively discovered that there were certain aspects that came into direct contradiction with accepted theological thought. The universe of Thomas Aquinas was finite, and bounded by the empyrean, the abode of God. Yet the God of religion was infinite and omnipresent, being found everywhere. Confining His presence to a single region amounted to casting doubt on His unlimited powers. If God was infinite, why should the universe not also be infinite? The seed of the idea of an infinite universe was thus sown. Moreover, was it not presumptuous of Man to believe that he occupied the center of the universe? Why should not God, who was everywhere, also be at the center of the universe?

In an even greater leap of imagination, the German cardinal Nicolas of Cusa advanced the idea in the fifteenth century that because God was infinite, omnipresent, and centered everywhere, every point in the universe must be the center. The starry sky should appear the same, no matter where an observer was in the universe. No single place was special, and there was an infinite number of centers. This idea, now known as the cosmological principle*, was to be adopted by Einstein 5 centuries later, when he developed his theory of relativity*. The geocentric universe was showing serious signs of cracking. The concept of the immobility of the Earth, which until then had remained sacrosanct, also began to be questioned. Was it not sacrilegious to suppose that God, who was omnipotent, could not overcome the fixity of the Earth, and cause it to rotate? After all, as the French prelate Nicole d'Oresme pointed out in the fourteenth century, all movement was relative. The motion of the stars across the sky could be caused either by the rotation of celestial bodies around an immobile Earth or by the rotation of the Earth relative to the fixed heavens. Human reasoning could not distinguish between these two possibilities. A man on a boat floating down a river, and watching the banks passing by, would gain the totally erroneous impression that the boat was stationary and that it was the landscape that was moving. Might we not be mistaken, like the boatman? asked Nicole d'Oresme. What if the Earth were moving?

Despite all those attacks against the geocentric universe, Ptolemy's model continued to reign supreme, because there was no better theory to replace it. More and more epicycles were added to account for the more and more precise observations of motions of the planets. The Ptolemaic structure became increasingly complex, moving each day further away from the simple harmony of celestial spheres so beloved by Pythagoras.

The Heliocentric Universe

It was another cleric, the Polish canon Nicholas Copernicus, who finally dislodged the Earth from its central place in the universe. In his book, *De revolutionibus orbium coelestium* ("On the Revolutions of the Heavenly Spheres"), which was published in 1543, just before his death, Copernicus completely altered the view of the universe, and began an intellectual revolution, the consequences of which we still feel today. The Aristotelian principles, which had been considered as self-evident for 2000 years, were called into question. In the Copernican universe, the center is located close to the Sun. The Earth was relegated to the ranks of the other planets. It lost its immobility, and was given motion, which carried it in its yearly revolution around the Sun, like the other planets. The order of the planets themselves became the one that we recognize today. In increasing distance from the Sun, they were: Mercury, Venus, Earth, Mars, Jupiter, and Saturn, the six known planets in the Solar System. Only the Moon retained the Earth as its center. It accompanied the Earth on its yearly journey around the Sun, while revolving around the former once a month (Fig. 7).

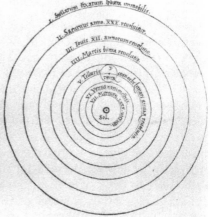

Fig. 7. *The heliocentric Copernican universe.* Copernicus (*left*), and the heliocentric universe shown in his work of 1543, *De revolutionibus orbium coelestium* (*right*). The rotating, planetary spheres, and the unmoving sphere of the fixed stars are centered on the Sun (photograph: Bibliothèque Nationale, Paris).

Because the Earth was in motion, the retrograde motions of the planets, which occurred whenever the Earth was overtaken by, or overtook, another planet, could be easily explained, without recourse to the epicycles proposed by Ptolemy. However, not even Copernicus was able to sweep away all the Aristotelian concepts. Ideas that have prevailed for 2000 years acquire a life of their own. In the Copernican universe, the planets continued to be located on crystalline spheres that were kept in motion by angels, and their orbits around the Sun remained the perfect circles so important to Greek thought. Like Aristotle's higher spheres, their motion had to remain perfectly uniform. Because the orbits of the planets are not perfectly circular, however, nor their motion completely uniform, Copernicus was also forced to have recourse to epicycles to explain their motions. Each planet moved on a small circle (the epicycle), the center of which moved itself on a circle on the corresponding crystalline sphere that rotated around the immobile Sun. The crystalline spheres were not perfectly centered on the Sun, but on a closely nearby point, between the Sun and the Earth.

The heliocentric universe dealt a severe blow to the human psyche. Man had lost his central place in the universe. He was no longer the center of God's attention. The universe no longer revolved around him, and the cosmos was not created for his sole benefit and use. Moreover, in the new universe, the Earth had become a higher sphere, like the other planets. According to Aristotle, everything connected with the higher spheres should be perfect, unchanging, and eternal, quite unlike the properties of terrestrial objects, which are imperfect, changeable, and ephemeral. Did this mean that the heavens themselves were imperfect? Confidence in the perfection of the heavens was severely shaken.

The Infinite Universe

In yet another strike against human self-esteem, the universe had been considerably enlarged, reducing the size and importance of the Earth with respect to the rest of the universe. The Copernican universe was still finite, bounded by the sphere of fixed stars, which was rigid, and did not move. Just as Nicole d'Oresme had thought, the apparent motion of the stars was a result of the Earth's daily rotation about its axis, not of the rotation of the heavens around the Earth. In the Aristotelian universe the sphere of the fixed stars was at a slightly greater distance than the sphere of Saturn. But even such a moderately small distance caused a problem: The circumference of a circle on the sphere of fixed stars was already so large that the stars would have to move at an inconceivably high speed to be able to cover such a path in a single day. Copernicus solved the problem by allowing the Earth to move and the stars to remain still. In doing so, however, he was forced to assume that the sphere of fixed stars was at a very great distance. From occupying almost the whole universe, the Solar System was reduced to just a tiny part. Copernicus had to ascribe a very great distance to the sphere of stars because the latter remained obstinately fixed relative to one another despite the annual revolution of the Earth around the Sun.

If the stars were relatively close and were observed at two different times during the course of the Earth's revolution around the Sun, they ought to have shifted their positions relative to more distant stars, as the simple, following experiment shows. Hold up a finger at arm's length and shut first one eye and then the other. The finger will appear to shift left and right with respect to objects in the distance. This effect is caused by the fact that the eyes are a certain distance apart. In an identical manner, the distance between two positions of the Earth in its orbit will cause a nearby star to appear to change its position relative to more distant stars. The angle that corresponds to this shift in position is known as parallax*, and the smaller this angle, the greater the distance to the star (see also Fig. 14). Because stars had parallaxes that were too small to be measured, Copernicus concluded that they had to be very distant.

In a single stroke, Copernicus had dethroned Mankind from its central place in the universe, sowed the seeds of doubt about the perfection of the heavens, and reduced Man to insignificance. It may seem remarkable that in proposing a heliocentric universe with such drastic consequences, Copernicus did not encounter the outright opposition of the Church, which supported the geocentric universe of Aristotle and Thomas Aquinas. This acquiescence on the part of the Church may be explained by several factors. First, Copernicus was himself an ecclesiastic. Then he only authorized publication of his book relatively late (3 years before his death; legend says that he only saw a copy on his deathbed). Above all, however, the book's preface maintained that the author did not believe that the proposed universe was necessarily real, but that it was a simple mathematical model, capable of predicting the motions of celestial bodies and of eclipses in a possibly more convenient manner than the Ptolemaic model. This preface, which is not signed, was probably written by Andrew Osiander, who undertook the publication of the book. In any case, the Church was happy with the interpretation of the Copernican universe as being simply a mathematical model. The Aristotelian universe of Thomas Aquinas could be retained, so the Church did not unleash its wrath on Copernicus.

The seeds sown by Copernicus began to germinate in the years that followed. Two men adopted the Copernican universe, which itself was much larger than any preceding universe, and cracked it wide open. The English astronomer Thomas Digges suggested, in 1576, that the outer sphere of fixed stars should be abolished. The universe was infinite and its stars were scattered throughout the unbounded abode of God. The Dominican monk Giordano Bruno went even further: He populated this infinite universe with an infinite number of worlds, inhabited by an infinite number of forms of life, all of which celebrated the glory of God. This was the last straw, and Giordano Bruno was accused of heresy, and condemned by the Church to die at the stake in 1600.

Imperfection in the Heavens

The Aristotelian concept of the perfection of the heavens continued to be subject to severe shocks. One of the decisive blows was delivered by Tycho Brahe, a Danish astronomer who had greatly improved the accuracy of astronomical observation—as far as was possible before the invention of the telescope—by constructing enormous instruments, from which the measurements could be more easily read, and also by taking account of variations in temperature, which affected the instruments by expanding or contracting their scales. In 1572 a new star appeared in the constellation of Cassiopeia, which was so bright that for about a month it was visible in daylight. Tycho, who was only 26, observed it day after day and night after night, establishing beyond doubt that it was very distant, well beyond the crystalline spheres of the planets. Unlike the planets, the star did not move at all relative to the distant stars. Tycho deduced from this that Aristotle was wrong, and that the heavens could change and were not immutable. We now know that Tycho was right, and that what he observed was actually a supernova*, an immense explosion that marks the death of a massive star in the Milky Way, and which, in its final death throes, releases the energy of 100 million Suns in a few days. Tycho's supernova, as it is often called nowadays, is one of the rare supernovae (seven in all) that have been observed in our own Milky Way galaxy.

The King of Denmark was so impressed by the discovery of the supernova that he gave Tycho an entire island, Hveen, just off the coast of Denmark, on which to build an observatory. There, for 20 years, Tycho accumulated observations of an unprecedented accuracy. In particular, his observations of the Great Comet of 1577 confirmed his doubts about the Aristotelian view of the perfection of the higher heavenly spheres. Until that time, comets had been viewed as phenomena occurring within the Earth's atmosphere, like rainbows. Tycho showed that this could not be true. The comet changed its position relative to the more distant stars, which indeed meant that it was much closer to the Earth than the supernova. Yet its movement was much smaller than that of the Moon, indicating that the comet must be farther away from the Earth. It was undoubtedly located somewhere among the crystalline spheres of the planets.

Once again, the Aristotelian immutability of the heavens had been shaken. A new object had appeared in the sky. Even more serious, Brahe's very accurate observations had enabled him to determine the orbit of the comet. This proved to be elongated, not circular. What had happened to the circular perfection of the heavens? There was another serious consequence of this discovery: If the orbit of the comet were elongated and closer than the most remote planets, then the comet must have passed through the solid, crystalline spheres of the planets, which, if the latter truly existed, was absurd. Tycho Brahe was forced to conclude that the crystalline spheres of the planets were not real, and that they only existed in man's imagination. Removal of the Aristotelian spheres raised a real difficulty: If the planets were

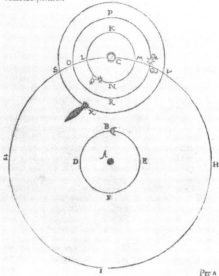

Fig. 8. *Tycho Brahe's universe.* An engraving from 1598 (*left*) shows Tycho Brahe, on the island of Hveen, just off the Danish coast. He is using a measuring instrument known as a quadrant, the telescope having not yet been invented. The diagram (*right*) illustrates Tycho Brahe's cosmology, which was a compromise between the Copernican heliocentric universe and the Aristotelian geocentric universe. The planets revolve around the Sun, but the Sun, with its accompanying planets, revolves around the Earth, as does the Moon (photograph: Bibliothèque Nationale, Paris).

not attached to such spheres, why did they not fall down? What held them up in the sky? Despite these questions, Tycho proposed a universe of his own that was a compromise between the Copernican heliocentric universe and the Aristotelian geocentric universe. In Tycho's universe, the planets revolved around the Sun, but the latter and the Moon revolved around the Earth, which occupied a central position (Fig. 8).

Galileo and His Telescope

Next on the scene was Galileo Galilei, Professor of Mathematics at Padua in Italy. For the first 18 years of his career, between 1591 and 1609, he studied the behavior of falling bodies. He was convinced that this would reveal the secret of the motions of celestial bodies, because he had rejected the Aristotelian idea that motion on the Earth was rectilinear, while motion in the heavens was circular. He had seen for himself that a ball thrown in the air fell back to earth along a curved trajectory. The same natural laws ought to govern the behavior of everything in the universe, and they could be discovered only by observation or by repeated, accurate experiments. Having, in effect, invented experimental physics in this way, Galileo investigated the motion of bodies on inclined planes, which slowed the rate of fall and allowed him to make more accurate measurements. He discovered that every falling object had exactly the same acceleration, whatever its weight. If there were no air resistance, a feather and a lead cannon-shot released simultaneously from the top of a tower would reach the ground at exactly the same instant. Although this result has been confirmed many times in an artificial vacuum created in the laboratory, some 360 years later, one of the American astronauts, in homage to Galileo, carried out the experiment with a feather and a geological hammer on the airless surface of the Moon.

In 1609, Galileo heard of the telescope that had been recently invented in Holland. He immediately constructed a device capable of magnifying 32 times, about the size of the small telescopes that can be found in the shops nowadays, and set out to explore the universe. As was to prove the case every time a new astronomical instrument was turned on the sky, Galileo discovered completely unexpected wonders. He saw many new phenomena and previously unknown objects, which only increased doubts about the Aristotelian universe. New imperfections appeared in the heavens. There were mountains on the Moon, and the Sun had dark spots on its surface. (Now called "sunspots," these appear dark because their temperature is lower than that of the surrounding surface.) With Tycho's supernova and the comet of 1577, the mountains on the Moon and sunspots were the final nails in the coffin of the Aristotelian concept of the perfection of the heavens. Turning his telescope toward Jupiter, Galileo discovered four satellites orbiting the planet. These satellites are now known as the "Galilean satellites," but to gain favor and financial assistance from the powerful Medici family, he called them "the Medicean satellites." As for the planet Venus, it exhibited phases, just like the Moon, going from full, when the planet was extremely brilliant, to new, when it was dark, with crescent and half phases in between. All these facts supported the Copernican world-view. The existence of satellites of Jupiter disproved the argument that the Earth was the center of the universe, and that everything revolved around it. The phases of Venus, which were the result of changing illumination by the Sun, could be explained only if the planet were in orbit around the Sun. Galileo became a champion of the heliocentric universe in his book *Dialogo de due massismi systemi del mondo* ("Dialogue on the Two Chief World Systems") of 1632, where he showed that those who defended the geocentric universe were simpletons (the character who upheld the traditional position was called "Simplicio"). This proved to be too much for the Church, which could no longer turn a blind eye to the pretext that the heliocentric universe was just a simple mathematical model. Thanks to Galileo's observations, this "model" was becoming too much of a reality for the Church's liking, and was likely to sow the seeds of doubt in the minds of the faithful about the ecclesiastical teachings. Galileo was brought to trial, and placed under house arrest until his death, in 1642. His book remained on the Index of books proscribed by the Catholic Church until 1835.

The Motions of the Planets

This unfortunate action by the Church had the effect of shifting the center of scientific activity to northern Europe. The next person to enter the scene as one of the architects of the universe was the German Johannes Kepler. As a young teacher of mathematics, he had become assistant to Tycho Brahe in 1600 at Prague in Czechoslovakia, where Tycho retired after having lost favor with the King of Denmark. Two years later, Tycho Brahe died, bequeathing to the young Kepler his innumerable, meticulous observations of celestial objects and, in particular, of the planets. Thanks to Tycho's magnificent obsession, Kepler was in possession of a treasure trove of observations of unrivalled accuracy, accumulated night after night over a period of some 20 years. This was long enough to follow the nearer planets over several complete orbits round the Sun. Kepler was convinced that among this wealth of information he would find the secret of celestial motions, and the laws that governed the universe. He believed in the heliocentric universe of Copernicus. He was also convinced by Tycho's reasoning that the crystalline spheres were only a figment of human imagination.

This attitude was all the more praiseworthy, because Kepler believed that the universe was governed by mathematics and that God was a geometer. For a long time, the astronomer thought that the number of planets (six were known in his time), or more precisely, the number of intervals between the planets (five in total), corresponded to the five regular solids known to the Greeks, which were called "Platonic" solids, the faces of which were regular polygons. The cube, for example, has square faces, and is one of these solids. After years of work, Kepler finally had to accept the evidence that planetary orbits with relationships determined by the Platonic solids just did not agree with Tycho Brahe's observations. The number of planets now known (nine) would in itself suffice to invalidate any theory based on the Platonic solids.

In his heliocentric universe, Kepler naturally considered circular planetary orbits, with uniform motions. Although belief in the perfection of the heavens had been well and truly shaken, the theory that orbits would be perfect circles, and motion perfectly uniform, had been accepted since the time of Plato. Copernicus, Tycho, and Galileo had all accepted this idea, without expressing the slightest doubt about its validity. In examining Tycho Brahe's observations of Mars, Kepler discovered that the planet's orbit was not symmetrical around the Sun as it would be for a circular orbit, but that it was slightly elongated in one direction. After 4 years of study, 900 pages of calculations, and contrary to his own prior convictions, Kepler became resigned to seeing the last bastion of Aristotelian thought collapse. The orbits of the planets were not circles, but ellipses. The Sun was not at the

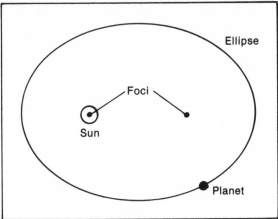

Fig. 9. *Kepler's ellipses*. Kepler (whose portrait is shown), discovered that the orbits of the planets were not circular, but elliptical, and that the Sun occupied one of the two foci.

center of the ellipse, but at one focus (Fig. 9). Epicycles were no longer required. The motions of the planets could be explained perfectly without them. Two thousand years of obsession with epicycles thus came to an end. The myth of the uniformity of planetary motions was also swept aside. The planets accelerated as they approached the Sun and decelerated as they moved away from it. There was a precise mathematical relationship between the time taken for a planet to complete its orbit around the Sun and its distance from it. If the Earth, by definition, took just 1 year to orbit the Sun, Mars, which was at 1.5 times the distance, required 1.9 times as long, while Jupiter, at 5.2 times the distance, had a "year" that lasted 11.9 Earth-years.

What keeps the Moon from Falling?

Although Kepler could describe the motions of the planets by precise mathematical laws, the problem that Tycho had raised by abolishing the crystalline spheres still remained. What maintained the planets in their elliptical orbits? What was the cause of their motion if there were no angels to push them around? Why did they accelerate when approaching the Sun and decelerate when receding from it? Kepler thought that it was magnetic forces originating in the Sun that kept the planets in their orbits. Such forces decreased far from the Sun, which would account for the slower movement of the planets, and they increased near it, accelerating the planets. But Kepler was wrong. This was brilliantly shown by the Englishman, Isaac Newton, who was born in 1642, the year that Galileo died, and 12 years after Kepler's death. Thanks to him, gravitation made a sensational entry onto the scientific stage.

In 1666, Newton was 23, and had just obtained his degree from the University of Cambridge. To escape the plague that was then raging, he went back to his home at Grantham in Lincolnshire. In the 2 years of his enforced rural retreat, Newton changed the face of the universe. He invented infinitesimal calculus, made fundamental discoveries about the nature of light, and discovered the principle of universal gravitation. Never before or since, except perhaps in 1905 when Albert Einstein simultaneously developed special relativity*, explained in the photoelectric effect (an effect concerning the interaction of light with atoms), and described the disordered motion of small particles suspended in liquids, has the world undergone such an intellectual upheaval in such a short time.

Legend has it that Newton conceived the idea of universal gravitation when he saw an apple fall from a tree. Before Newton's time, Galileo had already tried to resolve the motions of falling bodies. The great conceptual leap that Newton achieved was to equate the fall of an apple in his orchard to the motion of the Moon around the Earth. Newton swept away the Aristotelian distinction between heaven and Earth. To him, the Moon and the apple were subject to one and the same universal gravitation. Just like the apple, the Moon is attracted by the gravity of the Earth and falls toward

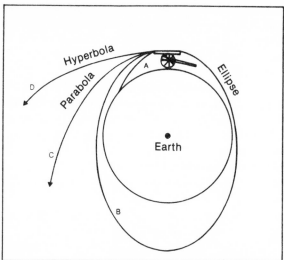

Fig. 10. *The motion of objects according to Newton.* The illustration (from an engraving by J. A. Houston, Mansell Collection) shows Isaac Newton conducting experiments on the nature of light. He was the first person to reproduce the colors of the rainbow by passing light through a glass prism. He was also the first to develop a reflecting telescope (which can be seen on the table in the illustration). But above all, he discovered the law of universal gravitation and was able to quantify the motion

it. But its orbit is such that it misses the Earth. As a result, it revolves every month around our planet in an elliptical orbit, with the Earth at one of the foci of the ellipse. No crystalline sphere is needed.

Moreover, there was no longer any need for angels to move the Moon around its orbit. Once in motion, the Moon would continue to move without the aid of any external intervention. The situation is similar to that of an apple thrown into the air. Once it has left your hand, it does not require any further intervention to continue along its path. But, you will say, the apple crashes to the ground, whereas the Moon continues inexorably in its orbit around the Earth. This is because you were unable to give sufficient impetus to the apple. If you throw it harder and harder, it stays longer and longer in the air, and falls farther and farther away. If you had superhuman strength, you could throw the apple so strongly that the distance by which it fell would be greater than the diameter of the Earth. It would then follow an elliptical path round the Earth, and would continue to do so indefinitely. You would have put an apple into orbit. If you were able to throw it even harder, the apple would follow a parabolic of hyperbolic path. It would escape from the Earth's gravitational influence and be lost in space (Fig. 10). In practice, of course, the Earth's gravity is too strong for objects to be put into orbit by any human effort, which is fortunate because we do not want space to be polluted with footballs! The energy released by tons of fuel is needed to put a Space Shuttle into orbit.

The Mechanistic Universe

Motions, once set in train, no longer require any divine or other intervention. The Newtonian universe is mechanistic. It functions like a clock powered by a coiled spring. Once it has been wound, the universe works on its own, obeying the law of universal gravitation. God had much more free time in the Newtonian universe than he had in the Aristotelian one. Instead of having to exert the constant vigilance that was required to oversee the army of angels that were keeping the planets and other celestial objects in motion, God only had to give a gentle push at the creation for the universe to continue on its own.

Universal gravitation allowed Newton to explain the planetary laws of motion discovered by Kepler. The force of gravity was transmitted by a

---

of objects. According to Newton, the path of a body (such as a shell fired from a cannon) depends on the initial velocity it is given. At a low velocity, it falls to the ground close to its starting point (path A). At a higher velocity, it would enter an elliptical orbit around the Earth (the center of the Earth would be at least one of the foci of the ellipse B). At an even higher velocity, the body would be lost to space, along either a parabolic (C) or a hyperbolic trajectory (D).

medium known as the ether (quite unlike the Aristotelian ether, however), a vague concept that Newton did not elaborate on. Being attractive, gravity caused two objects to fall toward one another, and it showed an inverse relationship to the distance between the bodies, because it was stronger close to the Sun, attracting a planet more strongly and causing it to accelerate, while it decreased away from the Sun, allowing the planet to slow down. In fact, Newton discovered that the force of gravity decreased as the square of the distance between two objects. If two people walk away from one another until the distance between them is ten times as great as originally, the gravitational force decreases in proportion to the square of ten, that is, to one hundredth of the original value. Moreover, the gravitational force is proportional to the mass of an object. Mass is a measure of an object's inertia, its resistance to movement. It is far easier to push over a person than an elephant, because the person has a far smaller mass. A pebble tossed in the air falls because of the gravitational force that the Earth exerts on it. This force is mutual, and the pebble exerts exactly the same force on the Earth. However, the motion of the Earth toward the pebble is imperceptible, because the Earth is far more massive, and resists any force applied on it to a far greater extent than the pebble. The pebble falls toward the Earth, and not the other way around. Similarly, the Moon orbits the Earth, not the opposite. It is important not to confuse mass with weight. You have weight because the Earth's gravity, by attracting you toward its center, pulls you to the ground. Weight is a word that implies "gravitational force." It varies according to the gravitational field. If you weigh 60 kilograms on Earth, you would weigh only 10 kilograms on the Moon, because the Moon's gravity is one-sixth that of the Earth. Your mass would remain the same, however.

Every object attracts every other object with a force that is proportional to the product of their masses and inversely proportional to the square of the distance between them. By enunciating this law of universal gravitation in his masterpiece, *Philosophiae naturalis principia mathematica* ("Mathematical Principles of Natural Philosophy"), published in 1687, Newton explained not only planetary orbits, but also the way in which the Moon governed the tides, the "elongated" orbits of comets, and many other natural phenomena. In fact, it was partly to further the study of cometary orbits that Edmund Halley, the Astronomer Royal, urged his friend Newton to publish the *Principia,* more than 2 decades after that eventful year of 1666. Newton, who was obsessed with the idea that others would rob him of his findings, delayed publication of his discoveries. For a long time he accused the German mathematician Leibnitz of having stolen the invention of the infinitesimal calculus from him, even though Leibnitz actually developed it quite independently. It was by applying the law of universal gravitation that Halley discovered that the comet now named after him was following an elliptical orbit around the Sun, and that it returned to visit humankind every 76 years.

The Deterministic Universe

Determinism now entered the scientific universe. The motions of terrestrial or celestial bodies were governed by precise, mathematically rigorous laws, which could be understood and used by human thought. If a stone is thrown in the air, it suffices to know its initial position and velocity to be able to predict precisely when, where, and with what velocity it will fall. The stone *must* follow the parabolic path determined by the law of universal gravitation.

The infinite universe, which was of a philosophical and theological nature for Thomas Digges and Giordano Bruno, acquired a scientific status with Newton. The universe ought to be infinite, because if it had boundaries, somewhere there would be a privileged central point. If this were the case, gravitational attraction would cause the whole universe to collapse toward that point, there to form a single massive body. This did not accord with the observed universe. On the other hand, in an infinite universe with no specific boundaries or privileged central point, where stars were evenly scattered throughout space, there would be no net gravitational force in any particular direction, and thus no risk of a collapse.

At the end of the seventeenth century, man looking at the sky saw an infinite universe, uniformly filled with stars, from the center of which he had been dislodged. Standing on the surface of an insignificantly small Earth, he was adrift in a mechanistic and deterministic universe, full of inanimate objects that behaved in accordance with rigorous laws, which he could determine with his reason. God, infinite like the universe, was still there, but in a far more remote manner. After having created the universe, and "set it going," he observed its evolution from afar, and stopped intervening directly in human affairs.

The Hypothesis of God Is No Longer Necessary

This new universe had two psychological effects, with diametrically opposed results. Some were overwhelmed by the concept of an infinite universe. Blaise Pascal cried out his anguish: "The eternal silence of infinite space terrifies me." He took refuge in Jansenism[4] in an attempt to come closer to a God that has become so remote. For most, however, the thought that human reason could understand God's intent and determine the laws that govern the universe was exhilarating. The eighteenth century was the Age of Enlightenment, the Age of Reason. God became more and more remote. The universe ("Nature," as it was termed in the eighteenth century) was more than ever a well-oiled machine that worked of its own accord,

4. A religious movement that reached its peak in France and Italy during the seventeenth and eighteenth centuries and which holds the pessimistic view that salvation of the soul is given or denied at birth, and that "acts in life" cannot change that fate.

without help from God. Reason reigned supreme, relegating faith to a secondary role. When the Marquis Pierre Simon de Laplace presented a copy of his work *Mécanique céleste* ("Celestial Mechanics") to Napoleon Bonaparte, the latter gently chided him for making no mention at all of the Supreme Architect. Laplace replied drily, "I have no need for such a hypothesis!" Confidence in human reason was unbounded. This optimism spilled over into all other spheres of human activity. The idea of progress made its appearance: Mankind could continually better and perfect itself. Nature could be domesticated to serve human needs. Social and political institutions could be perfected. It is no coincidence that the end of the eighteenth century saw not only the Industrial Revolution but also the American War of Independence in 1776, and the French Revolution in 1789.

The nineteenth century brought a reaction against Newton's mechanistic and deterministic universe in the shape of Romanticism. The scientific universe was strongly entrenched, however, and could no longer be rejected. Human reasoning continued to prevail. It was capable of anything, even the discovery of a new planet. Uranus, the seventh planet, which was discovered by the English astronomer William Herschel in 1781, showed irregularities in its orbit around the Sun that could not be explained by the law of universal gravitation if there were only seven planets. On the other hand, if the attraction of a yet-undiscovered, more distant, eighth planet were assumed, then the motion of Uranus could be understood. Two astronomers, the Frenchman Urbain Le Verrier and the Englishman John Couch Adams, independently calculated the position of the hypothetical planet, and indeed in 1846, the eighth planet, named Neptune, was discovered close to the predicted position. Neptune was not discovered by scanning the sky with a telescope, but with pencil, paper, and human reason. It was yet one more proof that the machinery of the universe was functioning correctly, in accordance with Newton's laws.

Dethroned from his central position by Copernicus, reduced to insignificance in an infinite universe, and remote to God in Newton's mechanistic universe, the western man of the nineteenth century could console himself with thoughts of his divine lineage. Was he not, after all, the descendant of Adam and Eve, themselves created by God? Even though no longer occupying the center of the universe, he remained the chosen child of God. Even that consolation was denied him after the publication in 1859 of Charles Darwin's *Origin of Species*. According to the naturalist, human origins were far less noble. Looking back through time, man's remote ancestors were, in turn, primates, reptiles, fishes, invertebrates, and, ultimately, primitive unicellular organisms. Biological evolution required an extremely long time in which to operate. The age of the universe, which Kepler and Newton had estimated at some 6000 years, was called into question. Thousands of millions of years were required, a time scale that the work of geologists appears to confirm. The universe, which had expanded in space, now expanded in time. God became even more remote. At the dawn of the twentieth century, Reason and Faith, Science and Religion, finally diverged.

| 2 |

From the Milky Way
to the Universe

The universe of the twentieth century is that of the Big Bang*. Nowadays, the majority of cosmologists believe that the universe began, approximately 15 billion years ago, with an enormous "explosion" from an initial state that was extremely tiny, hot, and dense. The rise of this new universe had been irresistible. In half a century, Newton's static universe with fixed, immobile stars was transformed into a dynamic, expanding universe, permeated with motion and violence. The "eternally silent infinite space" that horrified Pascal is now filled with sound and fury. From being immutable, it has been transformed into a state of perpetual evolution.

The rapidity with which the new universe was accepted is all the more extraordinary, because at the beginning of the twentieth century, the true extent of our Milky Way*, the Galaxy* (from the Greek *galaktos,* meaning "milk"), which contains our Sun and the Solar System, was completely unknown. Talk of other worlds in other galaxies was no more than pure science fiction. Yet in the last 50 years cosmology has acquired the status of an exact science, in other words, a discipline founded on accurate, rigorous observations, and not on vague, philosophical, or metaphysical speculations. This rapid development has arisen, above all, because of the giant advances in technology that have been made in the past 2 centuries in the methods of capturing and recording light from celestial objects. This light carries precious information and is the most effective communication link between man and the rest of the universe. Endowed with unshakeable confidence in the power of human reason, and in its capacity to discover and understand the laws of nature, the descendants of Newton and Voltaire threw themselves with a passion into inventing and perfecting the tools required to observe the universe in detail, and thus satisfy their curiosity.

Capturing Light

First of all, larger "eyes" were needed. The human eye, whose pupil has a maximum diameter of less than 1 cm, is too small a light detector. Yet it is an extremely efficient detector, because the brightness of the faintest star visible to the naked eye, on a moonless night far from the blinding artificial light of the cities, is 25 million times less than that of the Full Moon. Telescopes help the eye in two distinct ways. On the one hand, they magnify images, giving us a more detailed view, and on the other, they collect more light, enabling us to see fainter objects. Galileo was the first to turn a telescope onto the night sky, in 1609. Even his small telescope, with a lens that was only a few tens of millimeters in diameter, revealed details of the mountains on the Moon and the satellites of Jupiter. By allowing him to see objects about 1000 times fainter, Galileo's telescope increased the number of stars visible in the Milky Way from a few thousand to several million.

The search for greater detail and fainter objects has continued unabated, and telescopes have continued to increase in size. One of the most famous builders of these cathedrals of the twentieth century was the American astronomer, George Ellery Hale, who constructed the largest telescopes of his time. He had the gift of being able to persuade some of the richest philanthropists of his epoch to finance his projects. Charles Yerkes, for example, who became rich from building tramways in Chicago, financed the construction, in 1897, of the 1-meter refractor at the Yerkes Observatory in Wisconsin. This telescope remains the largest refractor in the world, and resembles Galileo's in that it uses a lens to gather the light. But the era of large refractors soon passed. Lenses larger than 1 meter (approximately 40 inches) in diameter become too cumbersome, because of both their weight and their thickness. Reflecting telescopes, which gather light by means of a large mirror in the shape of a paraboloid, came on the scene. This time, Providence appeared in the form of Andrew Carnegie, the steel magnate. With his financial backing, Hale built two telescopes on top of Mount Wilson, under the clear southern California skies. These, the 60-inch (1.5 m) telescope, completed in 1908, and the 100-inch (2.5 m) telescope, installed in 1922, were soon to change the face of the world. But Hale was not content to stop there. He dreamt of an even larger telescope, 200 inches (5 m) in diameter, which would be able to detect objects so faint that they were at the very edge of the universe. Legend has it that one fine day in the 1930s, Hale received a telephone call from the head of the foundation established by John D. Rockefeller, founder of Standard Oil and a petroleum magnate. He had read a popular article in *Harper's* magazine where Hale described his dream, and was interested in financing the project. The 200-inch telescope was commissioned in 1948 on top of another mountain in southern California, Mount

Palomar, and until the 1970s it remained the largest telescope in the world. With modern instrumentation, it is able to detect objects that are 40 million times fainter than the faintest star visible with the naked eye.

Nowadays, there are more than a dozen telescopes larger than 3 meters in diameter, scattered all over the world on mountain tops far from civilization and city lights. From Arizona to Hawaii, California to Chile, and Australia to the Canary Islands, their domes are opened every clear night so that they can capture light from the distant universe. But the quest is not yet over. Already plans are being drawn up for yet bigger telescopes, 10-15 meters in diameter. The mirrors of these mammoths can no longer be made in once piece. A monolithic mirror of such a size would be so heavy that it would be impossible to maintain the paraboloidal shape required to give good images. These new telescopes will be of the multimirror type. They will consist of several small mirrors, like the Keck telescope whose construction was recently completed on top of the extinct volcano Mauna Key in Hawaii, which is composed of 36 hexagonal mirrors, each measuring 1.8 m across, and whose combined effect is the same as that of one mirror 10 m in diameter, like the Very Large Telescope built in Chile by the European Southern Observatory. These telescopes of the future will allow us to see objects ten times as faint as does the Palomar telescope (Fig. 11).

Conserving Light

After having gone to so much effort to gather the light, the next problem was to trap as much of it as possible. The image had to be recorded so that it could be retained and studied. It was not enough to look through the telescope, letting the image fall on the retina, and be transmitted by the optic nerve to the brain, which would merely be left with the fleeting impression of a beautiful sighting. Yet this was all that the earliest observers could do. To capture his images, Galileo had to call on his not inconsiderable talents as an artist to make remarkable drawings on his observations. The situation was only solved when the Frenchman Nicéphore Niepce invented photography in 1826. Subsequently, the images of thousands of stars could be captured on a single glass plate. By allowing light to accumulate over a period of hours, and by doing so over a considerable area, the photographic plate extended the sky to be studied systematically. It remained paramount in observatories until the early 1970s, when it was supplanted by electronic detectors. The latter devices are much more sensitive and can accumulate as much light in half an hour as a photographic plate can in a whole night.

Splitting the Light

We are all familiar with the rainbow, that multicolored arch that is sometimes visible during or after a rainfall on the opposite side of the sky to the Sun. Its secret was unlocked by Newton in 1666, that fateful year when he took refuge in the country to escape the plague. It is white sunlight passing through rain-

Fig. 11. *The large telescopes of the future.* The large telescopes of the future will no longer be a single unit. They will have multiple mirrors, because a large monolithic mirror will sag under its own weight, deteriorating the quality of the image. (a) This photograph shows the Keck Observatory on the summit of the

drops that produces, for our greatest delight, the wonderful play of red, orange, yellow, green, blue, indigo, and violet colors. The raindrops act like glass prisms and spread out the light that their telescopes have captured from celestial objects, so that it may be analyzed. The instrument, developed by the German Joseph Fraunhofer at the beginning of the nineteenth century, was soon to reveal the chemical composition and motion of the stars, and later of galaxies. An astronomer's work may thus be summarized: Light is captured with large telescopes, recorded with either photographic plates or electronic detectors, and analyzed by splitting it into its components with a spectroscope.

The New Forms of Light

But does this definition really correspond to modern astronomy? In fact, what has been described covers only visible astronomy, the astronomy that uses the light to which our eyes are sensitive. There is, however, a whole range of "light," or radiation, of which visible light is only a small part. First there are gamma rays and x rays, which are so energetic that they go easily through your body. You may well have seen x-ray photographs of your lungs taken to check for tuberculosis. Then there is ultraviolet light, which is less energetic, but which can still cause burns, destroy cells, and even cause skin cancer if you receive too much. In order of decreasing energy, then come visible light, with which we are all familiar; infrared radiation, which gives us the sensation of heat; microwaves like those in a microwave oven; and finally radio waves, which are the least energetic form of radiation. This is the radiation that carries sounds and images from the transmitter to your radio or television set, and enables you to enjoy your favorite program (see Appendix A).

With such a wide gamut, is it reasonable to suppose that the universe chose only visible light in which to emit its signals? That would seem most unlikely, and indeed anti-Copernican. After all, the fact that the human eye is sensitive to just "visible" light is only a consequence of Darwinian bio-

---

dormant Mauna Kea volcano on the island of Hawaii. It is composed of two 10-meter diameter telescopes, the equivalent in light gathering of a monolithic 14-meter telescope. The mirror of each telescope is composed of 36 hexagonal smaller mirrors that work in concert as a single piece of reflective glass (Photograph: Courtesy W. M. Keck Observatory). (b) One of the world's most advanced research facilities, the Very Large Telescope consists of four 8.2-meter telescopes, the equivalent in light gathering of a monolithic 16.4-meter telescope. It was built at the Paranal site in the dry Acatama desert in Northern Chile by the European Southern Observatory (ESO), which is a consortium consisting of eleven European countries: Belgium, Denmark, Finland, France, Germany, Italy, the Netherlands, Portugal, Sweden, Switzerland and United Kingdom (Photograph: European Southern Observatory).

logical evolution. Most of the Sun's radiation that passes through the Earth's atmosphere is visible light, and nature has provided us with eyes that are sensitive to it, thus helping human evolution. But we cannot expect the universe to pay any regard to the way in which our eyes function. In fact, even the Sun emits all the other forms of radiation, although in lesser quantities than visible light. The Earth's atmosphere protects us from the most energetic radiation from the Sun, such as x rays and ultraviolet light, and has therefore been of considerable importance in allowing life to develop. (The protection against ultraviolet light is by no means total, as anyone who has ever suffered from sunburn can attest.)

It is obvious that if we restrict ourselves to visible light, we shall possess an incomplete and impoverished view of the universe. If you need convincing, just imagine for a moment what it would be like if, suddenly, your eyes were sensitive to just a single color, say blue. Your view of the world would be quite incomplete. Naturally, you would still be able to see the blue sky and the blue of the sea. But the green of trees, the red of roses and poppies, and the varied hues of butterflies would all disappear. Life would be much the poorer.

Modern astronomers have therefore acquired new eyes. The development of radar during the Second World War gave rise, in the 1950s, to radio astronomy. Advances in astronautics and the conquest of space in the 1960s enabled us to get above the Earth's atmosphere, so that we were finally able to obtain images of the universe in gamma rays, x rays, ultraviolet light, and infrared radiation, with telescopes carried by balloons, rockets, and satellites.

In 1990, the Hubble Space Telescope (Fig. 12) was put in orbit by the Space Shuttle. With a mirror of 2.4 m, the largest ever put in space, it functions in the ultraviolet, visible, and infrared and was supposed to look seven times further and obtain images with ten times more detail than the largest ground-based telescopes. Hubble has not fulfilled at first its promise because an unfortunate defect in its mirror has made it myopic, making the observations of faint and distant celestial objects very difficult. Fortunately, in a spectacular space shuttle repair mission in December 1993, NASA astronauts have succeeded in giving back a clear vision to Hubble by bringing to it a set of correcting lenses. But even before its myopy was repaired, the Space Telescope was already sending us superb images of the planets in the Solar System and of nearby bright celestial objects. These images can be corrected for the defect of the telescope by a sophisticated computer treatment on Earth.

The Limits of the Milky Way

Armed with our telescopes, photographic plates, and spectroscopes, let us now begin to find out the extent of the universe. The dimensions of the Solar System were well known by the end of the nineteenth century. The distances between the planets had been determined by the principle of parallax, de-

Fig. 12. *The Hubble Space Telescope.* This telescope, with a mirror of 2.4 m in diameter, is named after the American astronomer Edwin Hubble, the discoverer of galaxies and of the expansion of the universe. Launched in 1990, it can peer 7 times farther into the universe, with 10 times more detail than the largest telescopes on Earth. Since looking far in space is looking back in time, the Hubble Space Telescope lets us look all the way back to the time of the birth of the first galaxies, 2 to 3 billion years after the Big Bang. During several years after its launch, it suffered from an optical defect in its mirror which rendered it myopic. Fortunately, Hubble has recovered its full sight thanks to a spectacular repair mission by the Space Shuttle astronauts in December 1993. With a weight of 11 tons, a length of 11 meters, and an orbit of several hundred kilometers above the Earth, the telescope makes one turn around the globe every 90 minutes and is controlled from a center located in Greenbelt, Maryland. It is expected to be operational for a period of at least 15 years, with repair and maintenance missions by the Space Shuttle every 30 months or so.

scribed earlier when we discussed the discovery of the Earth's revolution around the Sun by Copernicus. Every nearby celestial object appears to change its position relative to distant stars when it is observed from two different vantage points. It was enough to photograph planets simultaneously from two different observatories. The latter needed to be as far apart as possible (preferably on two different continents), because the shift in position is larger, and thus easier to measure, the longer the distance (known as the baseline) between the two observatories (Fig. 13). Once the parallax

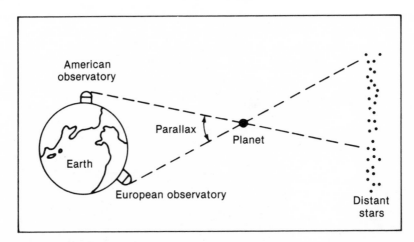

Fig. 13. *Measuring the distances of the planets.* The positions of the planets with respect to the distant stars, measured simultaneously by two separate observatories, differ by a small angle, known as the parallax. The greater the distance between the two observatories, and the closer the planet, the larger this angle becomes. Measurement of the angle and knowledge of the distance between the two observatories therefore give the distance of the planet.

(the angle corresponding to the shift in apparent position) has been measured, the distance of the planet is easily obtained by simple trigonometry, knowing the distance between the two observatories.

This work revealed that light from the Sun took about 8 minutes to reach us—in other words, the Sun was about 8 light-minutes (147 million kilometers) away from the Earth—and that Neptune, the planet discovered by the power of calculation, was at a distance of 4 light-hours. But at what distances were the "innumerable" stars that Galileo saw in the Milky Way, the numbers of which could be multiplied almost at will in the following centuries by the use of photographic plates on the increasingly larger telescopes? Was there any limit to the Milky Way? Or did it extend indefinitely, filling the whole of the infinite, Newtonian universe with an evenly distributed population of stars? What was the shape of the Milky Way? Was it spherical or flattened? The answers to these questions were by no means self-evident. The universe appears as a two-dimensional image on the celestial sphere, like a landscape painted on some vast canvas, where the artist has neglected all the rules of perspective. Direct observation does not provide any information about the third dimension, that is, about the depth of the cosmic stage.

A Thin, Flattened Disk

Naturally, there were always some brave souls ready to speculate about the nature of the Milky Way, which, until the revolution in ideas in the twentieth century, constituted the entire universe. To the Englishman Thomas Wright, in 1750, the universe was a thin layer of stars, sandwiched between two concentric spheres, God being located at the very center of the spheres. Wright's universe was mystical and philosophical rather than scientific, and he proposed several contradictory versions. But he discovered one essential aspect of the Milky Way, which will be recognized by anyone who, on a fine summer night, contemplates the yellowish band of stars that stretches overhead. Wright quite rightly observed that the distribution of stars in the sky is not uniform, and that the Sun and the Earth must lie within a very thin layer of stars. Anyone on Earth whose line of sight is tangent to this thin spherical shell of stars would see a multitude of stars, giving the appearance of a broad, yellowish band across the sky. Looking at right angles to the thin shell, however, would reveal only a few stars.

In 1775, the German philosopher Immanuel Kant took up Wright's ideas, but realized that it was not essential to have a spherical shape to account for the appearance of the Milky Way. The universe changed from a thin spherical shell to a flattened disk. Drawing his inspiration from the motion of the planets around the Sun, Kant suggested that the stars circled in the plane of the disk, and around its center. The stars appeared to be fixed in the sky, because they were so far away that their motions were imperceptible. The disk of the Milky Way was not infinite, but had limits. Beyond those limits, there ought to be other worlds, similar to our own. These "island universes," as Kant called them, might be the nebulous patches that the English astronomer William Herschel had just discovered. Such speculations were quite remarkable, because they are so close to the universe that we recognize today.

But inspired intuition was not enough. The discoverer of Uranus, William Herschel, a musician who abandoned his calling to study the music of the spheres, was, in the 1780s, the first to try to measure scientifically the extent of the Milky Way. He did not succeed, being unable to determine the distances of stars, but he struggled valiantly to determine the shape of the Milky Way by counting the number of stars in different directions on the sky. He reasoned that the Milky Way would extend farther in the directions where there were larger numbers of stars than in directions where there were fewer. This method could give valid results only if the stars all had more or less the same brightness, if they were uniformly distributed throughout space, and finally, if they could be seen right out to the edge of the Milky Way—assuming that there was nothing to absorb their light. We now know that none of these conditions is met. The brightness of stars varies between one-tenth and 100,000 times that of the Sun. They are not uniformly distributed throughout the Milky Way, and, above all, grains of interstellar dust absorb light from the stars and prevent them from being seen to the edges of the Milky Way. Herschel found a Milky Way that was flattened, centered on the Sun, but very irregular in shape. We now know that this apparent irregularity is not real; it is caused by the absorption of starlight by dust.

The Sun Is Just an Ordinary Star

The unraveling of the secret melody, of the cosmic fugue, could only begin with knowledge of the true depth of the universe. Measurement of distances beyond the Solar System began with the nearest stars. Parallaxes were again the key. Because stars were much farther away than the planets, the separation between the two points of observation also needed to be far larger. Otherwise the difference in position of a nearby star relative to more distant ones would not be measurable.

The Earth's annual voyage around the Sun was put to good use. The two simultaneous observations from two different observatories were replaced by two successive observations from different positions in the Earth's orbit. To maximize the distance between the two positions of the Earth, the photographs were taken at an interval of 6 months (Fig. 14). The baseline therefore changed from the diameter of the Earth (about 10,000 km) to the major axis of the Earth's elliptical orbit around the Sun (about 300 million km). This method enabled the distances of a few hundred of the nearest stars to be measured. Unfortunately, the parallaxes became smaller and smaller, and thus much more difficult to determine, as the distances increased. The parallax method no longer gave any useful information for stars beyond about 100 light-years.

Detailed examination of even this small corner of the universe, however, already revealed the insignificance of the Solar System and the utter void of space. The size of the Solar System could be measured in light-hours. Pluto, discovered in 1930, at the edge of the Solar System, was found to be 5.2

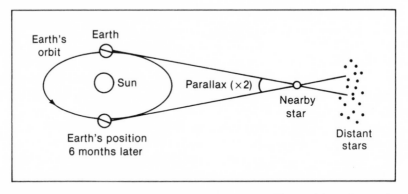

Fig. 14. *Measuring the distances of the nearest stars.* The position of a nearby star relative to more distant stars appears to change when it is observed at different times of the year. This change is not caused by any real motion of the nearby star, but is an optical effect produced by the change in perspective, itself the result of the Earth's annual revolution around the Sun. The amount of parallax is actually half the angle given by the change in position of the star relative to the distant stars, determined at an interval of 6 months, when the Earth is on opposite sides of its orbit. This angle is smaller, the greater the distance of the star. From a knowledge of the angle and of the Earth–Sun distance, the distance of the star can be derived.

light-hours away from the Earth, whereas the distances between the stars are measured in light-years (there are 8760 light-hours in a light-year). Space was extremely empty. The distance to the star closest to the Sun was no less than 4 light-years; to Sirius, 8 light-years; and to Vega, 22 light-years.

The distances to the nearest stars could be used in conjunction with their apparent brightnesses to derive their absolute, or intrinsic, brightnesses, thus settling an old argument between Kepler and Newton. Kepler thought that Sirius and the other stars were much fainter than the Sun, whereas Newton thought that they were similar, a hypothesis that Herschel adopted when he attempted to determine the shape of the Milky Way. In fact, among the nearest stars, some were brighter, and some were fainter than the Sun. Copernicus said that there was nothing special about the Earth. The same applied to the brightness of the Sun. Our star was quite unremarkable.

The Stars in Motion

The spectroscope came next to help in expanding the limits of the known universe. The stars, which until then had been thought to be fixed in the sky, acquired motion. Their movement was revealed by the spectroscope, thanks to a discovery by the Austrian physicist Johann Christian Doppler. In Prague, in 1842, he discovered that the sound emitted by a moving object is higher in pitch when the object is approaching the observer, and lower when receding. We have all probably experienced the Doppler effect*, when hearing the high pitch of the siren of an approaching ambulance, and its lower pitch as it speeds away from us. Just like sound, light is also affected by the Doppler effect. When a luminous object moves away from us, its light becomes "lower" in frequency, being shifted toward the red end of the spectrum, and losing energy. Similarly, when the object is approaching, its light becomes "higher" in frequency, being shifted towards the blue, and gaining energy. The change in color is greater the higher the velocity of approach or recession. To uncover the motion of the stars, the spectroscope splits up their light into their components, which allows the change in color to be measured, and the velocity of approach or recession to be obtained (see Appendix A). If you have ever been convicted of speeding on the freeways, you have probably been the victim of the Doppler effect. All the policeman has to do to measure your speed accurately with his radar is to bounce the radio waves off your car. The change in the "color," or frequency, of the radio waves reflected from your vehicle enables the speed at which you were driving to be determined exactly.

Stars That Converge in the Sky

Doppler measurements of the nearest stars showed that relative to them, the Sun was moving at a velocity of 20 kilometers per second. It had finally lost the immobility that it had retained ever since the time of Copernicus. All the other stars also acquired motion. In particular, the study of the motions of stars within what are known as galactic clusters* played a large part in expanding the boundaries of the known universe. These clusters of stars are groups of a few hundred stars that are not bound together by gravity (they disperse over a period of a few hundred million years), but that happen to lie close to one another through an accident of birth: they were born from the same cloud of interstellar gas, which collapsed and fragmented into a myriad of protostars (Fig. 15).

Like the Sun and other stars, galactic clusters are in motion. Stars in a cluster follow parallel paths in space, but perspective produces the illusion that they are converging towards a single point in space, known as the "convergent point*." This perspective effect is familiar to anyone who has ever looked at a set of parallel railway lines running away into the distance, where they appear to converge at the horizon (Fig. 16). This parallel motion of the stars in a galactic cluster produces a change in their position relative to more distant stars. This change is so minimal, however, that great patience is needed to detect it. For the motion to be perceptible, the stars in a galactic cluster have to be photographed at intervals of at least a decade.

Fig. 15. *A galactic cluster in the Milky Way.* This photograph shows the brightest stars in the Pleiades galactic cluster. (The six brightest stars are visible to the naked eye.) This cluster contains a few hundred young stars. (They are a few million years old, and about 250 have been counted.) These stars are not permanently bound together by gravity, and will disperse over a period of a few hundred million years. They are together because they were born from the same cloud of interstellar gas, which collapsed and fragmented into protostars. The nebulosity visible around the brightest stars is the remnant of this original cloud (photograph: Hale Observatories).

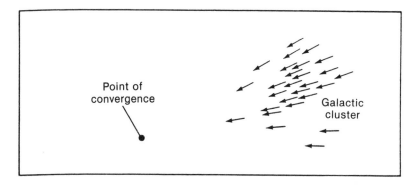

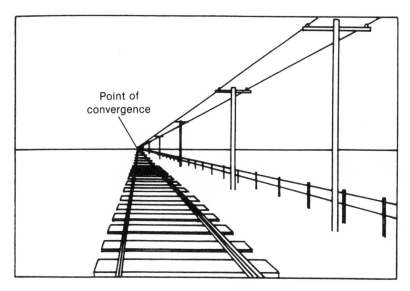

Fig. 16. *Measuring the distances of stars in a cluster.* The stars in a galactic cluster (like that shown in Fig. 15), although actually following parallel paths in space, appear to converge at a single point in the sky—the convergent point (a), just as parallel railway lines appear to converge at the horizon (b). Knowledge of the convergent point and of the motions of the stars enables the distance of the cluster to be determined.

We now have the problem of determining the distance of each star, having measured its change of position on the sky. To do this, we need to know the motion of the body from which we are observing, the Earth, at right angles to the line of sight to the cluster. The principle is simple, and you apply it unconsciously when you look at the countryside as you are driving along. The posts marking the side of the road pass very rapidly; farther away, a line of trees along the edge of the field moves slightly more slowly, and the mountains in the distance hardly appear to move at all. Instinctively,

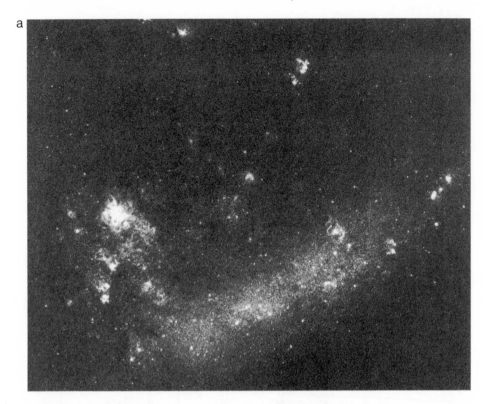

Fig. 17. *The Magellanic Clouds.* These photographs show: (a) the Large Magellanic Cloud and (b) the Small Magellanic Cloud, so called because they were "discovered" by Ferdinand Magellan, the navigator. (They are only visible from the Southern Hemisphere, and were already mentioned in the myths of the Australian aborigines and the various peoples of the South Pacific.) The Clouds are actually two dwarf, irregular galaxies, which orbit the Milky Way at a distance of some 150,000 light-years. The Magellanic Clouds played a key role in the story of the discovery of the

you rank the details in the landscape according to the speed at which they pass across your field of vision, and construct a mental image of the landscape's perspective. The posts flick past quickly; so they are in the foreground, close to the car, while the mountains, which appear almost stationary, must be the most distant. If you want to estimate the distance of the posts or of the trees, all you need to do is to measure their apparent motion and to know the speed at which you are traveling. In the same way, the motion of the stars in the cluster relative to more distant stars, and the motion of the Earth relative to each star in the cluster, are essential in determining their distances. This relative motion may be deduced easily if the convergent point is known, together with the motion of approach or recession of each star, which is determined from the Doppler effect.

The distance to the cluster is then the average of all the measured distances of individual stars. Use of galactic clusters—in particular, the

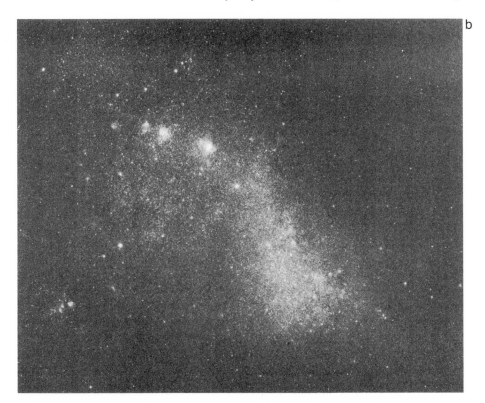

b

galaxies. It was by studying the brightness of Cepheid stars in the Magellanic Clouds that the American astronomer Henrietta Leavitt discovered the relationship that allows us to measure the distance of galaxies. Recently, the Large Magellanic Cloud was in the news again, because on February 23, 1987 it was the site of a supernova explosion, which enabled astronomers to examine the death of a star in great detail (photograph: Royal Observatory, Edinburgh).

Hyades* cluster, one of the nearest clusters—revealed the next region of the universe, out to about 1600 light-years. Beyond that distance, the motions of stars in clusters relative to more distant stars are so tiny that they cannot be measured, even by waiting several decades. Thanks to this method—called the "moving cluster" method—however, the known universe has expanded to be 2.5 million times as large as the Solar System. The Earth was becoming increasingly lost in the cosmic immensity.

Variable Stars and the Key to the Universe

But the limits of the Milky Way had still not been reached. The stars seemed to stretch out to infinity, and defied man's ability to measure their distances. The doors of the universe were only thrown open by the work of a young astronomer, Henrietta Leavitt, who worked at the Harvard College Observatory in 1912. She had been given the project of studying the two diffuse

nebulae that adorn the sky of the Southern Hemisphere, just as the Milky
Way arch graces the northern sky, and that filled Magellan with wonder
when his ship crossed the equator. These nebulae, called the "Magellanic
Clouds*," we now know to be two dwarf galaxies that are satellites to our
own Milky Way galaxy, and lie at a distance of some 150,000 light-years (Fig.
17). In fact, Miss Leavitt was ignorant of the nature of these nebulae, but
this was not relevant to her task, which was to locate stars in the Magellanic
Clouds that showed variations in brightness, by examining photographic
plates taken at different times. Although the majority of stars, like the Sun,
live most of their life uneventfully, and hardly vary at all in luminosity for
millions, or even billions, of years, some stars—known as "Cepheids*" after
the name of the constellation in which the first one was discovered—vary in
brightness periodically with the very short time scale of a few days or
weeks. (Yet another blow to Aristotle's immutability of the skies!) Leavitt
discovered that the time between two consecutive maxima of the brightness
(known as the star's "period") was longer, the brighter the Cepheid. There
was a direct relationship between the period of the variations and the apparent brightness of the Cepheid (Fig. 18).

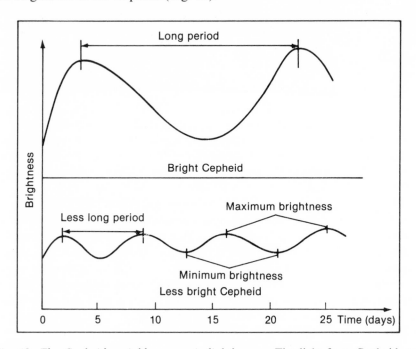

Fig. 18. *The Cepheid variables, cosmic lighthouses.* The light from Cepheid stars
varies periodically in a very specific fashion: The time between two maxima, or between two minima—known as the period—is longer, the more luminous the Cepheid.
This unusual property means that the Cepheids may be used as "standard candles"
to measure the distances of the galaxies in which they occur. A measurement of the
period gives the intrinsic brightness of the Cepheid. That intrinsic brightness may be
compared with the apparent brightness to give the distance.

A pair of car headlights shine through the darkness of the night. The apparent brightness of the headlights enables you to estimate the distance between you and the car. This is only possible because you have an idea of the intrinsic brightness of the headlights, having seen its full brilliance from close at hand. Similarly, the apparent luminosity of a star gives no information at all about its distance. A star may appear faint because it is intrinsically bright, but also very distant, or because it is intrinsically faint, but relatively nearby. The apparent brightness decreases as the square of the distance: a star at twice the distance appears four times dimmer. Just as you need to know the intrinsic brightness of the headlights to be able to estimate the car's distance, an astronomer needs to know the intrinsic brightness of a star to be able to deduce its distance from its apparent brightness. For Leavitt's relationship to be useful in deriving the distance of stars, the period-apparent luminosity relationship needed to be converted into a period-intrinsic luminosity relationship. To be able to determine the distance of a Cepheid, it would then suffice to measure its period (which indicates the intrinsic luminosity) and its apparent luminosity.

Transforming apparent luminosity into intrinsic luminosity requires the knowledge of the distances of some nearby Cepheid variables. The traditional methods for determining distances, the parallax and moving-cluster methods, were used. There were, however, no Cepheids close enough in the Milky Way (less than 100 light-years) for their distances to be determined by the parallax method. Similarly, there were no galactic clusters containing Cepheids close enough for the moving-cluster method to be used. An intermediate step was therefore required. The distance of the Hyades cluster was determined by the moving-cluster method. The distance of other galactic clusters that did contain Cepheids could then be deduced by assuming that the intrinsic luminosity of the stars within them was the same as that found in the Hyades. Once the distances of these clusters were known, the intrinsic luminosity of Cepheids that they contained was also determined.

This step was taken in 1916, and astronomers finally held the key to the universe. Cepheid stars are intrinsically bright and could be seen at great distances, out to about fifteen million light-years. In other words, they were visible to distances that are 500 times as great as those measured for galactic clusters. Like lighthouses spread throughout the vast cosmic ocean, they were to guide our journey to the edge of the Milky Way and far beyond. The universe would finally be revealed in all its immensity and fantastic extent.

The Sun Loses Its Central Place

The Cepheid cosmic lighthouses were to be put to good use by the American astronomer Harlow Shapley. In the 1920s, he was working with the 60-inch (1.5-m) reflector built by Hale on Mount Wilson in California, trying to discover the secrets of another type of grouping of stars, the globular clusters*. Unlike galactic clusters, which are irregular in shape, and which contain only a few hundred stars not bound together by gravity, globular clusters are

Fig. 19. *One of the globular clusters in the Milky Way.* The photograph shows the globular cluster known as 47 Tucanae, which is a spherical agglomeration of about 100,000 old stars, bound by gravity. The diameter of the globular cluster is about 300 light-years. The density of stars increases gradually toward the center, and it is so high close to the center that the individual stars can no longer be seen separately. It was by studying the spatial distribution of globular clusters in the Milky Way that the American astronomer Harlow Shapley came to realize that the Sun could not be at the center of our Galaxy (photograph: Cerro Tololo Observatory).

Fig. 20. *Anatomy of a spiral galaxy.* (a) The various parts of a spiral galaxy (and in particular those of the Milky Way), as they would appear to an observer located in the plane of the disk. (It therefore represents a galaxy seen from the side.) Young stars are found in the thin, flat disk. In the Milky Way system, the Sun also lies in the galactic disk, about two-thirds of the way out to the edge. The oldest stars are found in a spherical region at the center of the galaxy, known as the bulge. They are also found in the halo, which, together with the globular clusters, surrounds the whole galaxy. (b) A spiral galaxy (NGC 4565), seen from the side. The dark band crossing the galactic disk from one side to the other is a band of interstellar dust. It appears dark because the dust absorbs stellar light. (c) Also a spiral galaxy (Messier 51), but this time seen face on (as it would appear to an observer above the galactic plane). It is easy to pick out the central bulge and the exquisitely shaped spiral arms, which are delineated by young stars (the bright regions) and dust (the dark areas). At the end of one spiral arm there is a dwarf galaxy, which is gravitationally inter-acting with Messier 51 (photographs: Hale Observatories).

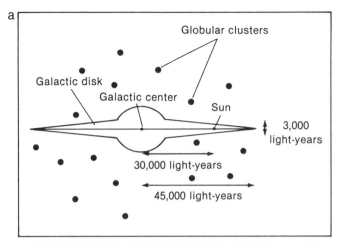

a

Globular clusters

Galactic disk

Galactic center

Sun

3,000
light-years

30,000 light-years

45,000 light-years

b

c

distinctly spherical, and consist of some hundreds of thousands of stars that are gravitationally bound together (Fig. 19). There are about 100 of them in the Milky Way. Using the Cepheids, Shapley determined the distances of the globular clusters and their distribution in space. The globular clusters were found to be spread throughout a large spherical region, but, surprisingly, the center of the sphere did not coincide with the position of the Sun. Instead it was some tens of thousands of light-years away in the direction of the constellation of Sagittarius.

Once again, the phantom of Copernicus appeared to haunt astronomers. Might the Solar System not be the center of the universe? Shapley decided that, to account for the distribution of globular clusters, this was the only possible solution. The Sun, rather than being at the center of the Milky Way, was relegated to its far-off suburbs. After having been assigned an ordinary brightness, the Sun was now found to occupy a perfectly ordinary position. Assuming that the system of globular clusters defined the extent of the Milky Way, Shapley deduced the Galaxy's diameter to be 300,000 light-years, and placed the Sun at a distance of 50,000 light-years from the galactic center. We now know that these values are exaggerated. Shapley, like Herschel, could not know that interstellar dust absorbed light from the Cepheids, causing them to appear fainter. What he thought to be an effect of distance was actually caused by the absorption of light by dust.

In the universe as we understand it today, the Milky Way galaxy is in the form of a disk, 90,000 light-years in diameter. This disk is very thin, its thickness being only about one-hundredth of its diameter. The several hundred billion stars of which it consists orbit the center of the disk, the galactic center (Fig. 20). The Sun lies about two-thirds of the way from the galactic center to the edge, at a distance of 30,000 light-years from the center. It carries the Solar System with it through space, at a velocity of 230 kilometers per second, in its orbit around the galactic center, which it completes once in every 250 million years. Since its birth, some 4.6 billion years ago, the Sun has circled the Galaxy 18 times.

The Extragalactic Universe

The limits of the Milky Way were thus finally established. The size of the Solar System had been reduced to one billionth of that of the Galaxy. The effort needed to do so has been prodigious, because measuring the Milky Way from our own small corner of the Earth is actually comparable to the task an amoeba would have in measuring the size of the Pacific Ocean!

But the work was far from finished. A fundamental question remained to be answered: Did the universe consist of only the Milky Way, or did it extend much farther? Were there other comparable systems beyond the latter's limits? Did Kant's "island universes" actually exist? Shapley thought that the universe consisted solely of the Milky Way. The nebulous spots that dotted the sky doubtless formed part of it. Paradoxically, the man who had dethroned the Sun from its central position in the Galaxy had forgotten the phantom of Copernicus. The center of the universe was now occupied by the Milky Way. Shapley had good reasons for this view, because he was convinced that the Milky Way was extremely large (150,000 light-years in radius). The distance of the Magellanic Clouds, obtained from Cepheid measurements, was also 150,000 light-years. The Magellanic Clouds and, by extension, all the other nebulous patches, must lie within the Milky Way. The latter was alone in the universe.

Shapley's universe was not unanimously adopted. Some workers thought that he had made mistakes in his distance estimates, that the Milky Way was much smaller, and that the spiral nebulae were not part of the Milky Way, but were galaxies like our own. The debate on the nature of the nebulae raged throughout the early 1920s.

The solution was found by Edwin Hubble, an American astronomer and former lawyer, who left the courtroom to devote himself to the study of the stars. In 1923, using the newly built telescope on Mount Wilson, he was able to resolve the light from the great nebula in Andromeda* into myriad stars, some of which were Cepheids. These flung open the doors to the world beyond our Milky Way. In fact, they gave a distance of 900,000 light-years for the nebula.[5] Even accepting Shapley's incorrect size for the Milky Way (300,000 light-years), the nebula was obviously far beyond it. The Andromeda nebula had been shown to be a galaxy, a twin to our own. Suddenly the universe became populated with innumerable galaxies. Kant's island universes were a reality. The universe grew more and more, and soon the Milky Way would become lost in the immensity of the universe, just as the Solar System had been lost within the vastness of the Milky Way. Copernicus' phantom had triumphed again. The Milky Way had lost its unique nature.

5. This distance is actually too small by a factor of about 2, because the Cepheids in the Andromeda galaxy are intrinsically 4 times as bright as those in the Milky Way that had been used to calibrate the Cepheid period–luminosity relationship. The true distance of the Andromeda galaxy is 2.3 million light-years.

| 3 |

The Actors in the Drama: The Galaxies and the Space–Time Couple

The Fleeing Galaxies

Just as a galaxy consists of stars, the universe consists of galaxies. To understand the universe, we need to study galaxies. Edwin Hubble set about this task with passion, using the 100-inch (2.5-m) reflector on Mount Wilson. His first task was to unravel the motion of galaxies. Making use of the spectroscope* and the Doppler effect to split and analyze the light from galaxies, he soon confirmed a strange fact, which had already been noted in 1923 by the American Vesto Slipher, working at the Lowell Observatory in Arizona. Out of 41 galaxies examined, 36 showed a shift toward the red: They were receding from the Milky Way. Only five showed a shift toward the blue. The evidence indicated that the motion of galaxies was not random. If that had been the case, on average, half of the galaxies would have been approaching our Galaxy, and the other half receding. Yet the majority of galaxies appeared to be fleeing away from the Milky Way, as if the latter had the plague.

In addition, this motion of recession by the galaxies was not completely random; it was ordered. After his success with the Andromeda galaxy, Hubble had continued to search for Cepheid variables, the cosmic "standard candles," in other galaxies, to determine the distances to these "island universes." In 1929, Hubble found a relationship between the velocities, deduced by measuring the shift in the color of the light, and the distances, a discovery that was to be a decisive step in our understanding of the universe: the velocity of recession of a galaxy is proportional to its distance. (This is known as "Hubble's Law*"). A galaxy twice as far away recedes at twice

56

the velocity. Moreover, the recession of the galaxies was the same in all directions. Wherever galaxies were examined, anywhere on the sky, they were fleeing in the same manner. This is known as an *isotropic motion*. The expanding universe had been born.

A Universe With a Beginning

Another profoundly important consequence of the fact that the recession velocity of a galaxy varies in proportion to its distance—the greater its distance, the faster it recedes—is that the universe had a beginning. Because of the proportionality between distance and velocity, each galaxy has taken exactly the same time (equal to distance divided by velocity) to reach its current position from its point of origin. If the sequence of events is reversed, the galaxies would all meet at a single point at a specific time in the past. This led to the idea of an initial, enormous explosion, the "Big Bang," which gave birth to the whole expanding universe.

Imagine yourself on the top of the Arc de Triomphe in the center of the Place de L'Etoile in Paris, where you can contemplate the magnificent vista of several of the most beautiful avenues in the world converging toward you. You notice several people jogging along the avenues, all heading away from the Arc de Triomphe and amuse yourself in estimating their distances and speeds. The first jogger, who is heading down the Champs Elysées, is 10 meters away, and is running at a speed of 1 meter per second. The second, on the avenue Foch, is 20 meters away, and his speed is 2 meters per second. The third, who is tearing down the avenue de la Grande Armée at a speed of 5 meters per second, is also the farthest away, 50 meters from the center. You deduce that the speed of each runner is proportional to his distance, and conclude that 10 seconds earlier all of them were at the same place, under the Arc de Triomphe. In just the same way, astronomers deduce that some 15 billion years ago, all the galaxies were together at one location. With the Big Bang, the universe lost its eternity and acquired a beginning. The notion of creation, which had been introduced into cosmological thought in the thirteenth century through religion by Thomas Aquinas, found an unexpected scientific support some 7 centuries later. After having taken diverging paths in the nineteenth century, religion and science seemed to be tentatively coming together again.

A Universe Without a Center

"But," you may say, "all the galaxies are rushing away from us, so the Milky Way must be the center of the universe." After having dethroned the Earth and the Sun from their central position, had the phantom of Copernicus finally failed in his task? Is Mankind the center of the universe after all? We must, alas, quickly renounce any such idea. The universe is arranged in such a way that the inhabitants of every galaxy would see exactly the same cosmic landscape; they would all see galaxies receding from them and have

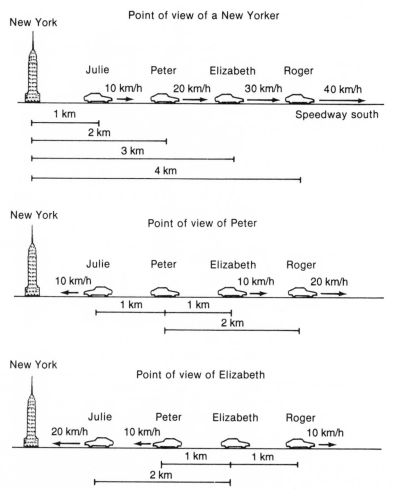

Fig. 21. *Hubble's Law on the freeway.* Heading out of New York, the cars behave in a very specific fashion: The farther from New York, the faster they move. (Galaxies behave in just the same way: Hubble's Law states that the more distant they are, the faster their recession velocity.) Any driver will then see exactly the same phenomenon: All the other cars will recede at a velocity that increases by 10 km/h for every kilometer separating them. Similarly, every galaxy in the universe sees all the other galaxies receding from it at a velocity that increases by about 25 kilometers per second for every million light-years between them. Just as there is no privileged observer among the drivers on the freeway, there is no center or privileged position in the universe.

the same illusion that they are at the very center of the universe. Because the center is everywhere, there is no true center. How does the universe create such an illusion? How does it perform this magical trick? Very simply, this is just another result of the fact that the recession velocity of a galaxy is proportional to its distance.

To illustrate this, imagine a stream of cars leaving a city, say New York, at dusk. The speed of each car is greater the farther it is from New York. Julie, only 1 kilometer away, is moving slowly, at 10 kilometers per hour (kph). Peter is twice as far away, at 2 km, and he is driving at 20 kph. Elisabeth, at 3 km, is moving at 30 kph, and Roger at 4 km, at 40 kph. Suppose also that all the drivers have a radar device and know how to use the Doppler effect to determine how fast other vehicles are approaching or moving away, and that they also know how to estimate distances of other cars from the brightness of the head- and tail lights. Peter sees that Julie's car, which is 1 km behind him, is receding at a velocity of 10 kph. Elisabeth, 1 km in front, is receding at 10 kph, while Roger, 2 km in front, is receding at 20 kph. Elisabeth observes exactly the same sort of motions as Peter: she sees Peter and Roger, both 1 km away, both receding at 10 kph, while Julie, 2 km away, is receding at 20 kph. Each driver sees the others receding at a speed which increases by 10 kph for every kilometer of distance (Fig. 21). In just the same way, the inhabitants of every galaxy see all the other galaxies receding from them at a velocity that increases by about 25 kilometers per second for each million light-years of distance. The galaxies are not fleeing from the Milky Way galaxy alone. They are all fleeing from one another. We do not occupy a privileged position. The phantom of Copernicus has triumphed yet again.

A Space in Creation

When you read "the universe is expanding," you must not imagine billions of galaxies hurled through an empty, static immutable space, which has existed for all time, and which was present before the big bang. You must not ask where, in this unchanging space, you can find the famous point from which everything was born, the location of the primordial explosion. This question, which could be asked in a Newtonian universe, is devoid of meaning in the Big Bang universe. Space did not exist before the Big Bang. Since the primordial explosion, it is constantly being created. Space in the Newtonian universe was static and unchanging. Bathed in an invisible substance named "ether," which serves to transmit forces, space was nothing more than a theatrical stage where the cosmic drama unfolded, with the planets, stars, and galaxies as actors. In the Big Bang universe, space is no longer passive. It relinquishes its role of spectator to become actor in the cosmic drama. From being static, it becomes dynamic. In the new universe, galaxies are not hurled through a static space; on the contrary, it is an expanding space that is pulling along galaxies fixed to it.

Think about a raisin cake that you put in the oven. As the dough expands, the surface of the cake increases and the raisins inserted in the dough move apart from one another. Or take the surface of a balloon on which are glued paper decorations. As air is blown into the balloon, its surface increases and the raisins move apart from one another. In the same manner as the raisins are encrusted in the cake or the paper decorations are stuck on the surface of the balloon, the galaxies are fixed in space. All the motions come from the expanding surface of the cake or the balloon, just as it is space which is in expansion. Just as the recession speed of galaxies increases in proportion with distance, the raisins and paper decorations will move away faster, the farther apart they are from one another.

As time passes, space, which was infinitesimal when created, grows larger and larger. The separations between galaxies increase more and more. After 15 billion years of evolution, from the epoch of galaxy formation to the present, the separation between two galaxies not bound together by gravity has grown a thousandfold.

A Dynamic Space

With the publication of his new theory of gravitation, known as general relativity*, in 1915, Albert Einstein announced the end of Newton's static space, and its replacement by a dynamic space. The Moon orbits the Earth. According to Newton, it serenely follows its elliptical orbit, thanks to the balance between two opposing forces: the gravitational force pulling it toward the Earth, and the centrifugal force, acting away from the Earth, that arises from its motion. The Newtonian universe is a world of forces, transmitted by a mysterious substance known as the ether, which fills a passive space. In Einstein's universe, the forces disappear. No longer is there any need for an ether. Space becomes active; it controls the situation and dictates the motions. The Moon follows its elliptical orbit, because that is the only possible trajectory in the curved space that has been created by the Earth's gravity. Einstein freed space from its rigid straightjacket. Space became elastic: It could be stretched, contracted, deformed, curved, or contorted, following the whims of gravity. And it is the final shape of space that determines the motion of objects or of light crossing it. The path of light passing near the Sun is curved, because the Sun's gravitational field distorts space around it (see Fig. 25). A black hole, which is the result of the collapse of a massive star (a few tens of solar masses), and which goes from an initial radius of several hundreds of millions of kilometers to a final one of less than 20 km, has such a powerful gravitational field that space is completely warped, preventing any light whatsoever from escaping.

The dynamical nature of space presented Einstein with a major dilemma. Space throughout the universe should be in motion. It should be either expanding or collapsing, just as a stone thrown in the air must either rise or fall. The concept of a static universe would be like having a stone hovering motionless in midair. At that time, however, all the observations seemed to indicate a static universe. Einstein therefore had to include an "antigravitational" force into his theory to cancel out the effect of the universe's own gravity, which made it expand or collapse. This antigravitational force is not observed in laboratory experiments nor in the motion of the planets, but it is not prohibited on the scale of the universe. Fourteen years later, in 1929, when he learned of Hubble's discovery of the expansion of the universe, Einstein undoubtedly regretted not having had sufficient confidence in his own equations, and not having defended his dynamic universe more vigorously in the face of observation. He described the introduction of that antigravitational force as "the greatest mistake of my life."

A Universe Without Limits

What would happen if the universe had limits and, standing near its edge, I were to throw a javelin over the border? Would the javelin return into our universe, or would it be lost somewhere outside? This question was asked by the Greek philosopher Archytas of Tarentum in the fourth century B.C. (Fig. 22). Twenty centuries later, the Englishman, Thomas Digges, and the Italian, Giordano Bruno, gave the only sensible answer to Archytas' question. The universe could not have any limits. The situation described by Archytas could not exist. There was no cosmic boundary, and nothing be-

Fig. 22. *Does the universe have a boundary?* This nineteenth-century drawing (by Camille Flammarion in his *Popular Astronomy*) of a medieval engraving wonderfully illustrates one of the oldest questions: Does the universe have a boundary? If we were able to look "through to the other side" (the sphere of the fixed stars formed the ultimate limit of the medieval world), what wonders would we discover? We now know that the universe, finite or infinite, has no boundaries (photograph: Bibliothèque Nationale, Paris).

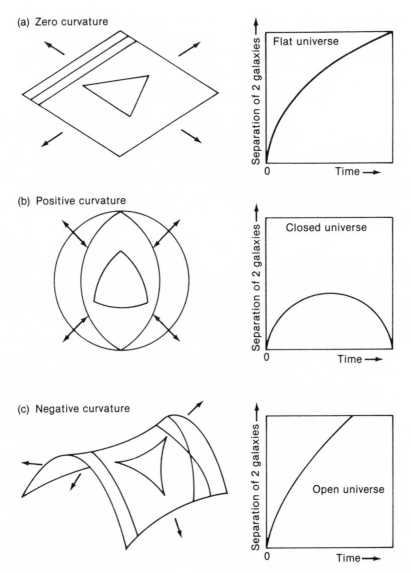

(a) Zero curvature

Flat universe

Separation of 2 galaxies

0   Time →

(b) Positive curvature

Closed universe

Separation of 2 galaxies

0   Time →

(c) Negative curvature

Separation of 2 galaxies

Open universe

0   Time →

Fig. 23. *The curvature of the universe.* This diagram shows, by analogy, the three types of curvature that the universe may possess. (The analogy is not perfect, because the universe's three-dimensional space is being represented by two-dimensional surfaces.) (a) The geometry of flat space with zero curvature is represented by a flat plane. This geometry was studied by Euclid: through any given point it is possible to trace just one line parallel to a given straight line, and the sum of the angles in a triangle is 180 degrees. The expansion of a flat universe would cease only after an infinite amount of time. (b) The surface of a sphere is used to represent the geometry of space with positive curvature. This geometry was studied by Bernhard Riemann: Straight lines meet at the poles; so it is impossible to draw a line parallel

62

yond, and any thrown javelin would necessarily remain in the universe, where its motion could be described. In Giordano Bruno's universe, Euclidean geometry reigned supreme. In this geometry, space was flat, two parallel lines never meet, and the sum of the angles of a triangle is always equal to 180 degrees. An unbounded Euclidean universe was, perforce, an infinite universe. Giordano Bruno paid for the discovery of the infinite universe with his life.

Giordano Bruno could have perhaps escaped from his obsession with infinity and saved his life had he known of the existence of non-Euclidean geometries. In these geometries, which were mainly developed by the German mathematician Bernhard Riemann in the nineteenth century, the universe could be without boundaries, and yet remain finite. If that seems impossible, just picture yourself as Magellan or as Jules Verne's Phileas Fogg, making several voyages around the Earth. You can go round and round as many times as you like, and you will never encounter any boundaries to prevent you from going any farther. Yet, despite that, the surface of the Earth is finite. This is possible because the Earth is not flat, but curved. Euclid is at a loss to describe a curved surface, and Riemann has to be called to the rescue. On Earth, parallel lines, such as the circles of longitude, no converge, in this case, at the North and South Poles. The tips of the Great Pyramid of Cheops in Egypt, the Eiffel Tower in Paris, and the Washington Monument, in Washington D.C. form the corners of a vast triangle on Earth. If you were to measure the three angles in this triangle and add them together, you would find a sum greater than 180 degrees. The curvature of the Earth is said to be "positive." Not all surfaces have "negative" curvature. On such a surface, parallel lines diverge, and the sum of the angles of a triangle is less than 180 degrees (Fig. 23).

These results may be applied to three-dimensional space. The universe may be flat (without curvature). But it may also have either positive or negative curvature. Imagine that in the dark of the night you shine a beam of light up into the sky from a powerful torch light. If the universe we live in is flat, the light will be lost in the infinite depths of space. If our universe has positive curvature, it would theoretically be possible for you to see the beam of light returning after having made a complete circuit of the universe, like

---

to a given line. The sum of the angles of a triangle is greater than 180 degrees. The expansion of a universe with positive curvature would cease at some time in the future, and it would then collapse on itself. Such a universe is said to be "closed." (c) The surface of a saddle is used to illustrate the geometry of space with negative curvature. This type of geometry was also studied by Riemann: Through any given point it is possible to trace a multitude of lines parallel to a given line (parallel lines being defined as lines that never meet), and the sum of the angles in a triangle is less than 180 degrees. A universe with negative curvature would expand forever. It is said to be "open."

circumnavigators returning to their point of departure after having circled the Earth. The universe would be finite, or "closed." But if our universe is negatively curved, the light would travel forever. The universe would be infinite, or "open." In each of the three cases, the universe has no boundaries, and Archytas can rest in peace.

Light Does Not Bring Fresh News

The propagation of light is not instantaneous, and it takes a certain time to reach our eyes. This statement, which seems perfectly evident nowadays, was for a very long period the subject of considerable discussion by some of the finest minds. Aristotle (350 B.C.) thought that objects instantaneously became luminous the moment they were exposed to a source of light. Twenty centuries later, Descartes still defended the instantaneous propagation of light through the ether. It was only in 1676 that light was found to have a finite velocity. The Danish astronomer Olaüs Römer, working at the Paris Observatory, was studying the eclipses of Io, one of the satellites of Jupiter. He noted that there was a variation between the observed and predicted times when Io disappeared behind Jupiter, and that the observed times were later when Jupiter was farther from the Earth, and earlier when it was closer. Römer interpreted these observations correctly as proving that light did not propagate instantaneously. The delay was caused by the additional time required for the light from Io to travel the greater distance to the Earth. The velocity of light that Römer calculated was very close to the value of 300,000 kilometers per second that is accepted today. This velocity is about one million times as fast as that of sound, which explains why during a thunderstorm you see a flash of lightning long before you hear the roll of thunder.

According to Einstein's theory of relativity, the velocity of light is the maximum velocity that can occur in the universe. It may seem immense. In 1 second, light can travel 7.5 times round the circumference of the Earth. Despite this, however, in terms of the overall scale of the universe, it creeps along like a tortoise. It takes 8 minutes to bring us news from the Sun; in other words, we see the Sun as it was 8 minutes ago. Similarly, the star closest to the Sun looks to us as it was 4.2 years ago, and the closest galaxy (the Andromeda Galaxy) as it was 2.3 million years ago, when the first men were appearing on Earth. We see the Virgo cluster of galaxies* as it was 40 million years ago; and the quasars*, those tiny specks of light in the very distant universe, as they were some 10 billion years ago, long before the Sun and the Earth formed. The news that came to us from farther and farther out into space loses more and more of its freshness. Waves from the distant past engulf from all sides our tiny island of the present. Astronomers are time travelers; with their telescopes, they can go back into the past of the universe.

Ten New Galaxies Every Year

Standing on the bridge of a ship in mid-Atlantic, a sailor looks out to sea: No land is in sight, and there is nothing but water out to the distant horizon. The sailor's sight is limited by the oceanic horizon, and in a similar way, astronomers' vision cannot extend beyond the cosmological horizon*. This fundamental limit is not determined by our telescopes, but by the twin factors of the finite velocity of light and the finite age of the universe.

The universe came into being, as we shall see later, about 15 billion years ago. The surface of a sphere, centered on the Earth and with a radius of 15 billion light-years, is our cosmological horizon. Regions of the universe beyond that sphere have not yet been able to communicate with us. The news carried by the light they have emitted has yet to reach us.

Whether it is finite or infinite, the universe only slowly reveals its immensities. As time goes by, our horizon sphere expands, and we are able to see a little farther. About ten new galaxies appear within our field of vision every year.

Why Is the Sky Dark at Night?

The simplest events are often the most revealing ones. They only need to be perceived with the right insight. The fall of an apple unlocked the secrets of universal gravitation to Newton. The fact that the sky is dark at night contains in it the beginning of the universe.

As early as 1610 Kepler puzzled over the secret of the dark sky at night. If the universe were truly infinite, he reasoned, the night sky, when the Sun is illuminating the other side of the Earth, should be just as bright as in the day. An infinite universe ought to contain an infinite number of stars that are as bright as the Sun. Just as in a dense forest where the view is blocked by the innumerable trunks of trees, any line of sight should, sooner or later, encounter a star, somewhere in the "stellar forest." The night sky should, therefore, be as bright as the Sun (Fig. 24). Kepler concluded that the dark sky meant that the universe was not infinite.

In 1687, when Newton again adopted an infinite universe so as not to have universal gravitation collapse everything into a single, central, enormous mass, the problem of the darkness of the sky at night reappeared. The German astronomer, Heinrich Olbers, taking up an idea proposed by the Swiss, Jean-Philippe de Cheseaux, suggested in 1823 that starlight was absorbed during its journey through space. The sky was dark because not all starlight could reach us. This explanation could not be the correct one, however, because all the energy that was absorbed should be reemitted. Light could not be lost. The enigma, which has now come to be known as "Olbers paradox," remained unsolved.

With the Big Bang universe, the mystery of the night sky finally found a solution. It is dark at night because there are not enough stars to fill the

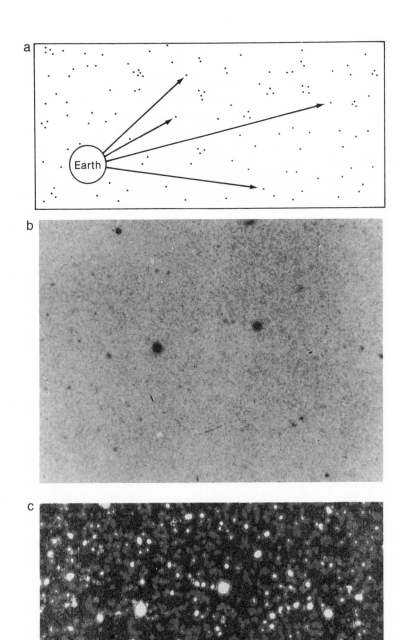

whole of the sky with light. The number of stars is limited, not because the universe has limits, as Kepler believed, but because we are unable to see the whole universe. Because the latter had a definite beginning, and because light does not propagate instantaneously, only the light from stars within our cosmological horizon reaches us. In addition, the number of stars is limited, because they are not eternal. The lifetime of very bright stars is short relative to the age of the universe. A few million years, or perhaps a few billion, and they are gone. Finally, the expansion of the universe adds its own small contribution. As the distances between the galaxies increases, light has to do more work to reach us. It loses energy and is shifted toward the red. The amount of luminous energy within our cosmological horizon decreases.

The next time you admire the starry sky on a fine, dark night, tell yourself that you are only able to enjoy this marvellous sight because the universe had a beginning, and because the lifetimes of bright stars are short.

Fig. 24. *Olbers' paradox: Why is the sky dark at night?* (a) In 1823, Heinrich Olbers asked himself the following question: Why is the sky dark at night? If the universe were infinite and filled with an infinite number of stars, our line of sight should always encounter a star, in whichever direction we look. The night should therefore be as bright as the daytime sky. Obviously, this is not the case. (The question would have been more accurate if it had been expressed in terms of galaxies, rather than stars. But Olbers did not know of the existence of galaxies.) We now know the explanation of this paradox: The universe had a beginning, and the number of galaxies from which light can reach us is not infinite. Large modern telescopes, equipped with the most sophisticated electronic detectors, are able to detect extremely faint, and therefore distant, objects. They are thus able to look far out into the universe. By pushing the observations to fainter and fainter limits, the telescopes reveal a sky filled with more and more sources of light. (b), (c) (by J. A. Tyson) The dramatic difference when the observational limit is extended. (b) A photograph of a particular area of sky, where the faintest objects have about one-millionth of the brightness of the faintest star visible to the naked eye. Only about a dozen stars and galaxies are visible. (The two circular images close to the center are both stars.) (c) The same area of sky, but this time photographed with a much greater sensitivity: The faintest object now has a brightness one-billionth of that of the faintest object visible to the naked eye. Once again, the brightest objects are the two stars near the center, but now the sky is almost entirely covered with sources of light. More than 1200 can be seen on this photograph, which would correspond to a figure of about 150,000 light-emitting objects over an area 1 degree square. Almost all of these sources are galaxies, each of which contains some 100 billion suns. A small number are probably not real galaxies, but galaxy images caused by nearer galaxies acting as gravitational lenses (*see* Fig. 54). Despite the fact that here light has invaded most of the picture, there are still, however, areas that are devoid of galaxies. It is these regions that cause the sky to remain dark. Conversely, the very fact that the sky is dark at night implies that the number of visible galaxies cannot rise indefinitely as the power of telescopes increases. If they were able to reveal yet fainter and fainter galaxies without limit, we would again be faced with Olbers' paradox.

Elastic Time

The greater their distance, the faster galaxies are receding. A telescope that was powerful enough to see as far as our cosmological horizon would discover objects receding at 80%, 90%, 95%, and even 99% of the velocity of light. Like a racehorse that exerts all its energy in trying to reach a winning post that is continuously getting farther away on a racecourse that keeps lengthening, the light from receding galaxies loses its energy to reach us.

Imagine that the inhabitants in a faraway galaxy send a radio signal in our direction every second. The signals, when they leave the galaxy, are separated by an interval of 1 second. During the course of their travel, the distance separating them increases because of the expansion of the universe. When they arrive at the Earth, the time interval between them is considerably longer than 1 second. The more distant the galaxy, the longer this time becomes. One second's duration in some far away galaxy becomes an eternity for us. The second may change into an hour, a year, or a century, depending on the recession velocity of the galaxy. Time has lost its universal nature. As with space, time has acquired elasticity. It lengthens or shortens depending on the motion of the observer who measures it. The unique and universal time of the Newtonian universe has given way to a multitude of individual times, all different from one another, in the Einstein universe.

The Fountain of Youth

Jules and Jim are twins. Jules, adventurous by nature, leaves on a powerful spaceship that can travel at 87% of the speed of light. Jim, more of a stay-at-home, prefers to spend his life quietly on Earth. Before parting, Jules and Jim take care to synchronize their watches. Jules leaves on January 1, 1998. Ten years later, he returns. When he lands on Earth, the calendar on the spaceship indicates that the date is January 1, 2008. Back home with Jim, he looks at the calendar on the wall, and finds that the date is January 1, 2018. Jim is 20 years older, whereas Jules has aged only by 10 years. The difference in ages is real. Jim has more wrinkles and grey hairs than Jules. His heart has beat more times, he has eaten more meals, drunk more wine, and read more books.

By abolishing absolute time, Einstein has provided us with a fountain of youth. This fountain does not bring rejuvenation, but it does slow down the inexorable passage of time. To age more slowly, all one has to do is to travel very rapidly. Speed is the secret of prolonged youth. If Jules had traveled at 99% of the speed of light, he would have slowed down his aging by a factor of 7. While 10 years elapsed on board of the spaceship, 70 would have elapsed on Earth. Jules would probably not have found Jim alive when he returned. And if he were able to travel at 99.9% of the speed of light, Jules would have returned to Jim's great-great-grandchildren. He would have slowed down his aging relative to that occurring on Earth by a factor of 22.4.

His 10-year voyage would have lasted 224 Earth years, and he would have returned in the year 2222 (see Appendix B).

Einstein would have been able to suggest another argument to those who advocate the benefits of jogging. Those who jog slow down their rate of aging when compared with others. Naturally, this slowing down is infinitesimal. One second for someone running at 1 meter per second is equivalent to 1.00000000000000005 seconds for someone who is stationary. The first figure other than 0 only occurs in the eighteenth decimal place. The difference is so small that it would not be detectable with even the most sophisticated atomic clocks. Even if a person spent half his life (50 years) running, he would only gain one hundred-millionth of a second relative to the rest of us lazy people. In everyday life, the speeds that we reach in a car, ship, or plane are minuscule by comparison with that of light. Thus we all experience the same time because the time differences caused by such speeds are imperceptible, which is just as well: Think how many appointments would be missed and what chaos would result if time were elastic in our everyday lives!

The Space–Time Couple

Time is elastic and changeable. Space, as we have seen, also possesses these properties. Both may expand, contract, stretch, or shorten at will. It is not by chance that these two principal actors in the cosmic drama are so similar. They do, in fact, form an intimately linked pair, the changes in one being complemented by modifications in the other. When time expands—that is, when it elapses more slowly—space contracts. Jim, on Earth, notices that Jules, racing through space at 87% of the speed of light, not only ages at just half his own rate, but also that his twin's space is contracted. To Jim, Jules's spaceship appears to have shrunk to half its length (see Appendix B). In Einstein's universe, space and time are indissolubly linked. The mutual changes in space and time may be considered as a transmutation of space into time. Space that contracts is transformed into expanded time, which runs more slowly. The rate of exchange given by the cosmic bank is not very favorable. You will only obtain 1 second of time for every 300,000 km of space, but you have no choice. Time and space cannot be separated, as they could be in Newton's universe. Henceforward, the universe has four dimensions. The time dimension is added to the three spatial dimensions. To describe your coordinates in the universe, it is not sufficient to give just your position; you need to specify the time as well.

The Speed of Light Is Costly

Space is immense. The closest star is more than 4 light-years away. From one end of the Galaxy to the other lies a field of stars nearly 100,000 light-years long. Our twin, the Andromeda Galaxy, is more than 2 million light-years away. On the other had, our human lives are about 100 years at most.

By forbidding us to travel faster than the speed of light, Einstein made interstellar, and even more intergalactic, exploration quite impossible. "But," you may say, "surely Einstein made amends by giving us the means to put a brake on the passage of time, to slow down the inexorable approach of death? All Jules would have to do is to press hard on the accelerator, increase the speed of his spaceship and go even closer to the speed of light. That way, he would age more slowly, and would have more time to travel the vast reaches of interstellar space." Unfortunately, that is easier said than done. Everything has its price, and the price of traveling close to the speed of light is extremely high.

If Jules were to travel at 99% of the speed of light, he would age seven times more slowly, but the mass of his rocket would also be seven times as great (see Appendix B). The spaceship would therefore need to burn more fuel. The faster the rocket, the more massive it becomes, and the more fuel it requires. This vicious cycle is inescapable. A rocket traveling at the speed of light would have an infinite mass, and would therefore require an infinite source of energy, which is inconceivable. Slowing down time by moving rapidly costs too much and is impractical. Interstellar and intergalactic exploration remains a distant dream.

My Past Is Your Present, and Their Future

In a railway station on a stormy evening, lightning suddenly strikes both ends of one of the carriages of a train that is coming into the station. Françoise, who is standing on the platform, sees the lightning strike the front and rear of the carriage at the same time. She knows that it took a fraction of a second for the light from both flashes to reach her, but because at the moment lightning struck she was exactly the same distance from each end, the delay was the same for both flashes. She concludes that they hit the carriage at exactly the same instant. Paul, who is sitting in the center of the carriage, is extremely startled. After his alarm has passed, he tries to recall exactly what happened. He saw lightning strike the front of the carriage before the flash at the rear. Because the train is moving forward, it is approaching the light from the flash that hit the front of the carriage, whereas the light from the flash that hit the rear has to catch up with him. So there was a tiny fraction of a second's difference in the times. Barbara, who is in a train passing in the opposite direction, also saw what happened. At the crucial moment she was also at equal distances from each end of the carriage. But to her, lightning struck the rear of the carriage first, and then the front.

Who is right? Einstein gave the answer. They are all right. The lack of absolute time has caused the disappearance of concepts such as simultaneity, and of a universal past, present, or future. For Françoise, lightning struck the two ends of the carriage at the same time. But the "same time" for Françoise is not the same time for Paul or Barbara, who are moving relative to her. If she defines the instant when she saw lightning strike the front of the carriage as her "present," she also saw lightning strike the rear during the same "present." Paul, however, saw lightning strike the rear later, in other words, in Françoise's "future," whereas Barbara saw it in Françoise's "past."

Motion dictates the sequence of events that are separated in space, and determines which occurs in the past, the present, or the future. There is no longer a universal "now." Jim, looking at his watch and calendar on Earth, may wonder "What's happening now on Jupiter?" Jules, racing through space at top speed in his spaceship may ask himself the same thing, at the same time and date as shown by his own watch and calendar. But the "now" on Jupiter for Jules is different from the "now" for Jim. The difference is greater, the higher the relative velocities. The quasars, those highly luminous objects in the distant reaches of the universe, are receding at more than 90% of the speed of light. The quasars' "now" when I am walking may differ by thousands of years from the quasars' "now" when I stand still.

Can We Make an Omelette Before We Break the Eggs?

Relativity has completely shattered our concepts of past, present, and future. Motion rearranges the order of events. Does this mean that causality is dead? Can the effect come before the cause; the result before the action? Can we make an omelette before we break the eggs? Does the bullet hit the target before you fire the gun? Will the light from distant stars reach us before it is emitted? Was I born before my grandmother? Luckily for our sanity, the reply to all these questions is always "No." The chronological sequence of two events may be altered only if they are very close together in time, or very distant from one another in space, so that it is impossible for light to travel from one event to the other during the interval of time that separates them. For the past, present, and future to lose their identity, the two events cannot be causally linked by light, to prevent them from influencing one another. Françoise sees lightning strike both ends of the carriage simultaneously. Light does not have time to travel from one flash to the other, and the order of events may be altered by motion. Light has ample time to travel from the egg when it is cracked to the pan where the omelette is to be made, or to go from a rifle to the target before the latter is hit by the bullet. The order of events is thus the same for everyone. Thank goodness! I cannot be born before my grandmother.

Remembrance of Things Past

"Time passes; it flows", we often say. We think of time as water in a river flowing past us. From our stationary boat, anchored in the present, we watch the river of time passing us, carrying away the streams of the past, and bringing the waves of the future. We attribute a spatial dimension to time, and it is this representation of the motion of time through space that gives us the impression of the past, the present, and the future. Only the present exists "now"; only it has a palpable reality. The past has gone and is lost among our memories. Marcel Proust, in his masterful *Remembrance of Things Past,* has enchanted us with his portraits of young girls in bloom, and his description of the lingering memory of the taste of long-gone, delicious madeleines. The future, yet to come, only exists in our hopes and dreams.

We all carry this subjective, or psychological time along with us. The distinction between past, present, and future rules our lives, and even forms a fundamental part of our language, with the different tenses of verbs expressing past or future action. We are convinced that bygones are bygones that cannot be altered, whereas we like to think that we can affect the future by our actions. Unfortunately, this idea of the passage of time, of its movement relative to our stationary consciousness (or, equivalently, of our moving consciousness relative to stationary time), does not agree with the findings of modern physics. If time moves, what is its velocity? The question is obviously absurd. On the other hand, the idea that only the present exists, and that it alone is real, is not compatible with the breakdown of a universal, rigid time brought about by relativity. The past and the future should be as real as the present, because Einstein showed us how one person's past may be another's present, and even the future for yet a third.

To physicists, time is no longer defined by a series of events. The distinctions between past, present, and future are no longer valid. Every instant is equivalent. There is no longer any "privileged" instant. If I throw a ball in the air, all I need to know are its initial position and velocity, and I can calculate its path. This path will always be the same, whether the ball is thrown at six o'clock in the morning or eight o'clock in the evening, on January 1, 1988, or December 31, 1998. Because the concepts of past, present, and future have been abolished, time no longer needs to be moving. It does not flow. It is simply there, stationary, like a straight line that extends to infinity in both directions. The flow of psychological time has been replaced by the quiet inertia of physical time. The question of the velocity at which time flows no longer arises.[6]

Why is there such a difference between the two times? Probably because physics is not yet capable of describing biological and psychological processes. It is our mental activity that makes us feel that time flows past us. The secret of passing time lies within our heads. It will only be unraveled when we understand the nature of sensation, thought, and creativity.

The Arrow of Time

A child is born, grows up, ages, and eventually dies. This sequence applies to all of us. The march of time is inexorable, and always takes place in the same sense, the same direction. Time carries us from the cradle to the grave, and refuses to operate in the opposite direction. The past has gone inevitably, while the future is yet to come. The past can never come after the future.

6. See also P. C. W. Davies, *God and the New Physics,* New York, Simon and Schuster, 1983.

Like an arrow that flies straight through the air once it has left the bow, psychological time always moves forward. It is irreversible.

This irreversibility of time, responsible for our fear of death, is nevertheless absent in the world of the particles that make up matter. At the subatomic level, time is no longer unidirectional. The arrow of time disappears, and time may flow in both directions. If you were to make a film of events in the subatomic world, and then projected it backwards, you would be unable to tell the difference. Two electrons fly toward one another, collide, and fly off again. Reverse the sequence of events, and you will still have two electrons flying toward one another, colliding, and flying off. The physical laws that describe these events, with a single small exception, are not subject to a particular direction of time. They apply equally in both directions.

The small exception concerns a subatomic particle with no electrical charge, called the K meson, or kaon. The subatomic world is a constantly changing, impermanent world. Most of the particles have a very brief existence. They come and go in a twinkling. The kaon has a lifetime of less than one millionth of a second. In disappearing, it turns in more than 99% of the cases into three other particles. This disintegration is time reversible. The three particles may recombine to give a kaon. But in less than 1% of the cases—and this is the nub of the matter—the kaon disintegrates into just two particles. This situation is not time reversible; it can occur only in one direction. The disintegration of kaons has defined a "tiny" arrow of time. Tiny, because the kaon is the only one of the thousands of different subatomic particles to show such a characteristic, because the disintegration into two particles is itself very rare, and also because kaons are not present in the matter that forms our bodies and makes up the galaxies and the universe. They appear only as a result of violent collisions between particles produced in large accelerators, moving near the speed of light. This tiny arrow of time does not appear to play any important role, but its significance remains unclear.

Although generally absent in the subatomic world, the arrow of time is all too apparent in the macroscopic world. Psychological time flows, as we have seen, from birth to death. Physical time acquires also a well-determined direction. Films of the macroscopic world cannot be projected in both directions.

A hot cup of tea cools down. An ice cube melts in the heat of the Sun. A drop of ink disperses into a glass of water. The walls of a ruined Gothic cathedral fall into disrepair and crumble into a thousand ruins. All these situations imply a direction for the passage of time. The opposite never occurs. A cup of tea does not heat up again on its own, and the water from the melted ice does not spontaneously turn into an ice cube. Neither do the particles of ink reassemble themselves into a drop of ink inside a glass of clear water, nor the pieces of stone gather to restore the cathedral to its ancient splendor.

In all these situations, the initial state is more organized than the final one. The structure in crystal of the ice is more ordered than the pool of water produced when the ice melts. The organization of a fine Gothic cathedral is far greater than the pile of rubble that it becomes. The information content decreases. I need far more words to describe a cathedral than I do to describe a pile of stones.

The cup of tea cools because the air surrounding it is cooler than the tea. There is a lack of equilibrium between the temperatures of the air and of the tea. The tea will cool and the air will grow warmer until there is no longer any difference in temperature between the two, until a balance exists between them. Imbalance decreases in favor of balance. The temperature of an object is governed by the agitation of the atoms and molecules of which it consists. The tea is hot because molecules of heated water are violently agitated. Their movements are disordered. The air is cold because the molecules of air are "quieter." Their motions are more ordered. The disorder of the molecules of water begins to affect the more ordered state of the molecules of the air, and their disorder increases until the temperatures are equal. The final situation contains less information than the initial state, because I can describe it with a single temperature rather than having to give two. Just as moving from the past to the future, from birth to death, defines the direction of psychological time, so does the transition from organization to disorganization, from more information to less information, and from imbalance to equilibrium, define the direction of physical time. Physicists encapsulate "disorder," "less information," and "less imbalance" in the term "entropy*." The principle that defines the direction of physical time may therefore be expressed as: "Entropy always increases." Disorder must increase, information must become lost, and any disparity tends to disappear. This principle is known as the Second Law of Thermodynamics, a science that studies the properties of heat. It was discovered in the previous century, during the Industrial Revolution, when attempts were being made to improve the efficiency of steam engines.

The Miracle of the Broken Plate

How did nature come to impose a direction of time on the macroscopic scale, when no such direction exists in the subatomic world? After all, macroscopic objects are made of subatomic particles. How could the whole acquire a property that its component parts do not possess? The answer lies in the large number of particles that make up the macroscopic world and in their mutual interactions.

A gram of water contains 1 million billion billion (10^{24}) atoms. Any attempt to relate the macroscopic world to the subatomic world always produces numbers of this order. It is absolutely impossible for us to follow the individual behavior of such a large number of atoms. We are able to gain an idea of their average behavior only by means of the laws of statistics and of probability. If I flip a coin into the air, I cannot predict whether it will land heads or tails. But the laws of statistics tell me that if I toss it a large number of times, it will, on average, fall heads or tails an equal number of times. The same applies to the law of entropy. It is a statistical law. On average, entropy should increase. But, in principle, entropy may decrease, with an increase in order, information, and lack of equilibrium, and with a reversal of the direction of time—just as, in principle, the coin could land heads up 100 times in a row. The arrow of time exists because the probability of such an event taking place is so small that it never occurs.

Let us imagine that I drop a plate, which breaks into 1000 pieces. The laws of statistics do not, in principle, forbid the opposite sequence of events from occurring. The statistical fluctuations of the molecules of air in the room could act in such a fashion that the pieces were pushed together to recreate the plate. Variations in temperature could occur along the breaks and weld the pieces together again. Other fluctuations could act to create a flow of air that lifted the repaired plate back into my hand. If you see such a sequence of events occurring, you would invoke a miracle—and you would be right. The probability of such an event is extremely small, so small that it is practically zero. It is about $1/10^{10^{25}}$. This number, $1/10^{10^{25}}$, is the figure 1 preceded by 10^{25} zeros. I could start writing that number: $1/10^{10^{25}} = 0.00000 \ldots$, but I should soon have to stop, because the number of zeros is so large that, even if I filled every page of every book in the whole world, I would never reach the end of the number.

A broken plate will never repair itself on its own. Plates are not made by miracles, but in factories. Although the atoms and molecules of which they consist are indifferent to the direction of time, their collective effect determines an extremely precise direction. The universe must flow from order to disorder, from a lack of equilibrium to equilibrium, from an abundance to a lack of information.

Light Cannot Travel into the Past

A stone falls into a pool of water and creates beautiful, concentric ripples that spread out from the point where the stone entered the water toward the edges of the pond. Once again, we are witnessing a phenomenon that is irreversible in time. We will never see the water organize itself in concentric ripples that converge on a point in the middle of the pond. The film cannot be run backwards. Like all undulatory phenomena, waves of light diverge from the source that produces them, and do not converge on it. Light propagates toward the future, not toward the past. The radio waves emitted by a police radar that are reflected from the back of your car return a fraction of a second later, not earlier. The Sun appears as it was 8 minutes ago, not as it will be in 8 minutes' time. Although the equations developed by the Scottish physicist James Clerk Maxwell describing the propagation of light are reversible with time, light propagates only into the future. We cannot communicate with the past. We cannot send a radio message to Eve to tell her not to eat the forbidden fruit, nor to our grandparents to prevent their meeting, and thus our birth. Causality is preserved. The electromagnetic arrow—light is an electromagnetic phenomenon—indicates the same direction of time as the psychological and thermodynamic ones. It must flow from the past toward the future. Time may slow down, but it can never change direction. Which is lucky for us, because it would be extremely difficult to communicate with people whose direction of time was reversed: they would know exactly what you were going to say before you said it, but would forget it immediately once the conversation finished.

Stars Are Machines for Creating Disorder

The fact that the three arrows of time all point in the same direction is probably not by chance. Physicists feel intuitively that their common direction has been imposed by the expansion of the universe, but the question is still far from being understood. So far, only the thermodynamic arrow has revealed its kinship with the cosmological time defined by the expansion of the universe.

At its origin, as we shall see, the universe was a homogeneous, uniform mixture of radiation and elementary particles. It possessed no structure. It was completely disordered, and its information content was very small. After about 15 billion years, hundreds of billions of galaxies have appeared, each containing hundreds of billions of stars. On a planet orbiting one of those stars, in one of those galaxies, human consciousness, capable of wondering about the universe, has appeared. Structure has emerged from a lack of structure, order from disorder, and complexity from simplicity. At first sight, this course of events appears to contradict the Second Law of Thermodynamics: Entropy seems to have decreased instead of increasing. Doesn't the universe have a thermodynamic arrow?

This is where the expansion of the universe and the cosmological arrow of time come in. The expansion cools down the radiation that bathes the universe. The temperature of a particle of light, a photon, is determined by the amount of energy it is carrying. As the distances between the galaxies increase, light loses more and more energy to reach us, and the universe cools down. From the temperature of several million Kelvin (K) that it had 3 minutes after the birth of the universe, the universal radiation now has a temperature of only 3 K.[7]

In contrast to this frigid temperature, the temperature in the cores of stars is several million K. There is a temperature imbalance between stars and the space surrounding them. Just as a cup of hot tea cools when in contact with the colder air around it, and communicates the disorder of the

---

7. In this book I shall use the Kelvin temperature scale* (named after the English physicist who established it). This scale is more convenient to use for measuring physical phenomena than the Celsius scale, because it embodies the idea that the temperature of a particle is described by its motion. The zero point of the Kelvin scale (0 K, equal to $-273°C$) is absolute zero, where all motion ceases.

molecules of water in the tea to the molecules of air, increasing the total amount of disorder, stars emit their hot radiation into the cooler radiation surrounding them and increase the total disorder of the universe. The stars are machines for creating disorder, and the overall amount of disorder they produce more than compensates for the deficit in disorder created by the organization of structures in the universe and by the rise in complexity. The net amount of disorder in the universe increases with time. However, disorder has not increased very greatly over the lifetime of the universe. The total contribution from stars over 15 billion years has increased the overall cosmic entropy by only 0.1%, but this is enough to satisfy the thermodynamic arrow of time. Thus the hot interiors of stars and the coldness of space caused by the expansion of the universe by allowing complexity to develop are directly responsible for our very existence.

The Second Law of Thermodynamics does not forbid pockets of order from arising in the universe, provided that, to compensate for that order, a greater degree of disorder is produced elsewhere. For example,[8] after having read this book, your brain will have acquired about one million pieces of information, and it will be more ordered by about one million units. But it does not do to be hungry if you want to read, and spiritual food is not sufficient. During the time that you spend reading this book you will have consumed at least 1000 calories in the form of food. You will have converted the ordered energy present in meat, vegetables, and fruits into a disordered form of energy, which you pass on to the space surrounding you in the form of heat, which your body radiates away, or is carried away by your perspiration. In doing so, you will have increased the total disorder in the universe by ten million million million million (10^{25}) units or 10 million million million (10^{19}) times more than the amount of order that your brain has acquired— assuming that you have read and fully understood everything! The amount of disorder is vastly greater than the amount of order. You need not fret about reading this book: You will not violate the direction of the thermodynamic arrow of time!

Certain questions arise. We do not, at present, know whether the expansion of the universe will be eternal, or if, at some distant future time, the galaxies will reverse their motion, and the universe will collapse back on itself. We have seen that the direction of thermodynamic time is linked to the expansion. Would it therefore reverse in a contracting universe? Would the pile of stones be transformed into a majestic cathedral? Instead of being radiated by stars, would light converge on them? Would psychological time also reverse its direction? If that is the case, then the inhabitants of a contracting universe would believe that they were in an expanding universe, because all their thought processes would be reversed. The misty veil that envelops such questions is still very far from being lifted.

8. Adapted from S. W. Hawking, *A Brief History of Time*, New York, Bantam Books, 1988.

Matter Slows Down Time

To slow down his rate of aging, all Jules had to do was to accelerate his spaceship and get as close as possible to the speed of light. But he could also have slowed down time by steering his spaceship close to a star. In his theory of general relativity*, published in 1915, Einstein showed that the gravitational field created by a star, or by any matter, slows down time. Just as space is curved by matter, so too does time lose its rigidity and become elastic in the presence of matter.

Time flows more slowly for someone who is at the base of the Washington Monument than for someone at the top, or for someone living on the ground floor of a building as against someone living on the tenth floor. Time would pass more slowly for an Eskimo living at the North Pole than for a native of Borneo, near the equator. Time is slowed down when gravity is higher. Gravity is the result of the attractive pull the Earth exerts on us, which varies inversely as the square of the distance from the center of the Earth. The persons at the bottom of the Washington Monument, living on the ground floor, or at the North Pole are all closer to the center of the Earth than their opposite numbers, and feel therefore a greater gravity. The Eskimo is closer to the center of the Earth than the native of Borneo, because the Earth is not perfectly spherical. The centrifugal force caused by the latter's rotation makes the radius of the Earth slightly greater (by about 30 km) at the equator than it is at the poles. However, the differences in time caused by the gravity variations are minimal. The cumulative effect is at the very most a billionth of a second over a whole human lifetime, hardly a heartbeat more. The slowing of time due to gravity goes completely unnoticed in our everyday life, which is just as well, otherwise there would be a housing crisis, because everyone would want to live at ground level and not on higher floors!

Black Holes Turn People into Spaghetti

The effect of gravity on the passage of time is more noticeable on a cosmic scale. Compared with time on Earth, time passes more rapidly in space, where the gravitational attraction of the Earth is much weaker. It flows slower on the Sun, where gravity is about 30 times what it is on Earth (you would weigh 30 times more on the Sun), and faster on the Moon, where gravity is only one-sixth of the terrestrial value. There are even places in the universe where gravity is so intense that time has come to a standstill. These places are the result of the deaths of massive stars (more than five times the mass of the Sun), which have collapsed on themselves after having exhausted their sources of nuclear energy. The large amount of matter in the collapsed, massive star is compressed into such a small volume that the resulting gravitational field is extremely strong, so strong in fact that space is folded back on itself, and light can no longer escape. The collapsed star has become a black hole*, and is no longer visible. Its existence may only be inferred from the gravitational attraction that it exerts on any object passing in its vicinity, and by the deformation of space–time that it creates.

Jules pursues his travels aboard his spaceship. He checks his instruments and realizes that the spaceship has gone off course, and is being pulled toward some object in space. Jules hurriedly looks through the porthole. There is nothing visible in the direction in which the ship is falling. A quick look at his charts tells him he is in the vicinity of a supermassive black hole of several million solar masses. Jules realizes that he is about to be engulfed by the black hole, and that he must act quickly, fire his engines immediately, and turn back. He knows that he needs to do this before the spaceship reaches the radius of no return of the black hole. Once inside that distance it would be impossible to turn back, regardless of how powerful his engines were, and he would be lost forever. The engines start, and the spaceship changes course. Jules heaves a sigh of relief; he is saved.

But let us suppose that Jules is an intrepid explorer who does not fear for his life, and who feels that this is a once-in-a-lifetime chance to explore the interior of a black hole. He decides not to turn back. As he gets closer to the black hole, he keeps communicating by radio his observations and impressions to his twin, Jim, back on Earth. But communication will be broken as soon as he crosses the radius of the black hole, because radio waves will no longer be able to escape. Whatever he might observe inside the black hole will remain secret forever because he will never be able to tell anyone whatsoever about it.

Let us follow Jules, in our imagination, as he approaches the black hole. The gravitational influence of the black hole gets greater and greater. Jules feels as if his head and feet are being pulled in opposite directions. This stretching is caused by the *difference* in the force of gravity that the black hole is exerting on his head and feet. His feet, which are closer to the black hole, experience a greater gravitational attraction than his head, which, because of his height, is 1.8 meters farther away. His feet are therefore falling toward the black hole more rapidly than his head, and his body is being stretched. These tidal forces—so called because it is also the *difference* in the gravitational attraction exerted by the Moon on the center of the Earth and on its surface that produces the tides of the Earth's oceans—stretch and elongate everything in the spaceship, turning it into long, thin spaghetti. When the gravitational forces become extremely high, the elongation becomes too great, and the electromagnetic forces that bind the atoms together and are responsible for the solidity and cohesion of the human body, are unable to resist the tidal forces any longer. Jules's body breaks up, and that is the end of him.

Black Holes Bring Time to a Halt

Let us not anticipate. Jules is still far from the black hole and his tragic end. He continues to radio back to Jim the pictures taken by the video camera inside the cabin. On Earth, Jim decodes the signals and watches on his television monitor the events happening inside the cabin. As Jules approaches the black hole and the gravitational field becomes more intense, the radio waves that are emitted have to do more and more work to escape from the gravitational field and reach Jim. They lose more and more energy, and the time interval between two successive waves, as received by Jim on Earth, becomes longer and longer. Each new picture takes longer to arrive and be updated. Events on board the spaceship appear to be unfolding at a slower and slower rate. From Jim's point of view, Jules now takes a considerable time to make each individual movement, or to finish any particular task. Well away from the gravitational influence of the black hole, Jules took 2 minutes to brush his teeth—the time being that measured by Jim's watch. As Jules approaches the black hole, the same tooth brushing takes 2 hours, 2 years, 2 centuries, 2 billion years. . . . Jim sees Jules's time becoming longer and longer relative to his own. Finally, just as Jules crosses the frontier of no return, his time, as shown by Jim's watch, freezes. From Jim's point of view, the black hole has brought Jules's time to a standstill. The picture on the television screen is no longer updated. Jules will have the same smile, the same posture, the same gesture, for all eternity. Jim will never see the spaceship and Jules disappear into the fathomless black hole. To him, the rocket will remain suspended for all time at the radius of no return.

All of Eternity in the Blink of an Eye

Jules sees events happen in quite a different manner. To him, the time that he reads from his clock appears to run normally. His spaceship approaches the black hole, and crosses the frontier of no return without problem. Jules is aware that he is heading straight for the center of the black hole, where the density and the gravitational field are infinitely great, and where the tidal forces will eventually tear his body apart. He continues to receive radio messages from Jim. As he approaches the center of the black hole, the increasing strength of the gravitational field makes those radio waves gain more and more energy, and arrive faster and faster. Jules sees Jim's time accelerate more and more, and at the instant that he crosses the frontier of no return, all eternity passes before his eyes in a single instant: Jim's old age and death; the death of the Sun after 11 billion years; the end of all the stars in the Galaxy, and of the universe. . . . Jules will not be able to return from inside the black hole to rejoin the universe, because, from his point of view, that universe has already lived its life, and ceased to exist at the very instant he crossed the frontier of no return. Rejoining the universe after having seen it come to an end would be equivalent to saying that Jules left the black hole before he entered it, which is absurd. Because his time has gone beyond that of the external world, Jules is condemned to remain and perish inside the black hole.

A Recipe for Making Black Holes

How can we make a black hole? In principle, any object may become a black hole. It is sufficient to compress it beyond a certain size for its gravitational field to become strong enough to bend space on itself, and prevent light from escaping. Let us assume for a moment that you have a mass of 70 kilograms. If two giant hands were to squeeze you down to less than 10^{-23} cm (1 divided by 100,000 billion billion), with a radius 10 billion times smaller than an electron, you would become a black hole. The radius of a black hole varies according to its mass. The Earth, which has a mass of 6×10^{27} grams (6 followed by 27 zeros), would become a black hole if its radius of 6400 km were reduced to 1 centimeter (smaller than a ping-pong ball). The Sun, with a mass of 2×10^{33} grams (2 followed by 33 zeros), would become a black hole if its radius of 700,000 km were reduced to 3 km (see Appendix C).

So you certainly should not imagine that a black hole is necessarily very small and very dense. It all depends on its mass. A black hole with a mass one billion times that of the Sun would have a diameter of 3 billion kilometers, that is, slightly less than that of the Solar System, and its average density would be no greater than that of the air we breathe. If, by bad luck, Jules had been captured by a low-density black hole, he would not have felt anything as he crossed the frontier of no return. He would not have realized that he had fallen into a black hole until quite a bit later, after having traveled billions of kilometers into its interior, when the tidal forces have begun to cause his bones to ache. It would then be far too late to do anything about it. All he could do would be to wait for the inevitable end.

In practice, black holes are not found just anywhere, because it is very difficult to compress objects. On Earth, the electromagnetic force, which binds atoms and molecules together and organizes them into crystalline lattices, gives rigidity to things and stubbornly resists such extreme compression. Neither you nor I, nor the Earth, will ever become black holes. To produce black holes, a powerful compressing agent is needed. Gravity comes to the rescue. It is attractive, and tends to make everything collapse. According to Newton, a very large mass is required for gravity to be effective. Such large masses are found in stars, but not just any star will do. The Sun, despite its enormous mass of 2 billion billion billion tons, will not end up as a black hole. In 6.5 billion years, when it has exhausted its sources of nuclear energy, gravity will cause it to collapse into a white dwarf, about 10,000 km across, which is about the size of the Earth, and far from the 3 km that is required. The Sun is not massive enough. To finish their lives as black holes, stars must have masses at least about five times that of the Sun. Such massive stars are relatively rare, which is why black holes are not found in great numbers in the Milky Way.

Against All Common Sense

At this point you are probably scratching your head and saying, "All this about time and space is very strange. Time has lost its universal character, and can dilate or contract, depending on my movements. And space does the same. Gravity is also somehow involved, and deforms both space and time. Then there are black holes where the whole of eternity passes in an instant. All this goes against my intuition and common sense. I just don't understand it!" Such a reaction is quite natural. All of us have an instinctive need to relate new, strange concepts to our everyday experience, reducing reality to familiar images. When this does not work, when our intuition is scorned and our dearest ideas and beliefs are spurned, and when our "common sense" is swept aside, then we throw our hands in the air and cry "I just don't understand!"

And yet there is nothing to understand. Nature is like that. We must accept it as it is. Our intuition and common sense based on everyday events are poor guides when it comes to the infinitely small or the infinitely large, to an atom or the entire universe. Einstein, by rejecting common sense, was able to build his theory of relativity, an immense monument to human thought.

A scientific theory is sound, not because it agrees with intuition or common sense, but because it correctly describes nature, because it predicts phenomena that may be observed and verified. There is no doubt now that time does dilate at high velocities. It does so whenever a subatomic particle reaches a velocity close to that of light in one of our particle accelerators, like the one at the European Center for Nuclear Research (CERN) at Geneva. This may be verified by accelerating particles that have very short lifetimes, those that disintegrate after a few millionths of a second. We find that the lifetime of these particles is increased by 10, 20, 100, or more times, depending on the velocity they reach—always precisely in accordance with the predictions of relativity. Time has slowed down for these particles. They do live longer, whether we like it or not.

Matter curves space. This is yet another of relativity's predictions that defies common sense, but was verified in 1919 by a solar eclipse expedition, which has remained famous in the history of physics. The idea, suggested by Einstein himself, was to profit from the absence of light from the Sun, which would be blocked by the Moon, to photograph distant stars, whose apparent positions in the sky were very close to that of the Sun. If space were curved by the Sun's gravitational field, the paths of light from these stars should also be curved. The curvature of the paths should be revealed by a slight shift in the positions of the stars' images when compared with a photograph taken 6 months later, when the Earth would be on the opposite side of the Sun, and the starlight no longer had to travel through the Sun's gravitational field to reach us. The angular deviation predicted by Einstein was minuscule (but nevertheless two times that predicted by Newton), and equal to the angle subtended by a person's finger at a distance of 1 km. But it could be measured (Fig. 25) and agreed with the amount predicted by relativity. This outstanding, observational confirmation established relativity as a theory to be reckoned with, and propelled Einstein to the height of fame. Since then, the deflection of light by the Sun has been measured many times, and with far greater precision, by radio astronomers observing the radio emission from distant celestial sources. Each time, relativity has proved to be correct. Space is curved by matter, whether or not our own common sense accepts it.

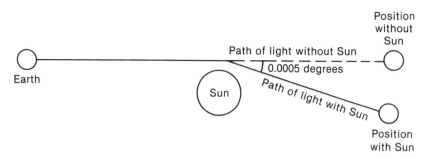

Fig. 25. *Matter curves space.* According to Einstein's general theory of relativity, the gravitational field of the Sun (or of any other massive object) causes a curvature of space, and of the path followed by light. This prediction has been confirmed by observing stars, the light from which passes close to the Sun, going through its gravitational field. We need to wait for a solar eclipse, during which we photograph the stars, and then compare the photographs with others taken when the light does not have to pass through the Sun's gravitational field, for example, 6 months later, when the Earth is on the opposite side of the Sun. Such comparisons always show a slight shift in the positions of the stars, in accordance with the theory of relativity. The deflection of light by the Sun's gravitational field was first measured by the British astronomer Sir Arthur Eddington during a solar eclipse expedition in 1919, propelling Einstein to instant universal fame. Newton's theory also predicts that light will be deviated in a gravitational field, but by only half as much.

Atomic clocks measure time with the most exquisite precision. Synchronize two atomic clocks, leave them side by side, and ask your distant descendants to check them in several billion years. They will find that the two differ by less than a fraction of a second. Using these fabulous products of modern technology, a team of physicists has shown that gravity slows down time. One atomic clock was placed in an aircraft and flown high above the Earth, where gravity is weaker than it is at the ground. After the flight, its time was compared with that of the clock remaining on the ground. The latter was several billionths of a second slow. For the clock that had remained on the ground in a stronger gravitational field, time had passed more slowly. Once again, relativity had triumphed over common sense.

Black Holes Are Hungry

Observations do not lie. You rather reluctantly accept the idea that space and time become elastic under the influence of motion or a gravitational field. In your defense of common sense, you have been pushed into a corner. So you now play your last card. You attack what appears to be the astrophysicist's Achilles heel: the concept of black holes. "Ah," you say, "It's all very well talking about black holes, but because they are invisible, observations can never prove or disprove their existence. Black holes will always remain a figment of astrophysicists' fertile imaginations. There's no need to reject common sense and agree with them there."

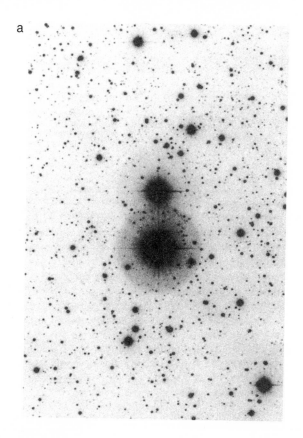

a

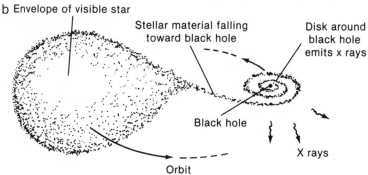

b Envelope of visible star

Stellar material falling toward black hole

Disk around black hole emits x rays

Black hole

X rays

Orbit

Fig. 26. *A black hole in the Cygnus Constellation.* (a) A supergiant star (the lower extremely bright star) in the constellation of Cygnus. There is an invisible object in orbit around it, which astronomers believe to be a black hole. The large size of the star's image is a result of optical effects caused by the star's brilliance, and does not reflect its actual size, which is far too small to be seen directly (photograph: Hale Observatories). (b) How the black hole, as it orbits the supergiant star, gravitationally pulls the latter's atmospheric envelope. The matter, in falling toward the black hole and forming a gaseous disk around it, heats up and emits copious amounts of x rays, which betray the black hole's presence.

84

But you would be wrong. Black holes are betrayed by their gluttony. Once one is born, it attracts everything in its vicinity and engulfs it, gaining mass and growing in size. This cannibalism has observable consequences.

Some stars, like humans, live in pairs. Many massive stars are members of binary stars, pairs of stars that are very close to, and orbit, one another. Suppose one of these stars collapses into a black hole. The other star will continue to revolve around its invisible companion as if nothing had happened. The gravitational field, which dictates the motion of the visible star, depends only on the total mass of the pair, which has not changed. Similarly, if giant hands were to compress the Sun into a black hole, daylight would disappear and there would be everlasting night, but the Earth would continue endlessly in its orbit, and the motions of all the other planets would be unchanged.

The powerful gravitational field of the black hole pulls the gaseous atmosphere of its visible companion toward it. The atoms of gas that form the atmosphere fall at extremely high speeds toward the collapsed star. In doing so they undergo collisions, heat up, and emit radiation. Because the atomic motions are exceptionally violent; the emitted radiation is very energetic and comes out in the form of x rays. This radiation, which is produced well outside the frontier of no return, may be detected.

In the constellation of Cygnus there is a very bright x-ray source, and a star is visible at the same position. By decomposing its light, and using the Doppler effect to investigate its motion, it is known that the visible star is orbiting another object, which has a mass about ten times that of the Sun. Yet this object is invisible. It is thought to be a black hole (Fig. 26). So even on this ground, nature seems to confirm the predictions of relativity. Common sense is defeated on all fronts.

| 4 |

The Big Bang Today

Astronomers Are Reluctant to Change Their Habits

Astrophysicists are conservative. They are not happy when new ideas or theories arise and change, from one day to the next, the knowledge that has been acquired with a vast amount of effort over the years. They take it badly when their representation of reality, their shared language, is suddenly altered in a drastic manner, when a new brushstroke suddenly alters the painting just finished, or when the notes of music are unexpectedly rearranged to give a new melody.

Yet this is just what happened with the Big Bang theory. In less than 50 years it has become the paradigm[9] of modern cosmology, a theory that is the starting point for planning and conceiving new projects and observations. The Big Bang has become the new common language, the newest representation of the world, the last painting and the latest melody. One of the major reasons for this new, rapid consensus is undoubtedly the theory's capacity for prediction, and the fact that its predictions have been confirmed in such a spectacular fashion. The Big Bang theory is the only theory that is able to explain observations that are apparently as disparate as the existence of a fossil background radiation that bathes the whole universe; the chemical composition of stars and galaxies—three-quarters hydrogen and one-quarter

9. T. S. Kuhn, *The Structure of Scientific Revolutions,* Chicago, University of Chicago Press, 1970.

helium—and the fact that the universe, the oldest stars, and the oldest atoms all have about the same age. Let us examine all this in more detail.

The Fossil Radiation That Bathes the Universe

With the advent of the Big Bang, the universe acquires a historical dimension. We can now speak of the history of the universe, with a beginning and an end: a past, a present, and a future. The static Newtonian universe, unchanging, and devoid of history, may be relegated to the rank of dead universes.

In the heart of Africa, paleontologists search for the bones of primitive hominids, which will enable them to trace human prehistory. Geologists burrow in the Earth's crust for fossils that will help them to reconstruct the history of the Earth. Similarly, astronomers turn their penetrating gaze on celestial objects, searching for cosmic fossils that will enable them to reconstruct the history of the universe.

The most important cosmic fossil, the one that has rallied the majority of the scientific community to the Big Bang theory, and the reef against which most rival theories have shattered, is a radiation that bathes the whole universe, and that originated at a time when the universe was just 300,000 years old. This background radiation* consists of photons* (about 400 per cubic centimeter) that are hitting my hand as I write these words, and your face as you read them. These particles of fossil radiation, which may be detected and captured by radio telescopes, are like scattered notes of music that nature sends us. It is up to us to discover their secret melody.

The existence of this fossil radiation was predicted by the Russian–American physicist George Gamow as early as 1946. Following the earlier work by the Russian mathematician Alexander Friedmann and the Belgian priest Georges Lemaitre, he used physical theory to retrace time back to the Big Bang, just as an explorer follows a river to its source. The laws of physics indicated that the universe must have been hotter and denser in the past. On the other hand, the ratio between the two components that make up the universe, matter (atoms, stars, galaxies, etc.) and radiation, must have been reversed when the universe was young. Einstein has shown that all matter is energy. In the universe today, matter is dominant; its energy equivalent is about 3000 times as great as the energy present in the form of radiation. During the first moments of the universe* the situation was reversed. Between 1 second and 300,000 years after the initial explosion, the temperature and density were so extreme that none of the structure that we observe today—the galaxies, stars, or even atoms—could exist. Radiation reigned supreme. According to Gamow, this radiation, which bathed the whole universe and was originally hot and energetic (its temperature was 10,000 K when the universe was 300,000 years old), should still be reaching us today, but greatly cooled. The amount of work that the fossil radiation has had to do over the past 15 billion years to overcome the expansion of the universe and catch up to the Galaxy has considerably weakened it. It has lost so much of its energy that its temperature now is a frigid 3 K ($-270°C$) above absolute zero*. This radiation is so cold and weak that it can be captured only with a radio telescope.

A fire leaves ashes in a fireplace. The background radiation is like the ashes of the fire of creation.

Telephones and Cosmology

No one bothered to search for this background radiation, this faint echo of creation, for the next 20 years. Gamow's work was forgotten. It was only in 1965 that the background radiation was discovered accidentally by two American radio astronomers, Arno Penzias and Robert Wilson, who were working at Bell Telephone Laboratories.

It is a good story that deserves to be told. Penzias and Wilson's concerns had nothing to do with cosmology. With the idea of improving telephone communications, they wanted to build as advanced an antenna as possible, to capture the signals from Telstar, the first communications satellite. Originally, this was to be done by a French antenna. But construction of the latter was behind schedule. The date for Telstar's launch was approaching, and Bell Telephone's directors became worried. In case the French antenna was not ready in time, they asked Penzias and Wilson to build another. As it happened, the French antenna was finished in time for the launch of Telstar, but Penzias and Wilson's telescope was still used. It not only detected Telstar, but also a mysterious radiation at a temperature of 3 K. Only later did Penzias and Wilson realize that they were hearing the music of creation. The French engineering delay had been the indirect cause of the discovery of one of the two cornerstones of the Big Bang theory! Without observations of the recession of the galaxies and of the background radiation, the edifice of the Big Bang would collapse.

You may well ask why it took 2 decades for observations of such importance to be made, and why, after such a long wait, it was only chance that brought about the discovery. The reasons are probably not technical ones. When Gamow's work on the Big Bang was published, several years after the end of the Second World War, radio astronomy was already expanding rapidly, thanks to the development of radar during the war. The real reason is probably psychological. The Big Bang provided a scientific basis for the idea of creation. Religion was "raising its ugly head," and the physicists, uneasy with the situation, subconsciously "forgot" Gamow's prediction.[10]

After Penzias and Wilson's discovery, astronomers frenetically observed the background radiation, as if to make up for lost time. Radio telescopes all over the world were pressed into service. The cosmic radiation was found to be present everywhere, with always the same intensity. It showed the same temperature of 3K, no matter in what direction the telescopes were pointed. (This is known as being isotropic.) This applies whether you are

---

10. See also S. Weinberg, *The First Three Minutes*, New York, Basic Books, 1977.

observing from your home or from the top of a mountain, and regardless of the frequency at which your radio telescope is operating. The Big Bang theory had passed its first major test with flying colors. The creation of the universe, as envisaged by the fertile imagination of a few physicists, did really take place. The universe did really begin its existence in a hot and dense state. It really was filled with radiation that reaches us now, greatly cooled. Numerically, the fossil particles of radiation dominate the universe. There are one billion of them for every particle of matter. But their energy is so weak that they represent less than one-thousandth of the universe's total energy.

Additional support for the Big Bang theory has come recently from a satellite launched by NASA in late 1989 named Cosmic Background Explorer (COBE) and designed to study exclusively the fossil radiation. COBE has studied the energy distribution of that radiation and found that it can only be understood if the universe had begun its existence in a very hot and dense state, as described by the Big Bang theory.

Moreover, by taking snapshots of the whole sky, COBE has also given us a picture of the early universe when it was only 300,000 years old, the time of birth of the fossil radiation. The COBE picture is extraordinary: It shows minute variations in the temperature of the fossil radiation, of the order of 30 millionths of a degree. These tiny temperature fluctuations are thought to be caused by variations in the distribution of matter. In a slightly denser region, gravity is a little stronger and light coming from that region would lose a little more energy, thus causing the fossil radiation to look cooler in that direction. Conversely, it would be slightly hotter in the direction of a less dense region. These tiny density fluctuations act as seeds of cosmic structure that will grow in time, thanks to gravity, to become the hundreds of billions of galaxies that make up the cosmic tapestry. The discovery of these cosmic seeds is fundamental to the Big Bang theory, since without them we would not be able to explain all the structures now seen in the universe, galaxies, stars, or men.

For Helium, the Die Is Cast After the First 3 Minutes

Our bodies consist of chemical elements. Our bones are made largely of calcium. Carbon, together with hydrogen, oxygen, and nitrogen, makes up the molecules of the genetic code that stores and transmits the information, ensuring that children resemble their parents. Zinc helps us to digest alcohol. Copper is involved in skin pigmentation. At least 27 chemical elements govern the correct functioning of our bodies. When just one of these elements is lacking, illness results.

Yet, the "heavy" elements—their atomic masses are heavier than those of hydrogen and helium, which are the lightest elements—that make up our bodies and are fundamental to life form only an insignificant fraction, about 2%, of the total mass of the universe. Stars and galaxies do not have the same chemical composition as men. They consist of 98% hydrogen and helium, the same helium that keeps children's party balloons bobbing at the end of their strings. In the 1960s, astronomers realized that the ratio of helium to hydrogen scarcely varied at all from star to star, or from galaxy to galaxy, quite unlike the heavy metals, which might vary by a factor of more than 1000. Cosmic objects always showed the same ratio: approximately one-quarter helium to three-quarters hydrogen by mass.

Such a remarkable regularity could not be the result of chance. The variation in the quantities of the heavy elements, and the lack of variation for hydrogen and helium, must reflect their origins. In 1939, the American physicist Hans Bethe had discovered that heavy metals could be manufactured in stellar interiors. These cosmic furnaces had temperatures of millions of degrees and fused the nuclei of hydrogen into the nuclei of heavier elements, liberating vast quantities of nuclear energy, which fed the fires of stars. Obviously, hydrogen had to be a primordial element, that is, one that existed before the stars were formed, because the latter consisted themselves of hydrogen. Helium had a peculiar status. It could not have been manufactured in sufficient quantity in the stellar furnaces, because such a process would have liberated an enormous quantity of energy, more than was actually observed in the universe.

The constant ratio of helium to hydrogen could be most simply explained if these two elements were produced exclusively during the first moments of the universe's existence, in other words, if they were primordial elements*. Their relative abundances would therefore be fixed once and for all, and would not depend upon the subsequent evolution of stars and galaxies, responsible for the wide variation in the abundances of the heavy elements. In addition, the current absence of the energy released during the formation of helium could be explained if that energy were diluted by the subsequent expansion of the universe. Astrophysicists frantically set to work, and the Big Bang theory enjoyed yet another triumph. If the temperature of the universe is now 3 K, calculations predicted that, about 3 minutes after the initial explosion, one-quarter of the mass of the universe was in the form of helium, and the remaining three-quarters were hydrogen, and also that these proportions subsequently would not vary significantly. These are precisely the values observed in stars and galaxies. Cosmologists became even more convinced that they were on the right track.

The Galaxies Are Slowing Down

The final argument in support of the Big Bang theory involves the age of the universe, that is, the interval of time between the instant when all the material in galaxies was together, and now. In theory, all that is required to obtain this age is to observe a galaxy, measure its distance and its velocity of recession, and divide one value by the other. If you pass a sign on the side of the freeway saying that you are 110 miles from Washington, and a glance at your speedometer shows that you are traveling at 55 mph, it is a simple mental calculation to find out that you left Washington 2 hours ago. The calculation is accurate, provided you have driven at the same speed all the time. The age of the universe, obtained by dividing the distance of a galaxy by its velocity, is correct, provided the galaxy's velocity does not vary as a function of time. However, every galaxy is affected by the gravitational attraction of all the mass contained in the universe, both the visible mass (such as that found in galaxies) and the invisible mass (such as that of black holes). The gravitational attraction slows down the rate of expansion; so the galaxies are subject to a slow deceleration. If, at the beginning of your trip, you had exceeded the speed limit, and driven at 70 mph, before slowing down to a legal 55, you would obviously have taken less than 2 hours to cover the 110 miles. Similarly, because of the deceleration of the galaxies, the age of the universe is slightly less than the age obtained by dividing their distances by their velocity of recession.

A Cosmic Ballet

Measuring the age of the universe therefore means measuring the distances and recession velocities of galaxies. The latter are easily measured. All we have to do is to split the light from the galaxies with a spectroscope, and then measure the Doppler shifts. But there is one problem, in that we must only measure the velocity caused by the expansion of the universe. All other motions, such as those caused by the gravitational attraction of nearby galaxies, must be excluded.

Galaxies do not like being alone. They are gregarious and prefer to be in the company of other galaxies. The best place for a galaxy to be is close to another. They therefore tend to occur in groups, which themselves cluster together on still larger scales. Our "home," the Galaxy, is part of a small "village" of galaxies, the Local Group*, that is bound together gravitationally, and that also includes the Andromeda galaxy, and about 15 dwarf galaxies, including our own satellites, the Large and Small Magellanic Clouds*. The Local Group extends over some ten million light-years. In the vicinity of our village there are other villages, other groups of galaxies. Farther out in the universe we come across "towns," clusters of galaxies*, consisting of thousands of galaxies linked by their gravitational attraction, with sizes of some 30 million light-years. Just as villages and towns make up part of a country, the Local Group is part of an enormous complex of some 10,000 galaxies that extends over some 200 million light-years. This is known as the "Local Supercluster" or the "Virgo Supercluster," the latter because the Virgo Cluster contains a significant proportion of its mass.

Gravity causes all the galaxies in these clusters to attract and "fall" toward one another. The motions thus caused are superimposed on the universal expansion. In fact, the Earth is involved in a fantastic cosmic ballet. First of all, it carries us through space at 30 kilometers per second in its annual travel around the Sun. The latter, in turn, carries the Earth with it in its trip around the galactic center at a velocity of 230 kilometers per second. The Galaxy is falling toward its companion, the Andromeda galaxy, at 90 kilometers per second. (From our point of view, the Andromeda galaxy appears to be approaching; it is one of the few galaxies whose light is shifted toward the blue.) But that is not all. The Local Group is moving at about 600 kilometers per second, under the gravitational attraction of the Virgo Cluster, and of the Hydra-Centaurus Supercluster, which is the nearest to our Local Supercluster. Current observations appear to show that things do not stop there, and that the Virgo Supercluster and the Hydra-Centaurus Supercluster are themselves falling toward another enormous collection of galaxies, which astronomers, for the lack of more precise information, have called the "Great Attractor*" (Fig. 27).

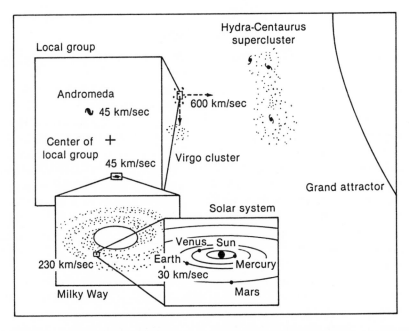

Fig. 27. *The great cosmic ballet.* On our terrestrial home, we take part in a fantastic cosmic ballet: The Earth carries us at 30 kilometers per second around the Sun, which is itself flying through space at 230 kilometers per second in its orbit around the center of the Galaxy. The latter, in turn, is falling toward the Andromeda galaxy at 90 kilometers per second. (In fact, each of the two galaxies is falling toward the center of gravity of the Local Group at 45 kilometers per second.) The Local Group, whose most massive members are the Galaxy and the Andromeda galaxy, is moving at 600 kilometers per second under the gravitational attraction of the Virgo Cluster, and of the Hydra-Centaurus Supercluster. The latter is itself falling toward the Great Attractor, the mass of which is equivalent to that of tens of thousands of galaxies, and whose exact nature is still unknown.

But in spite of all the aesthetic emotion which such a marvelous cosmic choreography inspires in us, we must exercise great care, when we use the motions of galaxies to determine the age of the universe, to include only those motions that are caused by the expansion of the universe. To suppress all motions due to the gravitational attraction of neighboring galaxies, we should only consider galaxies far beyond the Local Supercluster, at distances larger than 200 million light-years, 100 times further than the Andromeda galaxy. If we must include nearer galaxies, motions other than the recession motion must be subtracted out.

The Age of the Universe

Our search for the age of the universe is far from over. The sheer size of the universe causes problems. It is extremely difficult to measure the distance of remote galaxies. All our current uncertainties over the age of the universe arise from the uncertainties in measuring the distances involved. The true scale of the universe still eludes us.

Using the Cepheids as cosmic beacons, Hubble was able to determine distances out to 13 million light-years, about four times farther than the Local Group. Beyond that distance, however, the beacons became too faint. Brighter objects were required to probe even deeper. Astronomers turned to the brightest stars in a galaxy (the "supergiants," so-called because they are 100,000 times as bright as the Sun, and 300 times larger), to globular clusters, which contain hundreds of thousands of stars, and to supernovae*, which are the explosions marking the death of massive stars and which, at maximum brightness, release as much energy in a second as a whole galaxy. These beacons enable us to reach out to a distance of about 300 million light-years, and thus beyond the Local Supercluster. Unfortunately, unlike the Cepheids, the exact luminosity of these objects is poorly known. Just as a motorist who is unfamiliar with the true brightness of a car's headlights cannot accurately estimate the distance of the oncoming car, astronomers, who do not know the real brightness of the various cosmic beacons, are unable to derive very accurate distances.

To determine the true brightness of these distant sources, astronomers need to carry out a series of measurements, all of which are interdependent. Measuring the extreme distances that occur in the universe requires a whole ladder of steps, where the reliability of each step depends on that of all earlier steps (see Table 1). At the bottom of the ladder is the Hyades galactic cluster, the nearest group of stars. As you will recall, its distance (120 light-years) was determined by the convergent-point method. The true brightness of the stars in the Hyades cluster was then established from their apparent brightness. The next step required finding more distant galactic clusters and, in particular, clusters that contained Cepheids. The distance of these more remote clusters was obtained by assuming that their stars had the same actual brightness as those in the Hyades cluster. Knowledge of those distances then enabled the true brightness of the Cepheids to be established. The Ce-

Table 1   How astronomers measure cosmic distances

Distance in light-years	Celestial object	Method of determining distance
10×10^9		
	Distant galaxies and quasars	The Hubble law is used; the red shift indicates the distance
300×10^6		
	Supergiant stars, globular clusters, supernovae	Cosmic lighthouses whose intrinsic brightness is determined using the methods at the base of the ladder
13×10^6		
	Cepheid stars in Local Group galaxies	Relationship between period and intrinsic brightness
1500		
	Galactic star clusters (Hyades)	Convergent point
100		
	Planets, nearby stars	Parallax
0		

pheid period-luminosity relationship in turn led to the third stage in the process, enabling the distances of galaxies out to 13 million light-years to be determined. The true brightness of supergiant stars, globular clusters, and supernovae in these galaxies could therefore be obtained. To take the next step, out to 300 million light-years, another assumption had to be made, namely that the brightness of supergiant stars, globular clusters, and supernovae was the same in distant galaxies and that it did not vary with time. Time intervenes, because, as we have seen, light takes time to reach us, so that looking far out into space is the same as looking back in time. In studying the depths of the universe, astronomers are also investigating the past.

Assuming that stars and other objects do not vary with space and time is very rash. It is probably a mistake to assume that their histories and identities do not differ from one galaxy to another. Hubble learned this to his cost in 1929, when, in determining the distance of the Andromeda galaxy, he assumed that its Cepheids had the same true luminosity as those in our galaxy. This was wrong. They were four times as bright. Hubble thought that the Andromeda galaxy was much closer than it really is, and found, as a result, the universe to be only 2 billion years old, approximately one-tenth of the age accepted now.

The other great source of inaccuracy in evaluating the distance of our cosmic beacons is inherent in the construction of the ladder itself. A ladder method means that any errors introduced into any one step are carried forward, and accumulate, so that the overall error in the fourth step, for distances beyond the Local Supercluster, is enormous, even if the error in the distance of the Hyades, in the very first step, is minimal. After tremendous effort, and despite all the ingenuity employed, the universe still refuses to reveal its true scale and age. For the present, all we can say is that it is between 10 and 20 billion years old.

Will the universe be like some aging coquette, and never reveal its true age? New instruments are looming on the horizon that will possibly force its hand. The Hubble Space Telescope thanks to its capacity to observe unhampered by the Earth's atmosphere, will be able to study objects at least 50 times fainter, and 7 times as distant, as those detectable with the largest telescopes on the ground (Fig. 12). It will be able to detect Cepheid variables in the nearest cluster of galaxies, the Virgo Cluster, a group of about 1000 galaxies that is part of the Local Supercluster, and that is at about 42 million light-years. Distances determined from Cepheids are very accurate, which will considerably improve the fourth step in the ladder of distances. The first step in the ladder will also be improved in another way. A European satellite called Hipparcos (named after the Greek astronomer who first derived the parallax* of the Moon, but also as the acronym for "high-precision parallax collecting satellite") was launched in 1990. Although it did not reach the desired orbit, it will be able to measure accurate parallaxes for the stars in the Hyades cluster, on which the whole ladder rests. The distance of the cluster will then be established with an exquisite precision. On another front, astronomers are busy developing techniques that will allow us to measure distances of faraway objects directly, without having to resort to a ladder method with its inherent inaccuracies.

The Age of the Oldest Stars

The recession motion of the galaxies has served as a cosmic clock, enabling us to derive the age of the universe. Is the value of between 10 and 20 billion years reasonable? The universe contains objects such as the Earth, or old stars in globular clusters, the ages of which may be determined quite accurately. These ages should be equal to or less than that of the universe as a whole, because it would be very embarrassing to have a universe (which, by definition, includes everything) that is younger than its contents.

The globular clusters are among the oldest objects in the universe. They were formed during the first billion years of its history. They can serve as a second cosmic clock. Imagine that you collect all the babies in the country that were born on a particular date—let's say January 1, 1988—into a single town. The children will grow and will all celebrate their birthdays on the same day of the year. Some of them will be fat, and some thinner. Let's assume that the fattest will have more illness and heart attacks. Their lives will be short, say 50 years. Those of medium weight will live 75 years, whereas the thinnest will live to 100. You visit the town and all you have to do is to look at the inhabitants to decide their ages. Let's assume that you pay your first visit in 2008, 20 years after the babies were born. The inhabitants are all still alive, and you see fat, middling, and thin adolescents; so you can deduce that their age must be less than 50 years. You go back 30 years later, in 2038. You see only people of medium and low weight. The fat ones have disappeared. You conclude that the inhabitants must be between 50 and 75. Your son goes back another 30 years later, in 2068. He meets only very thin people, and concludes that their ages are between 75 and 100.

Similarly, astronomers can deduce the age of stellar "inhabitants" in a globular-cluster "town" by examining their physical characteristics, such as their mass, luminosity, or temperature. Like the children, all the stars in a globular cluster were born at the same time, as a result of the collapse of a cloud of interstellar hydrogen and helium. Like the human beings, some stars are born "fatter" (more massive, brighter, and hotter) than others. These massive, luminous, and hot stars consume prodigally their reserves of nuclear energy and rapidly exhaust them. They die after just a few million years. In the history of the universe they are like short-lived firecrackers. On the other hand, the "thinner" stars, less massive, less luminous, and less hot, are parsimonious with their energy reserves. They survive for billions of years by scrimping and saving. The Sun is one of these parsimonious stars. It has already lived for about 4.5 billion years, but is still only less than halfway through its existence. Stars that are less massive and less luminous than the Sun will live even longer—as long as 20 billion years. Astronomers, looking at globular clusters, see only low-mass, low-luminosity, and cool stars. The massive, luminous, and hot stars have died. They conclude that the ages of stars in globular clusters, the oldest stars in the universe, are between 12 and 20 billion years old. The supporters of the Big Bang theory are jubilant. The second cosmic clock also gives about the same age.

The Age of the Oldest Atoms

A third cosmic clock is available. It involves the lifetimes of some atoms that do not last forever. After a certain time, they disintegrate and change into other atoms. The best-known example is carbon-14. (The more common form of carbon, carbon-12, is more stable.) Carbon-14 has a half-life of 6,000 years; that is, half of the atoms of carbon initially present disappear after 6,000 years. If you start with 10,000 atoms of carbon-14, 6,000 years later only 5,000 remain. In another 6,000 years (12,000 years after the beginning), only 2,500 atoms will be left. Another 6,000 years (18,000 years from the beginning) and the number of carbon atoms has dropped to 1,250. And so on. All one needs to do is count the number of carbon-14 atoms in an object to date it.

This cosmic clock delights archaeologists but terrorizes art forgers. It permits them to date precisely any object that contains carbon atoms, whether they are old manuscripts or Impressionist paintings by Monet.

On the cosmic time scale, the carbon-14's lifetime is a mere flicker. Its half-life is too short to be used as a cosmic clock. We need atoms whose half-lives are comparable with the 10 to 20 billion year age of the universe. The atoms of uranium come to our aid—the same uranium that fuels our nuclear generating plants and provides us with electricity, and that was responsible for the devastating power of the bomb dropped on Hiroshima. Like some other heavy elements in the universe, this uranium was born in the explosive death throes of massive stars.

There are actually two types of uranium atoms: uranium-235, which has a half-life of 1 billion years, and uranium-238, which lasts longer, having a half-life of 6.5 billion years. Because the uranium-235 disappears faster than the uranium-238, the ratio of the number of atoms of uranium-235 to the number of atoms of uranium-238 continually decreases. This ratio indicates elapsed time, and can be used as a cosmic clock. The age that it indicates for the oldest atoms is again between 10 and 20 billion years.

There is, a priori, no obvious link between our three cosmic clocks, between the recession of the galaxies, the evolution of stars, and the disintegration of atoms. The fact that they all give the same answer cannot be accidental. Unless there is some immense cosmic conspiracy to fool us, this must be seen as another triumph for the Big Bang theory.

So far, we have considered only the Big Bang theory's successes, admittedly very spectacular. But does this mean that there are no clouds on the horizon, and that this theory accounts for everything? We shall see that this is not the case, but also that all is by no means lost. Recent developments in our understanding of the physics of the infinitesimally small, that of elementary particles, have already begun to disperse the great majority of those clouds.

Why Is the Universe So Homogeneous?

From the very beginning, there have been dark clouds that have cast a shadow on the Big Bang theory and that have not been dispelled until recently. The first concerned a remarkable property of the universe: its homogeneity*. The 3 K temperature of the fossil radiation that bathes the whole universe is the same in whichever direction we look: up, down, in front of us, behind us, to the left or right. It does not vary by more than 0.001% from one point in the sky to another. The universe therefore appears to be perfectly uniform in all directions. This background radiation was produced, as we shall see shortly, when the universe was only 300,000 years old. Because light is the fastest means of communication between different regions of space, this implies that only regions separated by less than 300,000 light-years could have had sufficient time to influence one another and equalize their temperatures. The cosmological horizon was then 300,000 light-years and there is no a priori reason for the temperature to be the same on a larger scale. Yet observations show that the temperature *is* the same in all directions, that is, in regions that were separated by far more than 300,000 light-years. The situation is comparable to that of two ships anchored at sea, so far apart that they cannot be seen from one another. Suddenly, and at exactly the same instant, they both weigh anchor and steer for the same island at exactly the same speed. As a spectator, you naturally think that such a series of similar and simultaneous actions cannot be accidental, and that the two boats must be in contact with one another by radio. Similarly, the extreme homogeneity of the universe implies that all its parts must have interacted. The Big Bang theory, at least in its initial form, asserts that this is impossible. Faced with this inexplicable puzzle, astrophysicists shrug their shoulders and hide their ignorance by invoking the "initial conditions": If the various parts of the universe are so similar today, it must be because they were like that from the beginning. The Grand Architect must have created them identical, an explanation that explains nothing.

Why Does the Universe Have Structures?

The second problem is, in some ways, the reverse of the problem of the homogeneity of the universe. Instead of wondering why the universe is so uniform, astrophysicists ask why there are irregularities, why the universe contains different structures.

The universe is rather like a gigantic pointillist painting by Georges Seurat. From a distance, you can appreciate the overall picture, but are unable to distinguish the thousands of individual tiny points of different, brilliant colors that create the figures in the picture. The contours appear perfectly uniform. It is only when you approach the picture that the figures break up into a myriad individual points. Similarly, the universe appears extremely uniform when viewed from a distance. All the details are suppressed, as evidenced by the remarkable uniformity of the fossil radiation.

Yet the universe is far from uniform. Happily for us, it contains many different structures, because a structureless universe would be like a desert without oases; life would neither be able to exist, nor to develop. If the universe were featureless, we would not be here to observe it. Just as for the painting of Seurat, the various structures in which matter is concentrated begin to become apparent when the universe is examined more closely. First the very largest structures, the superclusters of galaxies, can be distinguished. They are hundreds of millions of light-years across and weave an immense cosmic tapestry on the sky. On a smaller scale appear the clusters of galaxies, which are one-tenth the size of the superclusters; then groups of galaxies, 20 times smaller; then galaxies themselves (including the Milky Way), 2000 times smaller; the stars (including the Sun), 10^{15} times smaller (1 followed by 15 zeros); then finally the Earth, 10^{17} times smaller, our tiny haven among the immensities of the universe.

The cosmic landscape presents therefore a two-faced nature: an almost perfect uniformity at large scales, and an unbelievably rich, small-scale structure. From close by, and with a detailed examination, the perfectly monotonous, banal, and uniform canvas has become a wondrous tapestry, full of patterns and motifs. How could the universe develop such a rich hierarchy of structures from a state that was originally so featureless? How did complexity arise from simplicity? This problem of the structure of the universe and of the formation of galaxies remains to this day without a solution. Once again, the astrophysicists admit their defeat and call upon the "initial conditions": The Great Architect has sown irregularities in the otherwise uniform universe, which grew to give birth to galaxies, stars, and human beings.

Where Is the Antimatter?

The third black cloud casting a shadow over the Big Bang theory* concerns the most fundamental of all questions: the origin of the universe itself. How did matter and radiation arise from the structure of space itself? What are the physical laws that determined the universe's content of matter, antimatter*, and radiation?

Antimatter, despite a name that evokes science fiction, is actually not very different from the matter that makes up our bodies and the world around us. Take the components of matter: the nucleus of a hydrogen atom, a proton*, for example. By convention, it has a positive electrical charge. Reverse the sign of the charge and you will have an antiproton. The electron has an electrical charge equal to that of the proton, but with the opposite sign (the charge is negative). Reverse the sign (which becomes positive), and you have an antielectron (or positron*). Antimatter appears the same as matter, but as seen through some magic mirror that reverses the sign of every electrical charge. Apart from the electrical charge, the physical properties of matter and antimatter are precisely the same. An antiproton may combine with an antielectron to form an atom of antihydrogen. Antiatoms may combine to form antimolecules, and life can arise from the development of long helical chains of anti-DNA. I can imagine an anti-me writing these words, and an anti-you reading them on some distant antiplanet revolving around an anti-Sun, in the depths of some antigalaxy. You and your antimatter counterpart live parallel lives, provided you keep apart. But should you ever meet, it would be a disaster. Shake hands, and both of you would disappear in a flash of light. Matter and antimatter annihilate one another as soon as they come in contact, turning into radiation.

The probability of finding an anti-you anywhere in the universe, however, is extremely low. The universe appears to lack antimatter. Cosmic rays*, those winds of charged particles that come to us from the distant reaches of our galaxy, consist almost entirely of matter (protons in particular). In addition, we do not observe the large number of photons that would be produced by the annihilation of matter and antimatter if these existed in equal quantities. Photons certainly exceed in number the particles of matter (such as the proton) in the universe today (by about one billion to one), but this value is far smaller than would be the case if antimatter were as important as matter. Why do we live in a universe that consists exclusively of matter? Where are the anti-you and the anti-me? What has happened to all the antimatter that should have been present when the universe began? Until recently, these questions have remained unanswered.

Why Is the Universe So Flat?

Finally, there is the problem of the flatness of the universe. Here, "flatness*" describes the geometry of space. The large-scale, cosmic landscape today has no relief. It is like driving through a monotonous countryside, with flat fields extending as far as the eye can see. This lack of relief implies that the universe cannot contain a lot of matter, because the latter, when abundant, curves the overall geometry of space. In the vast expanse of the universe, it is very rare that we encounter galaxies whose gravity sculpts valleys in the cosmic landscape, interrupting its wearisome monotony. Why does the universe lack curvature?

Saying that the universe is flat means that an almost miraculous balance exists between the two opposing forces, which have been locked in a ferocious combat ever since the universe began: the explosive force of the Big Bang, which caused the universe to burst on the scene and gave it its expanding motion, and the gravitational attractive force of its matter and energy content, which slows down the expansion and tries to reverse it. If there had been more matter and energy, if the universal landscape had been more curved, in other words, if the gravitational force had been much greater than the explosive force, the universe would have collapsed on itself long ago, forming a giant black hole—and we would not be here to talk about it. On the other hand, if there had been less matter and energy, if the explosive force had far outweighed the gravitational force, matter would never have been able to come together to form galaxies, stars, Sun, and Earth—and we would still not be here to talk about it. The balance between the two forces must be remarkably precise. Change the universe's velocity of expansion when it was only 1 second old, by an infinitesimal amount, one billion billionth (10^{-18}) of the actual value, and you would have completely altered the fate of the universe. What physical mechanism enabled the universe to be so finely tuned? Once again, astrophysicists shrug their shoulders and admit their ignorance.

Until recently, these dark clouds were hovering over the edifice of the Big Bang, casting shadows that were tarnishing its luster. The theory, pushed into a corner, was beginning to show cracks in its structure. But recent successes in the field of subatomic physics, aimed at developing a unified theory of nature, where a single "superforce" would be responsible for the behavior of the universe at its beginnings, and also the idea of an extremely fast expansion phase (known as "inflation*") in the first few fractions of seconds of the universe's existence, promise to provide the cement necessary for repairing these cracks. These new developments have allowed us to go back to the very beginning of the universe and to sketch out a preliminary version of its history. But before opening the history book of the universe, we need to get acquainted with the four fundamental forces. We then need to familiarize ourselves with the quantum effects that govern the subatomic realm.

The Four Forces

Night falls and lamps light up in the houses. A storm descends, and flashes of lightning illuminate the sky. The wind rises, sending dead leaves whirling through the air before letting them flutter to the ground. All the forces of nature are in evidence in this familiar scene. Every change in our surroundings is a result of the interplay of these forces. Just four fundamental forces are responsible for the extremely wide variety of changes and movements in Nature. The Earth's force of gravity* causes the dead leaves to fall to the ground after being swept along by the wind. The electromagnetic force* is responsible for the lights in the houses and the flashes of lightning. The "weak" nuclear force* governs the disintegration of atoms and radioactivity, and is therefore responsible for the nuclear energy that powers the electrical appliances in the house. The "strong" nuclear force* is responsible for the existence of the atomic nuclei that form the houses, leaves, trees, and soil.

The Glue of the Cosmos

The force of gravity reigns supreme in the macroscopic world. Its role on Earth was noticed from the very dawn of humanity: Things fell down toward the ground. In the Aristotelian universe, in the fourth century B.C., this vertical movement was characteristic of only the imperfect world of the Earth and the Moon. The perfect world of the other planets, the Sun, and the stars, possessed perfect, circular motion, not governed by gravity. The concept of universal gravitation, which acts on everything in the universe, did not appear until Newton in the 17th century. Gravity is the "glue" of the cosmos. It causes everything to attract everything else. It binds us to the Earth, the Moon to the Earth, the planets to the Sun, the stars to the galaxies, and the galaxies to the clusters of galaxies. Make gravity disappear and we would float in space. The Moon, the planets, and the stars would disperse into the immensity of space.

Nothing escapes the grip of gravity. Everything that has mass or energy is subject to its effects. Paradoxically, this omnipresent influence is coupled with an extreme weakness. Gravity is the weakest of the four forces. On the scale of elementary particles, it is negligible. The atom of hydrogen, the simplest and lightest of the elements, consists of an electron and a proton. The gravitational force between the electron and the proton is 10^{40} (1 followed by 40 zeros) times smaller than the electrical force between the two particles. The atom of hydrogen is so small (10^{-8} centimeter = 0.00000001 cm) because the electrical force is so strong that it can hold the electron extremely close to the proton. Suppress the electrical force, and the hydrogen atom, left under the sole influence of the gravitational force, will swell up to fill the whole universe. The force of gravity is so weak that it cannot hold the electron closer to proton than a distance of a few tens of billions light-years.

The intensity of the gravitational force depends on the mass of the objects concerned. The gravitational force between a proton and an electron is weak because the mass of the electron is so extremely small (10^{-27} grams; the first figure differing from 0 is preceded by 27 zeros), as is that of the proton, despite the latter being about 2,000 times more massive than the electron. Because of its weakness and insignificance on an atomic scale, gravity follows the dictum of "strength through uniting" to gain in importance. Since its effect on a single particle is negligible, gravity will act on large, massive objects that contain an extremely high number of particles to manifest its influence. We can appreciate the incredible size of that number when we realize that even 1 gram of water contains about 10^{24} particles. Even on the scale of everyday objects, gravity is insignificant. You may perhaps weigh 70 kilograms (154 lbs), but you never feel the gravitational attraction exerted on you by someone (weighing 50 kilograms or so) you may be talking to. You may feel some "attraction," but it will not be caused by gravitation! If you walk past a large building, which weighs many tons, gravity does not pull you against the wall. Extremely sensitive instruments are required to measure the gravitational attraction of a massive building. It is only on the astronomical scale that gravity becomes truly apparent, and plays a real role. The enormous mass of the Earth (6×10^{27} grams) stops us from floating in the air—like astronauts in orbit—and prevents the Moon from flying off into space. The Sun (10^{33} grams), stars (10^{33} grams), galaxies (10^{45} grams), groups of galaxies (10^{46} grams), clusters of galaxies (10^{48} grams), and finally the universe (? grams) form a hierarchy of increasing mass, in an ever-larger realm over which gravity holds undisputed sway.

The Glue of Atoms

As we have seen, the electromagnetic force is far stronger than the force of gravity. Its strength is such that a magnet will easily pick up a nail despite the gravitational attraction exerted on it by the entire mass of the Earth. It is responsible for creating atoms by binding electrons (which have negative charge) to atomic nuclei. An atomic nucleus consists of a collection of protons* (with positive charge) and neutrons* (with a mass similar to that of the proton, but without any electric charge, as suggested by the name) that are bound by the strong nuclear force. It suffices to add up the number of protons to determine the positive charge of any nucleus.

In the electromagnetic world, only things that have a positive or negative charge are of any importance. Unlike gravity, which affects any mass or energy, the electromagnetic force is very selective. Anything that has no electric charge, such as a photon or a neutron, is ignored. As for charged particles, the electromagnetic force imposes very strict rules on their behavior: Opposite charges attract, and like charges repel. A proton and an electron attract one another, but two protons or two electrons repel each other. Unlike gravity, which attracts everything, electromagnetism attracts or repels according to charge.

The electromagnetic force's range of action is not restricted to the atomic world. It is involved in the construction of more complex structures. It binds atoms together into molecules by forcing them to share their electrons. For example, the electromagnetic force binds two atoms of hydrogen together with one atom of oxygen to form a molecule of water. It also causes molecules to be bound, in turn, into long chains, the most complex expression of which is the DNA helix, which is the basis of life itself. The electromagnetic force, by being the glue of atoms, is thus responsible for the cohesion, solidity, and beauty of everything around us. Without it, the Earth would not be solid, your skeleton would not support the weight of your body, and your hand would pass straight through the pages of this book. The beautiful shapes of a Rodin sculpture, the perfect lines of a female body, or the fragile and delicate outline of a rose are just a few of the aesthetic pleasures that would not exist without the electromagnetic interaction. Without it, the world would be devoid of any shapes and would be extremely dull. If it were left to gravity alone, atoms would be gigantic, and stars would be little more than giant nuclei of protons and neutrons.

Like gravity, the electromagnetic force decreases inversely as the square of the distance between two electric charges. But unlike gravity, which is able to compensate, on large scales, for its weak nature by acting on more and more mass, the electromagnetic force, depending as it does on the amount of electric charge, has a hard time increasing that charge. Although the positive charges may add up, the negative ones act to cancel them out, so that the majority of objects in the universe are electrically neutral; they do not carry a net charge. A book, a chair, a house, the Sun, the stars, the galaxies, and perhaps even the universe itself, are all neutral. They are not governed by the electromagnetic interaction. Its power is generally limited to the atomic world. It leaves the management of the universe at large to gravity.

As its name indicates, the electromagnetic force has a dual nature. It not only attracts or repels electric charges, but also aligns a compass or causes a nail to be attracted to a magnet thanks to its magnetic powers. These two aspects are inextricably interwoven, and one cannot be dissociated from the other. A moving electric charge creates a magnetic force. A magnetic field that varies creates an electric current. The Earth's magnetic field, which causes an explorer's compass to point approximately toward the North Pole, is the result of the motion of charged particles (protons and electrons) in the Earth's core. That region is so hot and under such a high pressure from the overlying layers of the mantle and crust that the Earth's core is no longer a solid, but a fluidlike magma where matter has been broken down into protons and electrons. Similarly, the magnetic field of the Sun, the stars, or the

Milky Way is the result of the motion of matter that has been broken down into electrical charges. This intimate connection between electricity and magnetism was discovered in 1864 by the Scottish physicist James Clerk Maxwell.

A Force for Decay

In general, matter is not eternal. Among the hundreds of "elementary" particles that make up matter, very few are eternal. The immortal few are the electron, the photon, and another neutral particle with zero (or a minutely small) mass, the neutrino. But all other particles eventually decay. Even the proton appears to be mortal, although its lifetime is extremely long, being at least of the order of hundreds of thousands of billions of billions of billions (10^{32}) of years. An elementary particle decays by disintegrating into other particles. The process continues until it has been completely changed into immortal, or stable, particles.

The force governing this disintegration and metamorphosis is the so-called "weak force," so named because it is not very vigorous. Although it is far stronger than gravity, it is about 1000 times weaker than the electromagnetic force. Its sphere of influence is minuscule. It is effective only on an atomic scale, over distances of 10^{-16} cm. It is so imperceptible on everyday scales that its discovery was the result of chance. One evening in 1896, the French physicist Henri Becquerel accidentally left a photographic plate in a drawer next to some crystals of uranium sulfate. The next day, he discovered that the photographic plate was covered with a slight, mysterious fogging. His studies showed that atoms of uranium were disintegrating into other particles that were affecting the plate. He called the disintegration process "radioactivity."

The weak nuclear force therefore plays a distinct role among the quartet of fundamental forces: It does not serve as any form of "glue" like the others. All it does is govern the disintegration of matter. It would not initially be missed if it were suddenly to disappear. The Sun would die out after a few million years (rather than after some ten billion years), because the weak interaction is responsible for certain nuclear reactions that occur inside its core, providing it with energy and prolonging its life. Above all, however, matter would last longer. The universe would be populated by all sorts of strange and exotic particles, which would coexist with our familiar protons, electrons, and photons. A new, strange form of chemistry would arise, and complex forms of life, different from our own (which is based on the chemistry of carbon compounds), might develop and thrive.

The Glue of Particles

Atomic nuclei are collections of protons and neutrons. Protons all carry the same positive electric charge. The electromagnetic force makes them all repel one another, yet they remain obstinately bound together in atomic nuclei. A force far stronger than, and opposed to, the electromagnetic force must be keeping the protons together and acting as a glue. This is the "strong" nuclear interaction, the strongest among the quartet of fundamental forces. It is 100 times as strong as the electromagnetic force. Like the weak force, its sphere of influence is minuscule, and its effects are only felt over atomic distances of the order of 10^{-13} cm. It is selective, and acts on massive particles only, such as protons and neutrons, completely ignoring light particles such as electrons, photons, and neutrinos. Here the concept of mass is all relative. The mass of a proton or neutron is insignificant (approximately 10^{-24} gram), but is nevertheless 1836 times as much as that of an electron. The mass of the neutrino is not known exactly, but is certainly far less than that of the electron. As for the photon, it is massless. In the kingdom of the blind, the one-eyed man is king.

The physicists' voyage into the heart of matter in the past 3 decades has revealed that the proton and the neutron are not the elementary and indivisible particles they were thought to be. They are now believed to consist of even more elementary particles, named "quarks*" by their discoverer, the American physicist Murray Gell-Mann. A great lover of poetry and literature, Gell-Mann had recalled a sentence by James Joyce, that great inventor of language, in one of his novels, *Finnegan's Wake:* "Three quarks for Muster Mark." Just as for "Muster Mark," three quarks are required to form a proton or a neutron. The glue binding the three quarks is the strong force. If the latter were to disappear, the world would consist of free quarks, without protons or neutrons, without atoms or molecules, Earth or Sun, stars or galaxies.

Now that we have met the quartet of fundamental forces, we need to become more familiar with the laws that govern the subatomic world. Such an understanding is essential if we are to understand the evolution of the universe, because the infinitely large came from the infinitely small, the whole universe arose from "practically nothing." Once again, as with space–time, our common sense will be severely tested.

Quantum Uncertainty

The nineteenth century bequeathed us a deterministic universe from which chance had been banished, and where everything could be described in a rigorous manner by mathematical and physical laws discovered by human reasoning. Every event had a cause—it could not occur by chance. Causality ruled the mechanics of a deterministic universe. The Marquis Pierre Simon de Laplace, the very one who rejected the hypothesis of God, wrote, in a burst of enthusiasm: "We should envisage the present state of the universe as being the effect of its previous state, and as the cause of the one that will follow. Consider an intelligence which, at any instant, knew all the forces governing nature, together with the respective positions of all the entities of which nature consists. If this intelligence were powerful enough to analyze all the data, it would be able to embrace in the same formula the movements of the largest bodies in the universe and those of the lightest atoms. For it, nothing would be uncertain, and the future, as the past, would be equally present to its eyes."[11]

The advent of quantum mechanics*, the physics of atoms, at the beginning of the twentieth century shattered the rigid shackles of determinism. Chance and extravagance came back in full force into a world where everything had been minutely regulated. Stimulating uncertainty replaced boring certainty. Quantum uncertainty replaced deterministic rigor, and the romantics had their revenge.

Imagine that you are playing tennis. The ball passes back and forth over the net. At any instant, you can, if you wish, accurately measure the position and the velocity of the ball in space. All you would have to do, for example, would be to film the match, and subsequently study the film. Now replace the two tennis players with two atoms in a molecule. Instead of hitting a ball back and forth, the two atoms exchange electrons. But if you try to do the same thing with the electrons as with the tennis ball—define both the position and the motion of an electron accurately—you will fail miserably.

The reason for this failure is the very act of observation itself. Light is the only means that we have for communicating with the electron, for knowing where it is and where it is going. To observe it, I need to bombard it with particles of light or photons. But each photon possesses a certain amount of energy, which is linked to its wavelength (see Appendix A). That wavelength determines the degree of accuracy with which light can define reality, and localize the electron. The lower the energy, the longer the wavelength, and the more indeterminate reality becomes. If the energy is increased, the wavelength is decreased, and reality sharpens. If I attempt to measure the electron's position with radio waves, I will only be able to say that it is located somewhere within a large region whose size is comparable with the wavelength of the radio emission, which is several tens of meters. If I illuminate the electron with the visible light from a lamp, I could pinpoint its position to within a few ten-millionths of a meter. Gamma radiation would permit the position of the electron to be defined with the extremely high accuracy of one billionth of a millimeter.

"So," you say, "what's the problem? All you need to do is to illuminate the electron with highly energetic radiation, like gamma rays, and you will be able to determine its position as precisely as you wish." But, and it is a very big "but," position alone is not enough to describe the electron's actual state. We also need to know its motion. In bombarding the electron with photons to determine its position, we will disturb it. The photons transfer their energy to the electron, and thereby alter its motion. The greater the inherent energy in the light, the greater the alteration. So we are face to face with a dilemma: The more we try to reduce the uncertainty about the electron's position by illuminating it with ever more energetic photons, the more

11. P. S. de Laplace, *Essai philosophique sur les probabilites,* Gauthier-Villars, 1921, p. 3.

we disturb it, and the greater the uncertainty in its motion. The very act of determining creates indeterminacy.

There will never be a way of getting around this dilemma. We must make a choice. Either we measure the position of an electron very accurately, in which case we will have to forego knowing its motion, or else we determine its velocity and accept that its position remains uncertain. We can never know both precisely and simultaneously. This indeterminacy does not arise because we are lacking in imagination in carrying out our calculations, or because our equipment is not sophisticated enough. It is a fundamental property of nature, and was discovered in the 1920s by one of the founders of quantum physics, the German physicist Werner Heisenberg. Nature obeys an uncertainty principle* (see Appendix D). The information we can derive about an electron can never be complete. Its precise future, which depends on such information, remains forever inaccessible. Laplace's earlier dream of a universe functioning like some perfectly oiled mechanism, where the past, the present, and the future of each atom could be understood by human thought, is reduced to tatters. There will always be an element of chance in the destiny of atoms.

The elements of chance and uncertainty that reign over the subatomic world fade away in the macroscopic world. We can, in principle, simultaneously measure the position and velocity, to whatever degree of accuracy we require, of a tennis ball, a ship at sea, an aircraft in the sky, or a star in a galaxy, and discover their past and future. The light that enables us to acquire this information has undoubtedly interacted with all these different objects, but its energy is so weak relative to that of the bodies themselves that the perturbations it induces are always negligible. It is just as if there had been no interaction. Because the act of observation does not disturb macroscopic objects, the physical laws describing their behavior are perfectly deterministic. They are perfectly adequate for describing everyday objects: the paths of aircraft, trains, or ships and the lives and deaths of stars and galaxies, which is lucky for our sanity. The uncertainty of our destiny is bad enough. We can certainly do without any additional uncertainty in the behavior of the everyday things that surround us. Like any tennis player, I like to know that if I hold my racket in a certain way, and hit the ball with a carefully applied force, it will land inside the court, and not randomly somewhere else. Limited to the macroscopic world, Newton and Laplace could only know determinism. The element of chance that rules at the very heart of matter was completely outside their experience.

Observation Defines Reality

An anthropologist turns up in the midst of the Amazonian rain forest to study the culture and customs of a tribe of Indians. His very presence will disturb the behavior of the natives. The Indians will not behave in exactly the same way under the eyes of a researcher as they would if they were on their own. The results of the anthropologist's observations will have been altered by his very act of observation. As for ourselves, who would not admit to having behaved differently, just because someone else was present?

Observation modifies reality and creates a new version of it. While this may only occasionally apply to human affairs, it constitutes a fundamental law of the subatomic world. It is pointless to speak of an electron's "objective" reality, that is, of some reality that exists regardless of whether it is observed or not, because that reality can never be seized. Any attempt to capture objective reality is doomed to failure, because it is irrevocably altered and turned into a "subjective" reality that depends on the observer and his instrumentation. We can talk about the reality of the subatomic world only in the presence of an observer. We are no longer passive spectators watching a majestic drama unfolding in the atomic world. Our very presence changes the course of the drama. The musical notes emitted by the electrons are altered by the very fact that we are hearing them. The shape of the melody created by the subatomic world is inextricably linked to our presence, and any equation describing the behavior of that world should explicitly include the act of observation.

The Dual Nature of Matter

Because an electron will never yield simultaneously its position and velocity, we can never speak of an electron's trajectory as we can speak of the Moon's orbit around the Earth. We can never say that the electron goes from point A to point B by a specific route, as we might speak of a vehicle going from San Francisco to Los Angeles on Route 1. But then how does it get from A to B? By simultaneously taking every path between A and B. All roads lead to Rome, and an electron takes them all. Within an atom, it is not confined to a single orbit around the nucleus, like a planet orbiting the Sun, but like some demonic dancer in a ballroom, twists and turns, pirouettes, takes a few steps here and a few steps there, and is everywhere at once. But how can an electron take every path and be everywhere at once? By stepping into its other persona, because electrons, photons, and all other elementary particles have dual personalities. They are both particles and waves.

The particle, when it adopts its wave character, is able to propagate across and fill the whole of the empty space within the atom, just as the circular ripples set up by a stone falling into a pond propagate to occupy the whole surface of the water. Provided I am not observing it, the electron can escape from the rigid, deterministic world—where the position and velocity of every particle must be accounted for simultaneously—and be everywhere at once. I will never be able to predict where it will be at a specific instant. At the most I can only estimate the probability that it will be at a certain point. To do this, I need first to calculate the wave form associated with it by following the procedure established in 1926 by the Austrian physicist Erwin Schrödinger. The electron's wave form, like ocean waves, has a large amplitude at certain points (corresponding to the crests of the waves) and a far smaller amplitude at others (corresponding to the troughs). To obtain the probability of finding the electron, I then need to follow the instructions first given (also in 1926) by the German physicist Max Born, which say that it will suffice to square the amplitude. Thus to maximize my chances of encountering the electron, I should choose the crests of the waves and avoid the troughs. But even at the crests of the waves, I can never be certain of encountering the electron. Perhaps, on two occasions out of three (corresponding to a probability of 66%), or four times out of five (a probability of 80%), the electron will be there. But the probability will never reach 100%. Certainty has been banished from the subatomic world, and chance has invaded it. If I flip a coin in the air, the laws of probability tell me that on average half of the time the coin will fall heads up, and the other half tails up. However, I cannot predict whether at the next toss, the coin will land head or tail, just as I cannot be certain of encountering an electron at a wave crest.

Chance is inherent in the subatomic world. Certainty is totally defeated. The great Einstein, an inveterate determinist, had great difficulty in accepting the major role that chance played in the subatomic world, despite his pioneering role in recognizing the wave–particle duality of matter. "God does not play dice," he said. But he was wrong, because God does play dice. The predictions made by quantum mechanics, which allocate a major role to chance, have always been confirmed by laboratory experiments. For "chance" does not necessarily mean "total chaos" or "lack of predictions." Predictions are the mark of a good scientific theory. Instead of predicting isolated events in the macroscopic world, such as the fall of an apple, the path of a tennis ball, or the motion of the Moon around the Earth, as in Newton or Laplace's classical mechanics, quantum mechanics describes statistically the average behavior of a vast number of events that occur in the subatomic world. Although the theory is incapable of predicting the exact instant when a single atom of carbon-14 will disintegrate, it makes up for it by telling us how many, on average, out of a large number of carbon-14 atoms will disintegrate after a wait of 1, 100, or 10,000 years. Here, causality does not apply to the individual, but is at work when large numbers are involved.

Quantum uncertainty is therefore an integral part of the life of an individual elementary particle. Before the observation, it is indeterminate, because it behaves like a wave, and is thus able to take all roads leading to Rome. After the observation, it appears as a particle. But the uncertainty persists. Because the observation perturbs it, the particle refuses to provide us with its position and motion simultaneously.

Again, common sense rebels. You ask: "How can an electron be a particle and a wave at the same time?" But there is no validity to your question. Nature is just like that. The dual nature of particles has been verified innumerable times in the laboratory. Our everyday experience is not a good guide when it comes to the infinitesimally small. Like Janus, every particle has two faces. They are two equally valid descriptions of nature, and they complement one another. In the atomic landscape, the principle of complementarity*, proposed by the Danish physicist Niels Bohr, came to supplement Heisenberg's uncertainty principle*.

Everything Comes to Those Who Wait

This raises a question. If chance rules the lives of individual atoms, why does it disappear on the macroscopic scale, yielding its place to determinism? After all, macroscopic objects are made of subatomic particles. Why doesn't the Moon suddenly leave its elliptical orbit around the Earth, and fly off into space to revolve round Jupiter? The laws of quantum mechanics say that, in principle, this is possible. But the probability of such an event occurring is so minimal that we would have to wait for all of eternity to see it happen. The key to the answer to our question is contained in the enormous number of atoms (10^{50}) that make up the Moon. In the presence of a large number of particles, chance gives way to determinism. But, and this is the important point, it is never completely absent. Quantum uncertainty does, in principle, permit the Moon to wander off to Jupiter, if all eternity were available. But our pitiful 100-year lives, the Solar System's 4.6 billion years, or even the universe's 15 billion years are only a brief instant compared to eternity. You will not wake up tomorrow and find that the Moon is orbiting Jupiter.

Similarly, the large number of atoms contained in everyday objects prevents chance from asserting itself. If you lay a book on a table, you don't run the risk of finding it in the bathtub. A robber outside a bank would not see the money in the vaults suddenly materialize in his pockets. The Mona Lisa is not likely to disappear from the Louvre and turn up in my home. Quantum mechanics says that there is a nonzero probability that such events can occur. Everything is possible, provided one is prepared to wait long enough. But the wait is likely to be very, very long, which is why tales of people who disappear from one point in space and turn up in another only belong to science fiction.

God's Dice and Genetics

But does this then mean that quantum uncertainty has no influence on our existence? Certainly not. Because, to take just one example, it is responsible for the energy of the Sun, which is the source of all life on Earth, particularly our own. The core of the Sun is a vast stellar furnace at some ten million degrees that produces energy by fusing protons and neutrons to form nuclei of helium. Driven by the immense heat, the protons and neutrons are hurled against one another. Protons carry a positive charge, however, and when they encounter one another, the electromagnetic force, which increases as the distance between the protons decreases, and repels particles of like charge, makes them turn back. This is what happens most of the time. The majority of protons cannot get closer than a certain distance before being repelled. Most of the collisions fail to take place. But from time to time, the code of conduct imposed by the electromagnetic force is violated. Quantum mechanics allows the protons to bypass the electromagnetic rules, and as far as it is concerned, anything goes; laws are made to be broken. It allows the protons to approach sufficiently closely to undergo fusion. Such quantum-mechanical violations of the law would be extremely rare if only a few protons were involved. But there are so many protons (10^{57}) in the core of the Sun that the combined effect of all the violations is enough to keep our star shining. Unlike the bank robber, the protons can quickly profit from quantum uncertainty, because of their extremely large number. Quantum uncertainty is therefore directly responsible for our existence. Without its capacity to overrule the rigid electromagnetic laws, stars and galaxies would not shine at night, and we would not be here to talk about it.

You are making use of quantum uncertainty when you put a Bach compact disc into your laser player, or watch the latest episode of your favorite soap opera on TV. These devices only function thanks to small electronic components known as transistors. They amplify the electrical current using an effect that can only be understood in the framework of quantum mechanics. And the very existence of every one of us is the result of quantum uncertainty. It affects the formation of the long chains of molecules in the DNA helix responsible for our genetic inheritance. God does play dice in determining our individual genes.

Quantum uncertainty is thus all pervasive. The next time you admire the pure and perfect lines of one of Rodin's marble statues, remind yourself that beneath that apparently solid, unchanging exterior and that apparent perfection lies a world in a continuous state of flux and of extreme uncertainty. Under extreme magnification, the statue, which seems such a solid object, dissolves into empty space—empty space that not only separates the atoms, impeccably aligned in crystalline lattices like rows of well-disciplined soldiers, but, above all, occurs within the atoms themselves. An atomic nucleus* occupies only one million-billionth (10^{-15}) of the volume of an atom, being completely lost in a vast immensity. It is like a grain of rice somewhere in the middle of a football field. The monotonous void inside the atom is interrupted only by the passage of trains of electrons that frantically crisscross it in all directions, whirling from one point to another, appearing everywhere and yet stopping nowhere. Neither are atomic nuclei tranquil places; the protons and neutrons within them are in constant motion like restless dancers in a vast ballroom. Their movements are less frenetic than those of the electrons, because protons and neutrons are far more massive, and also because the nucleus-ballroom is far smaller than the atom-ballroom occupied by the electrons. On a subatomic scale, Rodin's statue appears as almost completely empty space disturbed only by the quantum fluctuations of electrons, protons, and neutrons.

Energy Loans from Nature's Bank and Virtual Particles

Quantum mechanics is a rebel theory that has a fondness for transgressing laws and defying the established order. Even the bastions of classical mechanics and relativity, which were thought to be absolutely solid and well protected, have fallen beneath its onslaught. One of the most spectacular of the fortresses that have fallen is the black hole. As its name indicates, a black hole should not allow anything, whether matter or radiation, to escape once a certain critical radius has been crossed. Like a glutton, it devours anything that passes close to it, whether that be stars within a galaxy or a spaceship, becoming larger and larger and more massive in doing so. According to relativity theory, the matter that has been engulfed is definitively and forever lost to the outside world. But to quantum mechanics, "definitively" and "forever" are words that do not exist. For it, anything can happen, provided one has the patience to wait. Using quantum mechanics, the British astrophysicist Stephen Hawking showed, in 1974, that black holes were not completely black, that they could lose mass and evaporate, emitting light as they did so. In an ironic turn of events, "black holes" could now "shine"!

Quantum mechanics achieves this extraordinary feat by resorting to a different form of Heisenberg's uncertainty principle. Nature not only prevents us from knowing the position and the velocity of an electron simultaneously, but also introduces indeterminacy in the energy of an elementary particle, an indeterminacy that depends on the lifetime of the particle. The shorter its lifetime, the more uncertain is its energy (see Appendix C). This energy uncertainty enables quantum mechanics to break the rules when it comes to the principle of the conservation of energy, which prevails in the macroscopic world. This principle may be summed up as "Life is not a free lunch" or "You get nothing for nothing." We have to work and expend energy to feed ourselves. A car is only able to run because we refill its fuel tank. If we add up all the energy expended by a car, we will find that the total amount of energy consumed is exactly equal to the energy content of the fuel that has been used.

Nature behaves differently in the quantum world. It no longer respects the law of the conservation of energy. Thanks to quantum uncertainty, it is able to create energy where there was none before. One could say that its dictum is "Energy may be free. It may be obtained from nothing." Nature can lend energy, without demanding anything in return, and this free energy can create elementary particles. But Nature's energy bank is subject to the uncertainty principle. Every energy loan must be repaid sooner or later, and the greater the amount of "borrowed" energy, the sooner it must be repaid. Although the amount of borrowed energy required to create a particle is minute (no matter how energetic, we would never feel a particle if it were to collide with our skin), it is still far too large for the energy bank, so a particle's lifetime is an infinitesimal fraction of a second. The energy loan is repaid, the bank recovers its energy, balances its books, and the particle disappears.

The particles that arise from quantum fluctuations lead an ephemeral existence. A brief, elusive appearance, and they are gone. Left to themselves they are never able to leave their shadowy world and enter the real world. We cannot detect them with our instruments. They are "virtual particles*," potential particles that do not come into being. For them to become real, energy is required, and Nature's energy bank refuses long-term energy loans. The space that surrounds us is therefore filled with an inconceivably large number of evanescent particles that appear and disappear at a frenetic rate. At any given instant, a cube of space no more than 1 centimeter on a side may contain some 1000 billion, billion, billion (10^{30}) virtual electrons.

These virtual particles are not restricted to matter: In fact their very existence is possible only because of the existence of similarly ephemeral antimatter particles. Although Nature may allow energy to be borrowed, it absolutely refuses to sanction loans of electrical charge. The law of the conservation of charge is strictly respected. Because the electrical charge of space is zero before any virtual particles appear, the creation of a virtual electron with a negative charge must always be accompanied by the creation of a virtual positron with the opposite charge. The seething activity of the virtual particles is accompanied by similar frenetic behavior of their virtual antiparticles.

Having got this far, you will probably be saying to yourself that physicists have extremely tortuous minds. What is the good of inventing virtual particles that cannot be detected? Is physics going completely mad? Luckily this is not the case, because although virtual particles may not be directly detectable, their presence may be deduced in an indirect manner. Without their ephemeral existence, certain aspects of the behavior of atoms could not be explained at all. Virtual particles are carriers of the strong nuclear force, which binds protons and neutrons together into atomic nuclei.

The other particularly interesting point about virtual particles is that, under exceptional circumstances, they may come into being, and enter the real world. If a virtual particle is able to discover some generous benefactor who will repay its energy debt to Nature's bank, it can leave the virtual world and materialize, together with its antiparticle, in the physical world. Gravity—and this is where we rejoin the story about black holes—is happy to act as benefactor, helping virtual particles to realize their potential.

Evaporating Black Holes

As we have seen, a black hole's gravity is extremely strong. Because it is very rich in energy, gravity is able to pay back the energy borrowed by the particles and antiparticles that are just outside the black hole's frontier of no return. Once their loan repaid, the particles and antiparticles leave their virtual existence to enter the real world. Let us follow the fate of one such electron–positron pair. There are several possible outcomes: The two particles may fall into the clutches of the black hole and be swallowed by it, in which case their existence in the real world would have been of extremely short duration; or the electron may escape, while its partner, less fortunate, is engulfed by the black hole. The electron goes on to encounter a positron that has also escaped from the black hole's gluttony, and the two annihilate one another in a burst of energy. If the electron and its antiparticle both escape, their mortal embrace also produces a flash of energy. Light is therefore escaping from the black hole, which is "radiating." The energy that gravity pays to enable the virtual particles to materialize comes, in the last analysis, from the energy associated with the black hole's mass. As gravity generously makes use of its energy reserves to allow the virtual particles to enter the real world and convert into radiation, the mass of the black hole decreases. The black hole literally "evaporates" into light. So our friend Jules, who was dragged down into and killed by the black hole, actually reappears in the form of particles of light. A rather poor consolation!

The evaporation rate is not the same for all black holes. Each is characterized by a temperature that controls the evaporation rate and is inversely proportional to the mass. The more massive a black hole, the lower its temperature, and the slower its evaporation rate. Unlike humans, the most massive black holes live longest. The lifetime of a black hole is proportional to the cube of its mass: Double the mass and it will live eight times as long. Black holes that are produced when stars die have masses several times that of the Sun. Such a mass is so large and the evaporation rate is so slow that black holes resulting from dying stars appear eternal. For example, a black hole with the mass of the Sun (2×10^{33} grams) would have a temperature of only one ten-millionth of a degree, and would take 10^{65} (1 followed by 65 zeros) years to evaporate. Assuming that it can. Because just as the water in a teapot can only evaporate when the temperature of the air surrounding it is lower than that of the boiling water, a black hole can only evaporate if the surrounding space is colder than it is. We know that this is not the case at present, because the cosmic background radiation that bathes the universe has a temperature of 3 K. We have to wait more than 10^{20} years, until the cosmic expansion has cooled the temperature to one ten-millionth of a degree (assuming that the universe has not collapsed on itself beforehand), for a black hole with the mass of the Sun to begin to transform itself into radiation (see Appendix C).

Stellar black holes are essentially eternal. Are there no black holes with much smaller masses that might show a far more significant rate of evaporation? Low-mass black holes are not very common. It is very difficult to compress low-mass objects to a size smaller than the critical radius and turn them into black holes. You will recall that if we want to turn a person weighing 70 kg into a black hole, we have to compress him to a size of one hundred thousand billion billionths of a centimeter (10^{-23} cm) to achieve this. No natural process and no technology would allow this to happen. In our present universe, gravity is able to convert massive stars containing 10^{33} grams into black holes, but not individual persons weighing 70 kg.

Nevertheless, there was a time in the universe's history, right at its beginning, when it was so small and so dense, and gravity was so strong, that low-mass, primordial, mini black holes could be formed. Stephen Hawking has suggested that during the first fractions of a second of its existence, the universe could have given birth to a multitude of mini black holes, with masses of around one billion metric tons (10^{15} grams). Here, the term "mini" is all relative. The mass of a mini black hole is roughly one billion billionth of that of the Sun, but still ten billion times that of a human being. A primordial, mini black hole has a temperature of 120 billion K, and is the size of a proton (10^{-13} cm). It emits energy at the rate of 6000 megawatts, equivalent to the amount produced by several nuclear power stations. As it evaporates, its mass decreases, its temperature increases, it emits even more radiation, and the mass loss accelerates. After 15 billion years (the age of the universe), the billion metric tons has shrunk to 20 micrograms (the mass of a speck of dust), and then comes a brilliant firework. The mini black hole ends its existence in an enormous explosion that releases as much energy as 10 million 1-megaton hydrogen bombs, and which is as luminous as 10 million billion galaxies. The gamma-radiation that these cataclysmic explosions produce is the most energetic of all. These cosmic fireworks have never been detected, so for the present, mini black holes remain in the realm of speculation.

We have now come to the end of our exploration of the strange and fantastic world of quantum mechanics, and are ready to open the history book of the universe.

| 5 |

The History Book of the Universe

The definitive history book of the universe has yet to be written. Many pages of such a book will need revisions at a later date, and only those that are concerned with the present epoch are likely to remain unchanged. Both the beginning, which describes the birth of the universe, and the end, which deals with the distant future, are based on extremely daring extrapolations of the laws of physics, as currently understood. Is it really permissible to do this? Only time will tell. However, the story of the Big Bang as we understand it today is already marvelously fascinating and deserves to be told.

As we have seen, in our Big Bang universe, time is elastic. It dilates or shrinks, depending on the motion of the observer, or according to the amount of mass in the latter's vicinity. Any history book describes a series of events throughout time. Which of the infinite number of possible time scales should we choose? To tell our story, let us adopt "cosmic" time, namely, the time experienced by someone who is carried along by the expansion of the universe, and to whom the majority of galaxies appear to be receding. This time is practically identical with that of the Earth, of the Sun, or of the Milky Way. All these objects do, of course, take part in the fantastic cosmic ballet that we described earlier (Fig. 27), in addition to the expansion, but these motions are so minute in comparison with the speed of light that the changes they create are negligible. A year of cosmic time, therefore, is the time it takes the Earth to complete its orbit around the Sun. Inhabitants of other galaxies would measure the same cosmic time. They would be able to read our story without making any changes.

We could have asked our astronaut, Jules, to recount the history of the universe as he was speeding away in his spaceship, and seeing some galaxies approaching with their light shifted toward the blue, and others receding with light shifted toward the red. His time would be slowed down. Similarly, he could have recounted the history of the universe as he was getting closer and closer to a black hole, when his time was slowing down progressively until it eventually came to a halt. The story would be the same, but the events would be more complicated to describe and the book would be far longer. To spare you the boredom of an unending story, we have adopted the simplest method, the use of cosmic time. The universe has not yet revealed its true age, so we will assume that it is 15 billion years. If, at a later date, the universe is found to be younger or older, all we need do is to add or subtract the appropriate number of years.

To tell the history of the universe in all its glory, the "pages" of our book will have a very specific property: Each page will take us 10 times farther along in time. For example, if we are at the page concerning the universe at an age of 100,000 years, the next page will bring us to the period of 1 million years after the Big Bang. On the subsequent page, the universe will be 10 million years old, and so on. We have to arrange our book in such a way because of the frenetic activity that occurs in the very earliest stages of the universe's existence. At that time, events were taking place at an unbeliev-able rate. To cover what was happening thoroughly and accurately, we need to discuss events that were extremely close in time, only separated by frac-tions of a second. As the universe aged, however, this initial, youthful exu-berance was replaced by the calm and serenity of maturity. New events and changes did not occur as frequently, so the reports on the state of the uni-verse can be more widely spaced in time, without missing anything essential. Time then comes to be measured in millions or billions of years.

Let us therefore open our book, and begin our voyage of discovery of the Big Bang universe, following in the footsteps of those intrepid explorers who are physicists and astronomers, with their only baggage the physical laws of the infinitely large (relativity) and of the infinitely small (quantum mechanics), together with astronomical observations (such as the recession of the galaxies, the background radiation, and the abundances of the ele-ments).

The Frontiers of Knowledge

Don't be disappointed if we begin our story not at "time zero," at the very instant at which space and time originated, but slightly later. But don't worry, our first account comes an unimaginably short interval, 10^{-43} second (0.000 . . .s) after the primordial "explosion." (The first figure to differ from zero comes in the forty-third decimal place.) This fraction of a second is so small that, relatively speaking, the flash of light from a flashbulb would last one billion billion billion times longer in the entire history of the universe, than 10^{-43} s would last in 1 second. Even though I am an astrophysicist, used to dealing with astronomically large or infinitely small numbers all day long, I still feel that there is something unreal about this figure. Yet the physical laws deduced from observation of the present-day universe appear to be capable of being extrapolated backwards to such an incredibly remote time, when the universe was unimaginably small, hot, and dense.

At 10^{-43} second after the Big Bang, what was to become the universe observable today was just 10^{-33} cm in diameter. That is 10 million billion billion times smaller than a hydrogen atom. The universe was so young that light had not been able to travel very far, and the cosmological horizon* was extremely close. It was unimaginably hot (10^{32} K) and dense (10^{96} times the density of water). Its energy was immeasurably great. If we wanted to attain such energies, we would have to build particle accelerators with diameters not of several kilometers, like the one at CERN (the European Nuclear Research Center) at Geneva, but light-years across. The construction of an interstellar accelerator that reaches as far as the nearest stars is not likely to be started just yet, which is why the particle physicists are extremely interested in the universe at its birth. It is free. It enables them to test their physical theories at energies that can never be achieved on Earth. Naturally, there are some disadvantages that go with this cosmic accelerator. The experiment is unique, and cannot be repeated. It took place in the dim and distant past. One does not have control over it as for an ordinary laboratory experiment, but the advantages outweigh the disadvantages.

Just 10^{-43} second after its birth, the universe was so compressed, and its density was so high, that gravity, which is normally negligible on the subatomic scale, was as strong as the other forces: the strong and the weak nuclear forces and the electromagnetic force. But, and here's the rub, we do not know how to describe the behavior of atoms and light when gravity is so intense. This problem was noted by the German physicist, Max Planck, at the beginning of the twentieth century, which is why the time of 10^{-43} second is known as the Planck time*. Physics runs out of steam and cannot go beyond the Planck time. Quantum mechanics accurately describes the behavior of atoms and light when gravity is negligible. Relativity can account for the properties of gravity on a cosmic scale, when the nuclear and electromagnetic forces no longer predominate. As yet, however, no one has been able to unify these two theories and describe the situation where all the forces are comparable in strength.

Physicists are currently struggling to find a "Theory of Everything." For some years they have suspected that the four forces that govern our present-day universe are only different aspects of a single force that prevails throughout the universe. In this respect, the situation would parallel that of the electrical and magnetic forces, which are simply two different aspects of the electromagnetic force. Under concerted attacks by the American Steven Weinberg and the Pakistani Abdus Salam, the electromagnetic and weak nuclear forces were unified as the electroweak force in 1967. There are hopes that the strong nuclear force can be unified with the electroweak force to form the electronuclear force, although the theory, known as the Grand Unified Theory, is still in its early stages of development. The combination of the electromagnetic force with the two nuclear forces, the weak and the strong interactions, requires very specific conditions. It can only occur in an extremely hot state, where the energies of the elementary particles are exceptionally high, as in the earliest instants after the Big Bang. Heat and energy are indispensable for the forces to become one. Once the universe cools below a certain temperature, the union is broken. This is what happened as the universe expanded, so that now the three forces are separate. The coldness of the universe today has dissolved their union.

Gravity still obstinately resists all attempts at unification. Ruler of the infinitely large, it refuses to have any links with the forces governing the infinitely small. It does not want to be "quantified." The unification of quantum mechanics with relativity presents barriers that are currently insurmountable. Even Albert Einstein, who worked relentlessly on the problem for the last 30 years of his life, failed to overcome it. As long as gravity resists unification, it will be impossible to go back beyond the Planck time, which therefore marks the limit of our knowledge. We have no means of tackling whatever lies beyond the barrier at the Planck time. Our universe's four-dimensional space–time may be completely different, or no longer exist. Physicists who have made brief sorties into the world beyond the Planck time barrier, say that they have caught glimpses of a chaotic universe with 10, or even 26, dimensions, where gravity is so strong that it has rearranged the very fabric of space itself, causing it to have 6 (or 22) more dimensions, where space has collapsed, as a result of its own gravitational attraction, into innumerable microscopic black holes, and where the past, the present, the future, and even time itself, no longer have any meaning. In fact, an infinitely long duration of time may be hidden behind the Planck time barrier. The figure of 10^{-43} second only results from extrapolating our physical laws backward towards time zero. But because those laws lose all relevance at the Planck time, this is of doubtful validity.

As for the microscopic black holes, they have a mass of 20 micrograms, the smallest mass that can exist. With a diameter of 10^{-33} cm, they are 100 billion billion times smaller than a proton. They have a temperature of 10^{32} K, and evaporate in 10^{-43} second, disappearing and reappearing (thanks to quantum uncertainty) in an infernal cycle of birth and death. Was the universe infinitely dense and hot at its origin? Relativity says that it was, but even here there is no certainty. In the past, the appearance of infinite values has always signalled a breakdown of our theories, rather than extreme behavior on the part of the universe. It has been a sign of a lack of imagination on our part, rather than a property of nature itself. It will take years of hard work to break through the Planck barrier. In the meantime, we shall have to adopt the Planck time as our "time zero." When we speak of the origin, the beginning, or the creation of the universe, we shall mean the Planck time.

An Awesome Expansion

Let us, therefore, open our book at its first page. The universe has the unbelievable age of just 10^{-43} second, and its temperature of 10^{32} K is far hotter than any hell that Dante could have imagined. The entire universe is contained within a sphere one-thousandth of a centimeter in diameter—the size of the point of a needle. Only an infinitesimally small fraction of that point of needle-universe can be seen. The cosmological horizon which limits the observable universe, and is defined by the distance traveled by light during the universe's existence, is a minute 10^{-33} centimeter. We are just on this side of the Planck barrier. Space-time as we know it has just appeared. Three-dimensional space is being created continuously as the universe expands. Front and back, up and down, left and right have all acquired meaning. The other dimensions of space that may have existed beyond the Planck barrier have shrunk to such an extent that they are no longer detectable. This loss of dimensions may be likened to what would happen if a two-dimensional sheet of paper could be rolled up so tightly that it became a one-dimensional straight line. The "quantum vacuum" prevails. But this is not the calm, quiet vacuum, devoid of all substance and activity, which we all imagine, but a dynamic vacuum, seething with energy, that hides turmoil and activity beneath its calm and placid appearance. We have already seen how, thanks to quantum uncertainty, this vacuum is full of a vast number of virtual particles and antiparticles that appear and disappear by borrowing energy from Nature's bank and paying it back. Two forces control the universe between them: the electronuclear force, resulting from the combination of the electromagnetic force and the strong and weak nuclear interactions, and gravity, which still resists any attempt at unification.

Let us read on. As we turn the pages, and as time passes, the expanding universe becomes less dense and grows colder. At the ninth page, at 10^{-35} second, the universe has cooled by a factor of 10,000, but still has the infernal temperature of 10^{27} K. An extraordinary event is about to take place. The drop in temperature causes the electronuclear force to split in two. The strong nuclear force breaks apart from the electroweak force that results from the combination of the electromagnetic force and the weak nuclear force. Gravity is still present, obstinately staying apart. Previously ruled by a pair of forces, the universe will be, from this time onward, governed by a triumvirate. This moment marks the beginning of a slow rise in complexity, which will ultimately result in our own existence. The universe loses its simplicity, its "symmetry," as it cools. It behaves like salt or water that is cooled. Heated salt and water are liquids and possess no structure. There are no preferred directions, just as the primordial, extremely hot universe was "symmetrical." But as they cool, both salt and water crystallize in the form of cubes. They lose their initial symmetry, because there are now preferred directions. The matter within them favors the flat surfaces of cubes. Just like salt and water, the universe "crystallizes" as it cools. Water is said to undergo a phase change* when it transforms to ice. Similarly, physicists describe the "crystallization" of the universe as a phase change.

This "crystallization" at 10^{-35} second will have far-reaching consequences for the universe and therefore for our existence. Just as water releases heat when it freezes, the universe's phase change will release the incredible energy present in the vacuum. Like the explosion of a bomb that hurls fragments of a building in all directions, the injection of the vacuum energy causes the universe to undergo a stage of extreme expansion, which the American physicist Alan Guth has called "inflation." Just as severe economic inflation causes money to lose its value and prices to escalate in a very short period, "inflation*" of the universe causes every part to undergo an awesome increase in its volume in an infinitesimal period of time. During the inflationary era, which lasts an infinitesimal fraction of a second, between 10^{-35} and 10^{-32} second (pages 9 to 12 in our book), the universe tripled its size every 10^{-34} second. There are 100 intervals of 10^{-34} second in 10^{-32} second—the duration of the inflationary era—so every region of the universe tripled in size 100 times in succession. Multiply $3 \times 3 \times 3 \ldots$ 100 times, and you will find that the universe has increased in size by a factor of 10^{50}, and its volume (which is proportional to the cube of its size) by a factor of 10^{150} (1 followed by 150 zeros) during the inflationary era. This frenzied expansion of the early universe contrasts markedly with the "monotonous languor" of the present-day expansion. During the past 10 billion years, the universe has increased its volume by no more than the relatively small factor of one billion (10^9).

This breaktaking expansion of the universe shortly after its creation allows many of the clouds that were casting shadows on the Big Bang theory to be dispelled. You will recall the problem posed by the extreme homogeneity of the present-day universe. How could diametrically opposite regions of the sky, which apparently have never been in contact, have exactly the same properties? The contact that is forbidden in an ordinary expansion becomes perfectly possible with inflation. To understand how this is so, note that after inflation the whole of the currently observable universe* was scarcely the size of an orange, about 10 cm across. (From now on, when I speak of the universe, this should be taken to mean the tiny portion of universe that can be observed by us, within the far larger entire universe.) Because the universe grew by a factor of 10^{50}, it began as an infinitesimally small volume of space with a diameter of 10^{-49} cm, that is, a billion billion billion billion times smaller than an atomic nucleus. In the beginning (at 10^{-35} second), and before the inflationary phase, our universe was so small that each infinitesimal part was in contact with every other. Light, which was the preferred method of communication between the various regions, and had a velocity of 300,000 kilometers per second, could have traveled a distance of 3×10^{-25} cm, even in the infinitesimally small period of 10^{-35} seconds at its disposal. The zone of possible communication was already 1 million billion billion times larger than our universe, and the various regions had no difficulty in achieving identical properties. After inflation, at 10^{-32}

second, the regions of the orange-universe were no longer in contact with one another, but they "remembered" that they had been.

The universe's flatness* problem also seems to be explicable within the context of an inflationary phase. The geometry of space flattens during expansion, just as a small region on the surface of a balloon flattens when the latter is inflated. We all know that the curvature of a sphere is less, the larger its radius. The universe, by growing so dramatically, became flat.

Myriad Universes

During the inflationary era, between 10^{-35} and 10^{-32} second, the universe therefore grew by a factor of 10^{50}, reaching the size of an orange. The cosmological horizon, which defines the region of space within which communication by light is possible, grew by the same factor. At 10^{-32} second, it reached 10^{26} cm, 1000 times larger than the observable universe today. Inflation not only increased the size of our tiny corner of the universe, but also did exactly the same for every other region of space. Our universe is just a tiny bubble, lost in the vastness of another bubble, a meta-universe, or super-universe, that is tens of million billion billion times larger. And that meta-universe is itself lost among a multitude of other meta-universes, all created during the inflationary era from infinitesimally small regions of space, all disconnected from one another. Although our observable universe is growing as light has more time to reach us, with ten new galaxies appearing over the cosmological horizon every year so that more of our meta-universe may be revealed, we can never communicate with any of the other meta-universes, or know what is happening in them. They are forever excluded from our sphere of observation. They may contain galaxies, stars, and other forms of life, or they may be inhospitable and empty; they may be paradise or hell, but we will never be in a position to know.

The phantom of Copernicus has had the final say. Not content with dislodging the Earth from its central position in the Solar System, with transforming the Sun into just a nondescript star near the edge of the Milky Way, and with reducing the Galaxy to insignificance in a vast universe containing 100 billion galaxies, it has even caused our universe to be lost within a meta-universe, and our meta-universe in a multitude of other meta-universes . . .

Everything Arises from Nothing

The release, during the inflationary phase, of the energy contained in the vacuum has another extremely important consequence. It gives rise to the matter contained within the universe. We have seen that the original vacuum was a seething foam, full of virtual particles and antiparticles that owed their existence to short-term energy loans. The injection of energy pays back those loans and allows the particles and antiparticles to leave their virtual world and become real. Quarks*, electrons*, neutrinos*, and their antiparticles are born from the vacuum. But as soon as they materialize, the particles and their antiparticles encounter one another and are annihilated, turning into radiation. The packets of radiation (or photons), in their turn, disappear, giving birth to particle–antiparticle pairs. There is a constant interaction between matter, antimatter, and radiation. The universe is bathed in a soup of quarks, electrons, neutrinos, photons, and antiparticles.

If there had been as many particles as antiparticles, that would be the end of our story. I would not be here to write these words, and you would not be reading them. Matter would destroy antimatter, and all that would be left would be photons. These photons would later lose their energy through the expansion of the universe, and would no longer be able to give birth to particles and antiparticles. All that would be left would be a universe full of radiation, from which elementary particles, stars, galaxies, human beings, and you and I would be absent. But, luckily for us, nature is not impartial when it comes to matter and antimatter. The Soviet physicist Andrei Sakharov discovered that it has a very slight preference for matter. For every billion antiquarks that arise from the vacuum, one billion and one quarks appear. Later, at 10^{-6} second (27 pages after the inflationary era), when the universe has cooled down sufficiently to allow the quarks to form protons and neutrons (known collectively as "baryons*") and the antiquarks to form the appropriate antiparticles (the antibaryons), most of the baryons and the antibaryons will annihilate one another and turn into radiation. But the fact that there is a slight preponderance of quarks over antiquarks means that some protons and neutrons will remain. For every billion particles and antiparticles that annihilate one another and turn into a billion photons, just one particle of matter will remain, which is exactly the proportion observed in the universe today. All the antimatter disappears.

One of the problems that had been overshadowing the Big Bang theory has thus been dispelled. Observations that had hitherto remained unexplained may now be answered. Matter can arise from a vacuum if a large enough quantity of energy is available. Everything, galaxies, stars, trees, flowers, you and I, all arose from the primordial vacuum. The idea of creation ex nihilo, from nothing, which even comparatively recently appeared to be a purely religious concept, now seems to be supported scientifically by cosmology. We live in a matter universe, and we do not run the risk of encountering an anti-you or an anti-me, simply because nature's preference is one-billionth greater for quarks than for antiquarks. Particles of radiation dominate the present-day universe by their sheer numbers, because the majority of matter and antimatter particles annihilated one another.

The soup of quarks, electrons, neutrinos, photons, and their antiparticles that emerges at the end of that unbelievably short interval of just 10^{-32} second is by no means uniform. Scattered throughout it are irregularities and "rough spots." Some parts are denser than others. These irregularities will later be amplified to give rise to the various structures that form the cosmic tapestry: galaxies, stars, planets, and the various oases that favor the increasing complexity of matter, and that will eventually permit life to emerge. The phase change in the universe did not just allow the first particles of matter, our far distant ancestors, to appear; it also sowed the seeds of the oasis needed for our appearance on the scene. This dispels another cloud overshadowing the Big Bang theory, the problem of the origin of the various structures in the universe.

The irregularities that arose were subject to very severe constraints. They could not be too small, because otherwise they would not have time, even in the 15 billion years that have elapsed to the present, to grow to the size of the majestic galaxies, hundreds of thousands of light-years across, that fill the universe today. Neither could they be too large, because the satellite COBE has measured no irregularity larger than 0.001% in the fossil radiation that came from the early universe. This double aspect of the universe, the almost perfect uniformity on large scales, accompanied by an incredibly rich variety on small scales, means that there is a delicate balance to be found for the irregularities. They could not be too modest nor be too excessive. Simple, preliminary calculations show that the irregularities that arose in the universe's inflationary phase are too large by a factor of 100,000 or more. We still have to write the page that describes how this problem was resolved. In the meantime, we should not fail to appreciate the enormous stride taken with the introduction of the concept of an inflationary phase of the universe; instead of having to invoke a "Grand Architect" who sowed irregularities into a regular universe, we are now even able to calculate what they might have been!

Another Phase Change in the Universe

Let us turn the page and leave the inflationary era, which has proved to be so beneficial for our existence. During its mad expansion, from a tiny region of vacuum, one billion billion billion billion times smaller than an atomic nucleus to the size of an orange, the universe cooled considerably. But the energy of the vacuum released when the first phase change occurred, when the strong nuclear force became separated from the electroweak force, caused the temperature to climb again to 10^{27} K. We are just 10^{-32} second after the primordial "explosion," on the twelfth page of our book. Our orange-universe continues to expand, but in a far less frenetic manner than previously, and at a pace more similar to the one occurring today, some 15 billion years later. Although distances increased exponentially during the inflationary era, subsequently (up to about 300,000 years) they increase only as the square root of time (see Appendix E). As we have seen, during the inflationary era, between 10^{-34} and 10^{-32} second, the universe grew by a factor of 10^{50}. At the rate prevailing after inflation, it would have grown only by a factor of 10 over the same period of time. The headlong expansion in the inflationary era drops to a snail's pace in the postinflationary epochs.

During the next 20 pages, from 10^{-32} to 10^{-12} second, the universe continues to expand, cool, and become less dense, but nothing unusual happens. It presents a monotonous and uneventful landscape (due perhaps to the physicists' failure of imagination?). All the pages look the same. The universe contains a handful of quarks, electrons, neutrinos, and their antiparticles. Because of the extremely high temperature, these particles are feverishly active, hurtling in all directions and colliding with one another. Collisions between particles and their antiparticles are fatal, turning them into photons. The latter are just as restless and add to the general frenzy. They, in turn, disappear, giving birth to new pairs of particles and antiparticles. Creation and destruction follow one another at a breathless pace.

By the thirty-second page, at one thousand-billionth (10^{-12}) second after the instant of creation, the universe has grown considerably. It is now not much smaller than the orbit of the Earth around the Sun. It is still very dense (1 million million times denser than an atomic nucleus) and hot (1 million billion degrees).

A very important event is about to occur, and the universe is about to enter a new stage of its progress towards higher degrees of complexity. It is, in fact, about to undergo another phase change. The triumvirate of forces governing the universe changes into a quartet. The union between the electromagnetic force and the weak nuclear force is broken, and henceforth the two forces will be separate. With the gravitational and strong nuclear forces, all four will rule the universe. This quartet of forces, which originated one thousand-billionth of a second after the primordial explosion, still persists in the universe today. As during the first phase change, the universe receives an energy boost, although far smaller this time. Instead of undergoing another inflationary phase, the universe continues to expand at a snail's pace.

The Quarks Are Confined

Let us turn over six pages. The universe is now 1 millionth (10^{-6}) of a second in age. Its volume is roughly equal to that of the Solar System. Its temperature has dropped to 10,000 billion degrees. The activity and the energy of the particles and antiparticles decrease as the universe cools. The motion of the quarks and antiquarks slows down sufficiently for the strong nuclear interaction to force them to combine into more familiar particles: protons, neutrons, and their antiparticles.

Like a shepherd who eventually manages to round up his sheep because they have become tired, the strong nuclear reaction rounds up the quarks and antiquarks. But it does not do this in a haphazard fashion. It gathers them together in groups of three. Both protons and neutrons result from three quarks being combined under the influence of the strong nuclear force. Quarks themselves are divided into two categories, depending on their electrical charge. One type has a positive charge equal to two-thirds of the charge on an electron, and the other has a negative charge equal to one-third of the electron's charge. A proton, which has a charge equal and opposite to that of an electron, consists of two quarks of the first type, and one of the second. A neutron, which, as its name indicates, has no charge, has the opposite composition: one quark of the first type, and two of the second. Because they owe their existence to the strong nuclear interaction, the proton, neutron, and their antiparticles are collectively known as hadrons, which means "strong" in Greek. Because hadrons are the principal players during this phase of the universe's existence, it is often known as the "hadron era*."

Henceforth, quarks and antiquarks will no longer be free. Throughout the rest of the history of the universe, we shall never see them in the free state again. To liberate them from their yoke, physicists have struggled to smash their proton and neutron prisons using tremendous bolts of particles that have been accelerated to very high energies in monstrous machines, several kilometers in diameter, like the one at CERN in Geneva. Despite this, no free quark has ever been glimpsed. The strong nuclear force, true to its name, has never relinquished its hold on quarks, despite repeated attacks. Until a new development occurs, quarks remain theoretical entities, born of physicists' fertile imaginations, but whose existence appear essential to explain the properties of the matter surrounding us.

Matter's First Victory

When the quarks become trapped, the universe enters the hadron era and matter takes another important step in its ascent towards complexity. The loss of liberty for the quarks and antiquarks is not the only remarkable consequence of the continuous cooling of the expanding universe, however. The perpetual cycle of particles and antiparticles being annihilated and changed into radiation, and of radiation being converted back into matter and antimatter, is about to be broken. The annihilation of matter can go on, but the metamorphosis of radiation into particles becomes more and more difficult. To understand why, we only have to remember Einstein's statement that all matter is energy. According to this famous dictum (which, unfortunately, thanks to human folly, led to the atomic bomb), a particle–antiparticle pair has a certain mass (equal to twice the mass of the particle), and thus a certain mass energy (equal to the mass of the pair multiplied by the square of the velocity of light). For a photon to be able to turn into a particle–antiparticle pair, it must have at least the mass energy of the pair, otherwise it would violate the law of the conservation of energy. A massive pair, such as a proton–antiproton pair, naturally requires more energy than a less massive pair, such as an electron–antielectron (or –positron) pair.

The photons are becoming tired, however, and are losing energy as the universe expands and cools. One ten-thousandth of a second after the Big Bang, the temperature of the universe has dropped to 1,000 billion degrees. The photons no longer have enough energy to be converted into proton–antiproton or neutron–antineutron pairs. The pairs that already existed annihilated one another. The reservoir of protons and neutrons and their antiparticles was rapidly depleted, but there was no new influx to replenish it. The vast majority of protons and neutrons were converted into radiation. But, because nature has that very slight (one billionth) preference for matter over antimatter, for every billion particle–antiparticle pairs that annihilated one another to become radiation, one particle of matter survived, because it was unable to find a corresponding antiparticle that would trap it in a mortal embrace. Matter has won its first victory over antimatter.

So, after the first ten-thousandth of a second of the universe's existence, the primordial soup turns into a mixture that consists mainly of photons, electrons, and neutrinos, with an extremely tiny scattering of equal numbers of protons and neutrons that have escaped complete destruction.

Although in this battle between matter and antimatter, antimatter has suffered a serious reverse, with its vast population of antiprotons and anti-neutrons being reduced to zero, antimatter is by no means beaten. It still has whole battalions of antielectrons (positrons) and antineutrinos, because, at a temperature of 1,000 billion degrees, photons still have enough energy to be converted into electron–positron and neutrino–antineutrino pairs. Because the conversion of radiation into particles and that of particles into radiation is in equilibrium, there are equal numbers of photons, of electrons and neutrinos, and of their antiparticles. Victims of the general slaughter, protons and neutrons, have been reduced to an infinitesimal minority among the particles in the universe. Only one proton or neutron remains for every 100 million or so of each of the other particles. The hadron era, which was dominated by protons and neutrons, comes to an end. Other, lighter particles now come to the fore, and these interact by means of the weak nuclear force. When the cosmic clock reaches one ten-thousandth of a second, the lepton* era (from the Greek *lepton,* which means "weak") begins.

The Neutrinos Remain Aloof

The cosmic cocktail in the lepton era consists of photons, electrons, neutrinos, and their antiparticles, with a small addition of protons and neutrons. Apart from the neutrinos, which remain aloof, the exceptionally high temperatures cause every one of these to be constantly interacting with others; no particle is able to move without immediately colliding with another. The photons that carry radiation and information are therefore quite unable to force their way through the thick forest of electrons, protons, and neutrons. The radiation, which in these first moments of the universe's existence is extremely energetic, and is in the form of gamma rays, is trapped. The universe is completely opaque, and not even the most powerful telescopes could have pierced through it.

Neutrinos, on the other hand, pass straight through this particle jungle as if it were not there, because they interact with the rest of the universe only by means of the weak nuclear force. As we have seen, the radius of action of this force is extremely small, and extends for just one ten million billionth of a centimeter (10^{-16} cm). After the first half-second of the universe's existence (four pages farther on in the book), and still in the lepton era, the density of the universe is already too low, and the particles are too far apart for the weak interaction to make itself felt. The neutrinos then behave as if the other particles did not exist, and no longer interact with any of the latter. This lack of interaction means that they are completely free to move about. Instead of being constrained, like the photons, they are free to race throughout the whole universe and, by virtue of their enormous number, to fill it completely. This population of neutrinos, which became separated from the rest of the particles towards the end of the first half-second of the universe's existence, is still roaming the universe today. Numerically, it is the second largest population in the universe, coming just behind the photons that make up the 3 K background radiation.

As I write these lines, hundreds of billions of neutrinos, born during the first instants of the universe's existence, cross every square centimeter of my skin, and pass straight through my body at nearly the speed of light. Each cubic centimeter of space contains hundreds of these fossil neutrinos. For every atom in the universe there are 100 million neutrinos (compared with 1 billion photons), and yet this incredibly large population of neutrinos has never been detected. It remains just a prediction of the Big Bang theory. The reason for this strange state of affairs is precisely the lack of interaction between the neutrinos and the other particles that populate the universe. Our telescopes and detectors are made of the particles with which neutrinos refuse to interact; so we have little chance of capturing and examining them. In addition, the primordial neutrinos have lost a lot of energy as a result of the expansion of the universe. Even if we were lucky enough to trap a couple of these neutrinos, they no longer have enough energy to trigger any nuclear reactions among the particles forming our detectors to reveal their presence. To detect these weak neutrinos, we need to build detectors that are millions of times more effective than those we have now, which is not likely to happen soon.

So, unlike the photons of the cosmic background radiation, which are easy to capture with radio telescopes like that used by Penzias and Wilson, because they interact more readily with matter, the primordial neutrinos remain elusive. Our only hope of being able to detect them one day is if they possess a mass. Even if their mass were only one ten-thousandth of that of an electron, neutrinos would be the predominant mass in the universe, thanks to their vast numbers. Because of their gravitational influence, they would profoundly alter the motion of stars and galaxies, and even change the future of the universe. We shall return to this point later, when we discuss the future of the universe. For the present, let us continue our history of the universe.

Antimatter Is Finally Routed

Apart from the separation of the neutrinos from the other particles, the lepton era is also notable for the final destruction of antimatter. The cosmic clock shows a time of 1 second. The temperature of the universe has dropped by a factor of 100, down to 10 billion degrees. The universe is still so dense that a single cubic centimeter weighs 100 kilograms. This is the instant at which a second major annihilation of matter and antimatter occurs. The sequence of events that occurred with protons and neutrons is repeated with electrons. The photons, weakened by the expansion of the universe, no longer have sufficient energy to turn into electron–positron pairs. The pairs that already exist are converted into radiation. But for every billion pairs that annihilate one another, one solitary electron remains without a partner, and escapes the massacre.

Matter always has a slight advantage. It continues to benefit from the slight preference that nature had for it during the inflationary era, and even if antimatter produces one billion antielectrons (positrons), matter always has one billion and one. The victory of matter, despite phenomenal losses, is now final, and antimatter's defeat is total. Ever since its first second of existence, the universe has been such that you and I can exist, and not an anti-you and an anti-I. On the other hand, because nature shows the same preference towards protons, which carry a positive charge, as towards electrons, which have a negative charge, there are as many positive charges in the universe as negative. This means that the overall charge of the universe is zero, and that we live in a universe that is electrically neutral.

The Decline of the Neutrons

The destruction of antimatter creates an imbalance between the populations of protons and neutrons, and this will have important consequences for the future chemical composition of the universe. We have seen that equal numbers of protons and neutrons emerged after the universe's first millionth of a second, because there were equal numbers of the two forms of quarks. When it comes to their lifetimes, however, protons and neutrons are fundamentally different. A free proton will live for at least several ten thousand billion billion billion (10^{31}) years—which is practically all eternity. It is a particle that is "almost" stable. The neutron, on the other hand, is quite unstable. It rapidly turns into other particles. A free neutron disintegrates into a proton, an electron, and a neutrino after just 15 minutes, this disintegration being controlled by the weak nuclear force. Left to themselves, therefore, neutrons would disappear from the face of the universe after just a quarter of an hour. But protons come to the rescue. They replenish the population of neutrons by combining with electrons, turning into neutrons and neutrinos. This transformation is also accomplished by the weak nuclear force.

Before the time of 1 second on our cosmic clock, as many neutrons were produced by the coupling of protons and electrons as the number that disintegrated; so the population of neutrons remained constant, and equal to the number of protons. There was an abundance of electrons with which the protons could combine to form neutrons. But as soon as the clock reaches 1 second, practically all the electrons are annihilated by their antiparticles.

The protons are no longer able to find enough electrons to produce neutrons. Henceforth, the birth rate of neutrons drops, and the population of neutrons progressively declines relative to that of protons. The numerical equilibrium is upset. After the first second, only two neutrons remain for every ten protons, a ratio that will be of paramount importance for the chemical composition of the future universe. Thus ends the first second, which has seen the birth of the universe, almost ex nihilo, the emergence of matter, and the setup of the physical conditions needed for the subsequent buildup of complexity. This 1 second, by virtue of all the various events that occurred within it, is of greater significance than all the other 10^{17} seconds in the 15 billion years that followed.

A Factory That Builds Helium

The end of the first second tolls the knell for the lepton era and ushers in the beginning of the radiation era*, the reign of the photons. As we have seen, these photons arise from the destruction of proton–antiproton and neutron–antineutron pairs in the first great annihilation event, and also of the electron–positron pairs in the second. By virtue of their number they predominate over the few protons, neutrons, and electrons that escaped from the wholesale destruction that took place in these two events. For every proton or electron that survived there were one billion photons, and this ratio has persisted to the present day. The universe is almost entirely radiation, being full of photons, which interact strongly with matter, and neutrinos, which are supremely indifferent to it. The initial particle soup has been almost entirely changed into radiation. The dominant form of energy in the universe is photons, and they control the rate of universal expansion. The energy of a particle is the sum of two forms of energy: its energy of mass (calculated by Einstein's famous equation, where energy is equal to the mass multiplied by the square of the velocity of light), and its kinetic energy, which arises from its motion. Because photons have no mass, all their energy derives from their frenetic motion. When the universe is 1 second old, the energy of the photons is 10 million times the sum of the mass and kinetic energies of the particles of matter (protons, neutrons, and electrons). Nevertheless, despite their numbers and energy, photons are quite incapable of going through the jungle of electrons and protons. The universe continues to be hidden within an opaque veil.

In the succeeding 100 seconds (the two following pages in our book), the universe will pass yet another important milestone on its slow ascent towards complexity, and will become a factory that builds atomic nuclei. These nuclei will later be indispensable in the construction of atoms and chemical elements. To build atomic nuclei, the universe uses protons and neutrons (known collectively as nucleons*), as building blocks. The cement that binds them together is the strong nuclear force.

The simplest structure is, of course, one that consists of a single building block. Nature naturally chooses the most stable block, and the proton is selected to serve as a hydrogen nucleus. A single neutron is rejected, because no mason worthy of the name would want to build anything that would fall apart after 15 minutes. The most complex structure that the universe then builds consists of two nucleons. One proton and one neutron, bound together by the strong nuclear force, form a deuteron, a deuterium* nucleus. But deuterons are ephemeral. The proton–neutron bond is fragile, and during the first few seconds of the universe, it is constantly broken by energetic photons. Deuterons disappear as rapidly as they are formed. Not being able to create structures from two nucleons, the universe has no means of constructing even more complicated ones with three, four, or five nucleons. Progress towards complexity is temporarily interrupted. Luckily, all that we have to do is wait, because the photons lose energy and become weaker as the universe expands. Soon they will not have sufficient energy to break the deuterons' bonds.

This happens when our cosmic clock shows that 100 seconds have passed, and the temperature has fallen to 1 billion degrees. Henceforth, deuterons have no trouble in forming, and the ascent toward complexity begins again. Each deuteron binds with a neutron to form a nucleus of helium-3* (the number represents the number of nucleons in its nucleus), which in turn captures a proton to make a nucleus of helium-4. This capture takes place despite the electromagnetic force, which tends to repel the proton from the nucleus of helium-3, both having the same positive charge. The temperature of the universe (which is responsible for the extremely high velocities of the particles) and quantum uncertainty overcome this resistance. Helium-4 is very familiar to everyone, because it is the gas that lifts children's multicolored balloons into the air. (The balloons rise because the helium inside them is much lighter than the surrounding air we breathe.)

Stopped in Its Tracks

Nuclei of helium-4, consisting of two protons and two neutrons, are, unlike deuterons, extremely stable and lasting. Once produced, they are there for the duration. This very durability has, nevertheless, one great disadvantage: It again paralyzes any increase in complexity. The very strength of the bonds linking the nucleons of helium-4 causes the latter to withdraw into their shells and reject any association with other nucleons. They are self-sufficient, and will not permit any additional nucleons to force their way into the nucleus. The universe certainly tries to turn the helium nuclei into larger structures, forming lithium-5, or else combining two helium-4 nuclei to form beryllium-8. But all its efforts are in vain. The resulting nuclei are not stable, and most disintegrate as soon as they are created. The cement is not strong enough. To make the situation worse, by the time that helium and deuterium are formed (scarcely 3 minutes after the birth of the universe), the universe is already so diluted by its expansion that the particles are no longer able to collide with one another, to coalesce, and to form more complex nuclei. Nuclear reactions come to a halt.

The construction of stable forms therefore stops with the nuclei of helium. On its route towards complexity, the universe has come to an impasse, and can go no further. It has been stopped in its tracks. Its first attempt on the long road to life and consciousness has ended in a serious reversal. If the universe had not managed to overcome this problem, it would be a very dull and dismal place today; it would be full of clouds of hydrogen and helium, and its landscape would be completely monotonous, with no other structures to enliven its drabness. Because helium is chemically inert, hydrogen would have had no partner to interact with, and the universe would be condemned to being simple and sterile. Such a universe would never produce the heavier and more complex chemical elements needed to bring about trees and flowers, Cézanne's apples or Monet's water lilies. And, above all, it could never create the intertwined helix of DNA molecules that governs all life, including you and me. Yet we are here. Much later, the universe will try again to invent methods for creating heavy elements. But we must not anticipate: Back to our story.

One Helium Nucleus for Every Twelve Hydrogen Nuclei

At 100 seconds, therefore, the universe has taken on a different aspect. It contains nuclei of hydrogen, nuclei of helium-4, and mere traces of lithium and beryllium nuclei, all in a bath of neutrinos and photons. The latter have lost a lot of energy because of the expansion of the universe. The gamma radiation present in the first instants has become ultraviolet radiation.

With a bit of simple arithmetic we can obtain the exact proportions of hydrogen and helium nuclei in the universe. Because hydrogen nuclei consist of just protons, and helium nuclei of both protons and neutrons, all we need to know is the number of neutrons relative to that of protons, and we can find the answer. As we have seen, when the universe was 1 second old, a decrease in the birth rate of neutrons considerably reduced their numbers, and that only two neutrons remained for every ten protons. The difference has increased, by 100 seconds: There are now only two neutrons for every 14 protons. Out of 14 protons, two will combine with two neutrons to give one helium nucleus, while the other 12 become hydrogen nuclei. So, at the end of the first 3 minutes, there is one helium nucleus for every 12 hydrogen nuclei. Because a helium nucleus consists of four nucleons, it has a mass about four times that of a hydrogen nucleus that consists of a single nucleon. The Big Bang theory therefore predicts that about one-quarter $[= 4/(4 + 12)]$ of the mass of the universe consists of helium, and three-quarters of hydrogen. (The heavy elements that form our bodies and are essential for life represent only about 2% of the mass of the universe.)

What a wonderful surprise! This is exactly the ratio that astronomers observe, wherever they look, in stars or galaxies. It seems clear that the two most abundant elements in the universe were formed in the first few minutes of its existence, and that its chemical composition has not changed much since. This agreement is one of the great triumphs of the Big Bang theory.

The Universe Lifts Its Veil

Once the period of hydrogen and helium nuclei formation is over—it is often called the period of primordial nucleosynthesis*—nothing much important will occur for a very long period of 300,000 years. It is as if the universe is trying to get its breath. Naturally, throughout this time, the expansion continues, and the universe becomes less dense and cooler as space increases in size. Just before the cosmic clock indicates 300,000 years (eleven pages further on in our book), the temperature of the universe, which was one billion degrees at 100 seconds, is only slightly greater than 3000 degrees, which is roughly comparable with that of the surface of the Sun. As the cooling proceeds, the photons continue losing energy. The invisible ultraviolet radiation that bathed the universe has now become yellow, visible light like that of the Sun. The universe continues to be hidden behind its opaque veil. The photons that carry the light and information still cannot go through the thicket of free electrons. The latter are unable to combine with the protons to form hydrogen atoms, or with the helium nuclei to form helium atoms. The surrounding photons still have too much energy. If any atoms do form, photons soon collide with them, breaking the bonds and freeing nuclei and electrons.

When the cosmic calendar reaches 300,000 years, a series of changes will occur in the universe that will mark it forever. Henceforth, photons will not possess enough energy to dissociate atoms. The electromagnetic force can exert its influence to make every proton capture an electron (you will recall that under the action of the electromagnetic force, opposite charges attract) to form an atom of hydrogen, and each helium nucleus (with two positive charges) capture two electrons to become a helium atom. For the first time, neutral matter in the form of atoms appears on the cosmic scene.

In another dramatic event, the universe lifts its veil to reveal its true nature. Because all the electrons have been trapped in atoms, because there are no longer any free electrons to impede the movement of photons, the latter can henceforth travel where they please. The universe, from being opaque, becomes transparent. Matter and radiation, previously intimately coupled with one another, separate and will subsequently live separate lives (Fig. 28). The photons that reach us from that fateful 300,000th year are the oldest that we can ever capture with our telescopes. They form the famous fossil radiation that bathes the whole universe, which, along with the expansion of the universe, is one of the two main cornerstones of the Big Bang theory. Naturally, this radiation no longer has a temperature of 3000 K, nor the yellowish color of the Sun. Neither are we able to see it with our eyes, as we would have done if we had been present when the universe was 300,000 years old. The temperature of the radiation has continued to decline inexorably as the universe has expanded. (It decreases inversely in proportion to the distance between any two points in the universe. If the universe expands by a factor of two, for example, the temperature drops to one-half.) Over the following 15 billion years, the cosmic radiation slowly passes

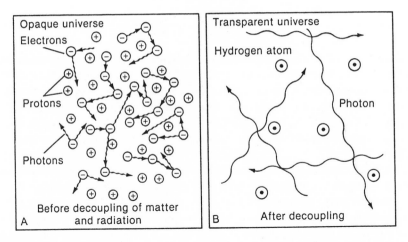

Fig. 28. *The divorce between matter and radiation.* (a) During the first 300,000 years of the universe's existence, the photons of radiation had so much energy that they prevented protons and electrons from combining into atoms of hydrogen. In their free state, the protons and electrons could not move without colliding with photons. Matter and radiation were intimately linked, and no density fluctuation could grow by attracting other fluctuations gravitationally, because the photons prevented them from moving. Conversely, the movement of the photons was impeded by the protons and electrons, radiation could not propagate, and the universe was completely opaque. (b) By the time the universe had an age of 300,000 years, the expansion had weakened the energy of the photons, and the protons and electrons could finally combine to form atoms of hydrogen. The latter could not be destroyed by the less energetic photons. The formation of hydrogen atoms was accompanied by the emission of photons, which today form the 3 K background radiation, the oldest fossil remnant in the universe. From that time onward, matter and radiation were decoupled, and had separate histories. Radiation no longer interfered with matter, and *vice versa*. The universe became transparent, and density fluctuations could at last grow by gravitationally attracting nearby fluctuations. The construction of large-scale structures in the universe could finally begin.

through the range of visible colors, from yellow to orange, to red, and deep red, before becoming invisible to the human eye. The sky, once as bright as the surface of the Sun, gradually darkens, and gives way to the black night sky, scattered with stars, which fills us with delight on a clear summer's night. The cosmic radiation has cooled down so much (to the temperature of 3 K, as we have seen), that now only radio "ears" are able to detect it.

The Reign of Matter

At practically the same time as the universe lifted its veil, matter took over from radiation the rule of the universe. The photons produced in the two great annihilations are still far greater numerically than the matter particles in the universe, there being one billion of them for each proton or neutron.

But when their energy is considered, the situation is about to be completely reversed. Before the 300,000th year, there was far more energy in the universe in the form of radiation than in the form of matter. It was radiation that called the tune, and controlled the rate of universal expansion. But matter is about to make up for lost time. Its energy becomes equal to that of the radiation. As time passes, the energy content of matter begins to exceed that of radiation, and the difference continues to increase. In the universe today, the energy content of radiation is only one-thousandth that of matter. The reign of matter*, which began when the universe was 300,000 years old, is still with us, and will persist into the far distant future, unless the universe collapses on itself.

This dramatic reversal of fortune is easily explained. The universe is involuntarily on the side of matter. Its expansion weakens radiation more than it does matter. In any volume of space that expands with time, the number of particles, whether of matter or radiation, remains the same. The ratio of one billion photons to every particle of matter does not change, but the energy ratio does alter. Although the total energy of the particles of matter (which is mainly the mass–energy, equal to the mass multiplied by the square of the velocity of light, according to Einstein) remains the same, the energy carried by the packets of radiation decreases (in inverse proportion to the size of the universe), so that eventually the energy content of matter exceeds that of radiation (see Appendix E).

At 300,000 years, therefore, the universe becomes transparent and matter dominated. It is full of hydrogen and helium atoms, and ready for the next act.

The Villages and Cities of the Universe

We now have to leave the firm ground represented by the period between 1 second and 300,000 years, where our knowledge is perhaps most reliable. Unlike the period before 1 second, where our attempts to derive a Theory of Everything that unifies the four fundamental forces are still hesitant and uncertain, we have not had to make a bold extrapolation of the known laws of physics, which have been checked innumerable times in terrestrial laboratories, to understand the properties of the soup of matter and radiation that filled the universe throughout that period.

We now have to venture across quicksand again, and tackle a period that is veiled in a dense fog, whose details are still very hazy. Its history has yet to be determined in detail, and we can only give a very rough sketch here. This mysterious period is the epoch of galaxy formation, which will occupy the first 2–5 billion years, and the next two pages in our book. In the past few years, the mists have cleared slightly, thanks to a flood of observations of tens of thousands of galaxies, which has helped to establish a more accurate map of the spatial distribution of galaxies in the universe, and thus give a more detailed view of the cosmic tapestry. This mapping has, in turn, helped to cast some light on the era of galaxy formation.

Well before the true nature of galaxies was recognized, their tendency to cluster together and form still larger structures was well known. Galaxies, like human beings, are very gregarious and occur in groups, avoiding solitude and isolation. Naturally, their bonds are not those of affection, like humans, but are created by gravity instead. The catalogs of positions of "nebulae*," extended fuzzy patches of luminosity later recognized to be galaxies, which were made by the British astronomer Sir John Herschel around 1864, already clearly showed that the best place to find one nebula is next to another. At the beginning of the twentieth century, in 1908, the Swedish astronomer Carl Charlier maintained, rather daringly, that these nebulae were extragalactic, that is, situated well beyond our Milky Way system, and he even proposed a hierarchical model of the universe, where the tendency of nebulae to cluster together was reproduced ad infinitum: two nebulae joined together in pairs; the pairs forming groups; the groups, clusters; the clusters, superclusters; and so on.

In 1925, Edwin Hubble definitively established the extragalactic nature of the nebulae and flung open the door to the world beyond our Milky Way. By photographing objects that were fainter and fainter, and therefore more and more distant, using the newly built telescopes on Mount Wilson in California, the two Americans Edwin Hubble and Harlow Shapley—the same man who dislodged the Sun from its central position in the Milky Way—showed that the Galaxy was part of a still larger structure, known as the "Local Group*." Apart from our Milky Way system, the latter contains the Andromeda Galaxy and some 15 dwarf galaxies*, including the Galaxy's satellites, the Large and Small Magellanic Clouds. Such groups of galaxies have diameters that average some 13 million light-years, about 130 times the diameter of a galaxy, and masses of some 10,000 billion solar masses (10^{46} grams) (Fig. 29). If we regard galaxies as being like houses, groups* are like small villages.

A new stage in the study of the hierarchy of structures in the universe began with the commissioning of the 1.2-meter (48-inch) Schmidt telescope on Mount Palomar in the late 1940s. This telescope, which was specifically designed to photograph large areas of the sky, took just a few years (between 1950 and 1954) to record the whole sky visible from the Northern Hemisphere on thousands of photographic plates. Copies of these plates are now found in every observatory around the world. They constitute a permanent record of the cosmos, and visual archives for astronomers. These photographic plates reveal structures that are even larger than groups of galaxies, the clusters of galaxies*. These are collections of a few thousand galaxies (Fig. 30), bound together by gravity, with an average diameter of about 60 million light-years, and a mass of some million billion solar masses (10^{48} grams). About 3000 clusters have been discovered in the Northern Hemisphere. They may be likened to fair-sized towns.

But clusters are not the largest aggregates of matter in the universe. They themselves assemble in superclusters*. Each supercluster contains five or six clusters, and is some 200 million light-years across, having a mass of

Fig. 29. *A group of galaxies*. The photograph shows the brightest galaxies of a group known as "Seyfert's Sextet" (after the name of the astronomer who discovered it), which lies at a distance of 195 million light-years. The diffuse, faint arms extending from the galaxies consist of stars that have been torn from their parent galaxies during gravitational encounters. A group of galaxies generally consists of 20-odd members, has a typical mass of 10,000 billion solar masses, and is some 13 million light-years across. This group of galaxies also illustrates the problem of accidental superimposition; the smallest galaxy, approximately circular in shape and appearing in the center of the picture, does not belong to the group, but is about 4.5 times as far away. It just happens to lie on the same line of sight as the group (see also Fig. 32) (photograph: Hale Observatories).

about 10 million billion solar masses (10^{49} grams). Superclusters are like large cities dotting the cosmic landscape. Our Local Group, for example, is part of a larger structure, which itself contains some ten other groups and clusters, and which is known as the Local Supercluster. This was discovered in 1960 by the French–American astronomer Gérard de Vaucouleurs (Fig. 31). Charlier's inspired intuitive guess appears to be confirmed, at least up to the scale of superclusters.

The Cosmic Tapestry: Pancakes, Filaments, Voids, and Bubbles

Those rapid advances in understanding the hierarchy of structures in the universe were based almost exclusively on the analysis of catalogs of the positions of galaxies. The tendency for galaxies to occur in groups was determined from their positions as they appear projected on the sky. For lack of knowledge about the distances of galaxies, astronomers were forced to ignore the third dimension. By doing so, they ran the risk of being misled by projection effects. Two galaxies may appear close to one another in the sky

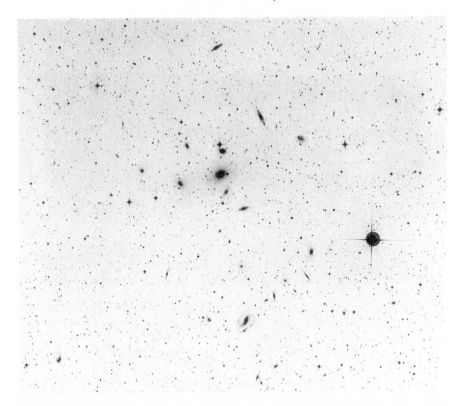

Fig. 30. *A cluster of galaxies.* This photograph shows the brightest galaxies in a cluster in the direction of Pavo, a southern constellation. The cluster lies at a distance of 325 million light-years. In the general hierarchy of structures in the universe, clusters of galaxies come after groups (Fig. 29). A typical cluster contains a total population of about a thousand spiral and elliptical galaxies. Elliptical galaxies predominate in the central areas of the cluster, while spiral galaxies are more common in the outer regions. (A giant elliptical galaxy, similar to that shown in Fig. 40, may be seen in the center of this cluster.) A cluster contains, on average, one million billion solar masses, and is approximately 60 million light-years across (photograph: Royal Observatory, Edinburgh).

but be, in fact, at completely different distances, if they happen to lie near the same line of sight (Figs. 29 and 32). Such projection effects are negligible on small scales, where lines of sight are short, but they become significant on the scale of clusters or superclusters, where lines of sight are very long. To obtain a detailed picture of the large-scale structure of the universe, knowledge of the third dimension is essential. Thanks to recent technical advances, it has finally become possible to measure the distances of large numbers of galaxies.

To determine distances in the universe, all we have to do is to make use of Edwin Hubble's great discovery in 1929, that the light from distant galaxies is red-shifted, with the amount of the red shift being proportional to

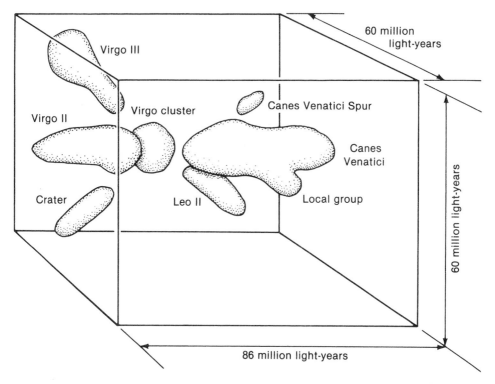

Fig. 31. *The Local Supercluster.* Superclusters of galaxies are the largest aggregates of matter known in the universe, and they are the next higher structural level after clusters of galaxies. Our Galaxy is part of the Local Group, which is itself a member of the Local Supercluster, a three-dimensional representation of which is shown here. The Local Supercluster consists of about 10,000 galaxies that are found in clusters, such as the Virgo Cluster (the Local Supercluster is also known as the Virgo Supercluster), or in smaller groups, all bound together by gravity. The Local Supercluster is in the form of a flattened disk, containing about 60% of the galaxies. The remaining 40% forms filamentary structures outside the plane of the disk that point toward it. There are also large voids. The galaxies in the Local Supercluster occupy only about 5% of the volume represented above. Our Local Group is at the edge of the disk, and falls, at a velocity of 250 kilometers per second, toward the Virgo Cluster, which lies in the center. This infalling motion is caused by the gravitational attraction of the Virgo Cluster acting upon the Local Group (diagram after R. B. Tully).

the distance of the galaxy. Spread out the light of a galaxy into a spectrum, measure the red shift, and we obtain the galaxy's distance. Progress was very slow at first, which is explained by the fact that even though the positions of thousands of galaxies may be recorded on a single photographic plate in one observation, measuring the red shifts of the light from galaxies requires as many observations as there were galaxies. Hubble had only about 30 measurements at his disposal when he made his great discovery. By the end of the 1970s, the number of galaxies for which measured red

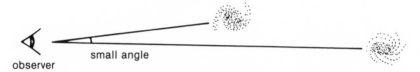

small angle

observer

Fig. 32. *Projection effects can lead to misleading conclusions.* The positions of two galaxies, as projected against the sky, may appear very close to one another: The angular distance between them is very small, and they lie on almost the same line of sight. In fact, they are separated by a considerable distance (*see* the specific example shown in Fig. 29). The only way of making allowance for this is to determine their distances by measuring the red shift of their light.

shifts were available did not exceed 2000. Luckily, the development of electronic detectors made the task of mapping the universe much easier. Such detectors are far more sensitive than photographic plates. They are able to detect the arrival of single photons one by one, and obtain, in just half an hour, information that took Hubble and his contemporaries a whole night's work.

Rapid progress was then made, and the cosmic landscape that emerged was most astonishing and unexpected. First, superclusters of galaxies, the largest known structures, were found not to be spherical, but to be in the form of flattened pancakes, or long, thin filaments. The thickness of the pancakes was some 40 million light-years, and approximately one-fifth of their diameter. The filaments, for their part, might run through space for hundreds of millions of light-years. But the greatest surprise was the discovery of large voids* in the universe, giant regions tens of millions of light-years across, that were completely devoid of galaxies. Galaxies, as we have seen, have a tendency to gather in "villages," "towns," and "cities," but they have taken this gregarious tendency to such an extreme that the "countryside" is completely empty. You could travel for tens of millions of light-years without encountering a single galaxy. The galaxies within the pancakes and filaments occupy only one-tenth of the volume of the universe. The remaining nine-tenths are empty of galaxies. Even more surprising, these voids appear to be in the form of gigantic spherical cavities, resembling giant soap bubbles, whose surfaces contain the flattened or filamentary superclusters (Fig. 33). These giant voids are not isolated in space. They are all interconnected, forming an immense network, where you could go from one void to another without crossing a pancake or filament. The topology of the universe resembles that of a sponge: If you started from any cavity in a sponge, you could move to any other cavity through what is admittedly a highly complicated labyrinth, but without ever having to cross one of the walls (Fig. 34).

What are things like on a still larger scale than that of the superclusters? Are the superclusters themselves organized into bigger structures? To an-

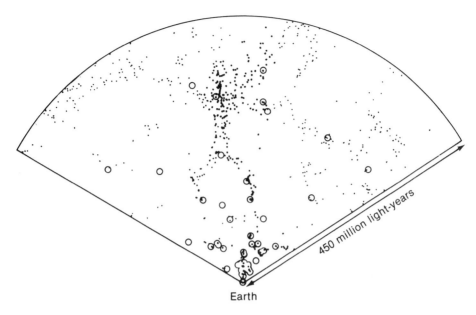

Fig. 33. *The giant voids in the universe.* By measuring the red shifts of the light from thousands of galaxies, thus obtaining their distances, cosmologists have been able to probe the universe in depth. An extraordinary picture has emerged. This diagram shows the distribution of the 1100 brightest galaxies in a small slice of the universe. These bright galaxies are shown by dots. It is immediately obvious that there are enormous voids that do not contain any bright galaxies, and that these voids are approximately spherical, with diameters of some tens of millions of light-years. Are these voids full of faint galaxies, however? Together with colleagues, I have determined the spatial distribution of much fainter dwarf galaxies (which are much too faint to be observed with optical telescopes) by measuring red shifts using radio methods. These dwarf galaxies (indicated here by open circles) are similarly distributed; the voids do not contain large numbers of faint galaxies (diagram after de Lapparent *et al.* and Thuan *et al.*).

swer these questions correctly, we need a large number of measurements oı the red shifts of more remote galaxies, which are much fainter and more difficult to observe. We shall have to wait for another 5 or 10 years. But, in the meantime, we can begin to find an answer by examining the apparent positions of a very large number of galaxies (about 1 million) in an extensive area of sky. What we find is like a marvelous cosmic tapestry, of which the network of pancakelike and filamentary superclusters constitute the texture, the clusters (which are regions of high density) the stitches, and the giant, almost spherical voids the space between stitches (Fig. 35). The universe, which from a distance appears as a dull, featureless cloth, becomes on closer inspection a fantastic patchwork quilt in which the galaxies trace an infinite variety of shapes and patterns. The task now facing cosmologists is to try to understand how such a complex pattern has arisen. Somehow they have

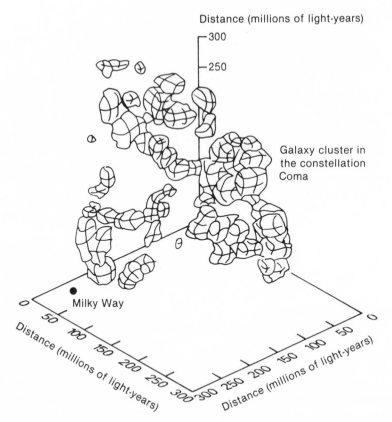

Fig. 34. *The spongelike structure of the universe.* This diagram shows the three-dimensional distribution of the brightest 1000 galaxies in the northern celestial hemisphere. The faintest galaxy is about 1600 times fainter than the faintest star visible to the naked eye. The individual positions of galaxies have been joined by smooth surfaces. The structures thus revealed form an extraordinary landscape: The galaxies form a completely interconnected network, intermingled with a network of large voids completely empty of galaxies. The topology of the universe resembles that of a sponge, where the structures formed by the galaxies correspond to the walls of the sponge, and the voids to its cavities (diagram after Davis *et al.*).

to fill in the middle of a story of which the beginning and the end are known. The beginning is a universe, whose properties, 300,000 years after the Big Bang, varied by only 0.001%, as revealed by the observations of the background radiation by the COBE satellite. The end are the structures observed in the universe today. How could the universe develop such a rich hierarchy of structures from such a uniform state? How did complexity arise from simplicity? We are now going to see that gravity holds the key to the answer.

The Seeds of Galaxies

During the first 300,000 years of the universe, gravity, although present, did not play a major role. Standing in the wings, it left the stage to the other

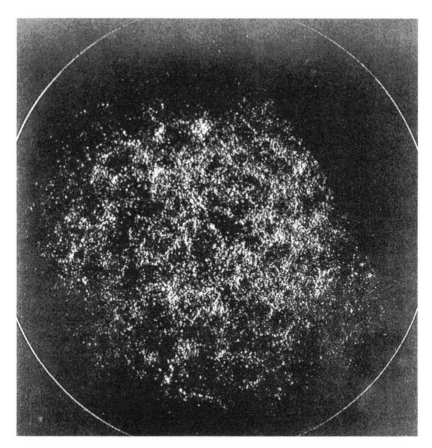

Fig. 35. *The very large-scale structure of the universe.* We can gain an idea of the very large-scale structure of the universe by examining the positions, as projected against the sky, of a large number of distant galaxies. The map shown here represents the million brightest galaxies in the northern celestial hemisphere. (The faintest galaxy represented is about 160,000 times fainter than the faintest star visible to the naked eye.) This map is based on galaxy counts made by the American astronomers C. D. Shane and C. A. Wirtanen, at Lick Observatory in California. The map does not represent the positions of individual galaxies, but was constructed by dividing the sky into small, square cells, and counting the number of galaxies in each cell. The counting was undertaken manually (and not with automatic scanning machines, as is done now): Shane and Wirtanen took 12 years to complete the work. An amazing landscape is revealed: Galaxies form an immense cosmic tapestry, of which the vast network of pancakelike and filamentary superclusters constitute the texture, the clusters the stitches, and the gigantic voids, tens of millions of light-years across, the space between stitches. Explaining how this wonderful patchwork quilt has arisen from a universe that was initially extremely uniform is one of the most fundamental, unresolved problems of modern astrophysics (photograph: P. J. E. Peebles *et al.*).

forces. The latter had helped the universe to take its first faltering steps towards complexity. The strong nuclear force built atomic nuclei from quarks, and the electromagnetic force helped to make atoms by binding nuclei and electrons together. But as we have seen, the manufacturing of the chemical elements essential for life ceased abruptly, because atoms of helium are too stable and prevent the formation of heavier elements.

Gravity comes to the rescue. It provides the universe with a second chance to continue its ascent towards complexity. It saves the day by creating in the cosmic immensity small oases that will avoid the continuous cooling caused by the universal expansion, and that will eventually allow life and consciousness to arise. These "oases" are called galaxies, stars, and planets.

But first let us go back a step and examine why gravity was essentially paralyzed, and did not begin its work until the year 300,000. We have seen that before the universe became transparent, it consisted of a soup of radiation and matter, and that this soup was not perfectly even and uniform. Scattered here and there within it were irregularities or density fluctuations*. You will recall that these irregularities arose during the universe's inflationary era, in its first second of existence, and during one of its "phase-change" episodes. Just as ice, which is the result of a phase change of water, may exhibit defects—cracks in its internal structure—the universe develops defects, irregularities in the soup, as it changes phase. Our very existence depends on these density fluctuations, since they are the seeds that will grow, with the help of gravity, into stars and galaxies. Without those seeds, galaxies, those oases where life can develop, would not have existed.

As we have seen, there are restrictions on the properties of these original irregularities. On the one hand, they had to be small enough not to cause temperature fluctuations larger than those observed by the satellite COBE in the 3 K background radiation and equal to about 0.001%. On the other hand, they had to be large enough to grow into the structures we observe today, within the time span of the 15 billion years that has elapsed since the appearance of the seeds. Although stars, galaxies, clusters, and superclusters differ in mass and size by a factor of more than one million billion, the actual values of their masses and sizes span only an infinitesimal range of all the masses and sizes that are possible. The experiment would have failed utterly if the largest structure that these irregularities could have grown into were of the size of a microbe!

Two types of fluctuations are known: There are those where matter and radiation vary in concert from place to place, and in such a fashion that the ratio of a billion to one between the number of particles of radiation (photons) and the number of particles of matter (protons and neutrons), is always preserved. The constancy of that ratio is called "adiabaticity," and the fluctuations are known as "adiabatic fluctuations*." It is also possible, however, for the radiation not to remain in step with matter. The latter can vary, but the radiation remains constant. Matter fluctuations then arise, superimposed on a perfectly homogeneous background of radiation, devoid of any fluctua-

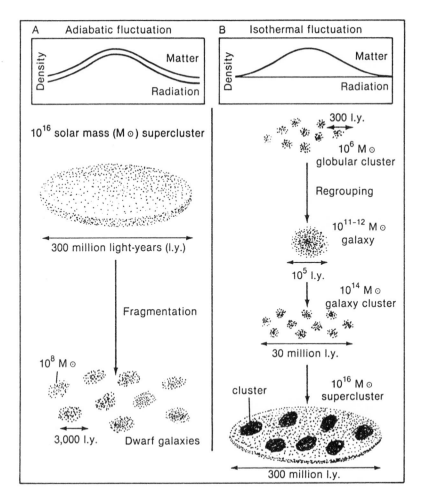

Fig. 36. *The seeds of galaxies.* To explain the structures observed in the universe, cosmologists have to invoke the existence of irregularities, or density fluctuations, in the early stages of the universe. These irregularities were observed in 1992 by the satellite COBE to be about 0.001%. They act as seeds for galaxy formation. There are two possible scenarios that can explain the formation of structures in the universe, and these correspond to the two possible types of density fluctuations. In the case of the so-called "adiabatic" fluctuations (a), matter and radiation vary in concert, so that the ratio of one billion photons for every proton or neutron is preserved throughout the universe. The radiation that is trapped in the fluctuations escapes, taking the compressed matter with it, and causing the destruction of small-scale structures. The first structures to emerge, after gravity has caused the fluctuations to grow, are superclusters, which are in the form of flattened pancakes (*see* Fig. 37), with a mass of 10 million billion suns. These superclusters then fragment to form smaller structures. In the case of what are known as "isothermal" fluctuations (b), only matter is compressed, and the radiation remains perfectly homogeneous. The smaller structures are not destroyed, because the radiation is not trapped in the fluctuations. The first structures to arise are small globular clusters, with masses of about one million Suns. The larger structures (galaxies, clusters of galaxies, and superclusters) form later as a result of gravitational attraction.

tion. In this case, the temperature of the radiation is constant, so the radiation is said to be "isothermal," and the matter fluctuations are known as "isothermal fluctuations*" (Fig. 36).

The Actors in the Drama: Gravity and Expansion

Gravity enters the scene to make these seeds grow into stars and galaxies. The excess gravity associated with the excess of matter or radiation in an irregularity attracts other, neighboring irregularities, which combine with the original irregularity, causing it to become larger. But before the year 300,000, all of gravity's efforts were in vain. Just as the photons could not propagate through the jungle of electrons and protons, the free electrons did not have much freedom of movement, because whenever they tried to move, they collided with the photons, which were far more numerous. The protons, which are 1836 times as massive as the electrons, had even greater difficulty in trying to force their way through the throng of photons. This lack of freedom of movement completely prevented any fluctuation from growing, because the protons and electrons could not cluster together, despite all the efforts of gravity to bring them closer together.

Making a bad situation worse is the fact that, during the first 300,000 years, the adiabatic fluctuations (those where matter and radiation varied in concert) were subject to wholesale destruction, which eliminated the smallest of them. The photons trapped in the fluctuations resisted their confinement, and tended to diffuse outwards. They were able to do so in the case of small fluctuations. By carrying along the particles that formed the excess matter, in such a way as to preserve the photon to baryon* ratio, they cause the destruction of the inhomogeneities. Only the largest fluctuations were able to escape that fate.

The magical moment of the 300,000th year finally arrived. The universe became transparent, and the electrons imprisoned in the atoms could no longer impede the motion of the photons. The irregularities were now allowed to grow. Matter could move freely without being halted by radiation. Gravity asserted its rights and pulled matter towards the density enhancements, making them larger. In addition, the separation of matter and radiation put an end to the destruction of the fluctuations. Photons were no longer able to cause protons and neutrons to diffuse out of the fluctuations. The latter could now grow without restrictions.

But gravity's task was not an easy one. It had to fight incessantly against the expansion of the universe, which tended to undo all its work. Between the age of 300,000 years and today, the distance between the galaxies has increased by a factor of 1000, and the average density of the universe has decreased by a factor of one billion. The effect of this expansion has been to reduce the inhomogeneities and lessen their power to attract one another. It gets harder and harder for the irregularities to grow as time passes. The situation is similar to that of a racehorse that is struggling valiantly to reach the winning post on a race track that is continually increasing in length. We do know the outcome of the struggle: Gravity finally won. The horse did actually reach the finishing post. The irregularities grew until they were so massive and their gravity so strong that they ceased taking part in the general expansion of the universe, and collapsed on themselves to give birth to stars, galaxies, clusters of galaxies, and superclusters. Because they no longer followed the expansion, these structures escaped the general cooling, thus allowing life to emerge. Let us examine this epic struggle in more detail.

The Invisible Mass in the Universe

Before discussing the fight between gravity and the expansion of the universe, and becoming involved in the story of how the initial seeds of galaxies grew into the range of structures seen now, we need to determine a fundamental quantity, the amount of matter in the universe. Because of its gravitational effects, matter slows down and regulates the rate of expansion and, as a result, the rate at which the seeds grow into galaxies. "Easy," you will say, "all you need to do is count the number of stars and galaxies." Unfortunately, the problem is not that simple. Astronomers can easily count objects that emit radiation. It does not matter whether this is in the form of gamma rays, x rays, ultraviolet or visible light, infrared radiation or radio waves, provided it can be "seen" by a telescope. (Henceforth we will use the term "visible," not in the conventional sense of "perceptible to our eyes," but in the extended sense of "perceptible with our telescopes.") But not everything in the universe is emitting radiation: An extreme example is a black hole, which may be massive, but which traps radiation within its radius of no return. Astronomers have had to resort to highly ingenious methods (which we will come to in the next chapter) to get around this lack of radiation, and take a census of the amount of matter present in the universe. The results are still not very definite, but one fact is certain: The universe contains 10 to 100 times as much dark matter as luminous material. The visible universe appears to be merely the tip of an iceberg.

The total amount of matter does not just govern the evolution of the fluctuations; it determines the fate of the universe itself. If the mean density of the universe—the mass, visible and invisible, divided by the volume—is greater than a critical density* of three hydrogen atoms per cubic meter, gravity is strong enough to halt the expansion of the universe and cause the latter to collapse back on itself at some time in the future. The universe is said to be closed*. If the universe contains fewer than three hydrogen atoms per cubic meter, the expansion will continue forever, and the universe is said to be open* (see Appendix E). At present we do not know the exact amount of matter in the universe and therefore whether the latter is open or closed. How, then, are we to discuss the growth of the seeds that were later to become galaxies? We have to turn to models.

Model Universes

Astrophysics (and particularly cosmology) is distinct from the other exact sciences, such as physics, chemistry, or biology, in that laboratory experiments cannot be performed. The experiment took place once and for all 15 billion years ago. We cannot create stars or galaxies in our test tubes. All we can do is observe them from afar.

The situation has changed considerably in recent years, however, with the development of very fast computers capable of carrying out billions of calculations per second. To make up for their inability to do laboratory experiments, cosmologists use powerful computers to carry out numerical experiments, which allow them to simulate and study the evolution of various astronomical objects. Stars, galaxies, clusters of galaxies, superclusters, or the universe itself, all can be investigated. The computer can produce a vast range of model universes, most of which having nothing to do with the "real" universe. To create a model, the cosmologist provides the computer with a set of conditions (known as initial conditions) that appear to be "reasonable" and that may have applied at the epoch when radiation and matter separated, 300,000 years after the Big Bang. Among the parameters to be specified are the total density of matter (both visible and invisible), the various components of that matter (protons, neutrons, electrons, neutrinos, etc.), and the nature of the initial fluctuations (adiabatic or isothermal). Then the matter is allowed to evolve in accordance with the known physical laws, such as the law of gravitation. The computer can thus follow the motions of hundreds of thousands of galaxies. After a simulated evolution of 15 billion years (which takes the computer just a few hours to calculate)—that is, when the distance between two points anywhere in the model universe has increased by a factor of 1000—the cosmologist stops the calculation and asks the computer to produce an image of the model, which may then be compared with the universe as currently observed. If the model disagrees with the actual universe, out it goes. The cosmologist goes back to consult his computer, like an ancient priest returning to consult the Delphic Oracle. The initial conditions are slightly modified, and the computer churns out another model, which can again be compared with the observed universe. And so on, until the model resembles the actual universe. The cosmologist then concludes that the initial conditions that were entered into the computer were probably similar to those that prevailed in the universe when it was 300,000 years old, and that some of the mystery surrounding that period has been dispelled. Let us look at some of these models in more detail.

Too Small Pancakes and Too Large Seeds

For our first model, let us have a taste of immortality by studying a universe that will expand forever. The mean density of such an open universe needs therefore to be less than the critical density of three hydrogen atoms per cubic meter. Let us set the density at one-fifth of that value. Why one-fifth?

Luminous objects (such as stars and galaxies) in the universe contribute only one-fiftieth of the critical density. However, as we shall see later, studies of the motions of galaxies in clusters seem to imply that there is 10 times as much dark matter of unknown nature as luminous matter in the universe; so the total amount is one-fifth of the critical density. This value is also in agreement with the quantity of matter needed to create in the first 3 minutes the primordial elements* (such as helium* and deuterium*), in the proportions observed in stars and galaxies.

We next introduce adiabatic seeds of galaxies, those where radiation fluctuates in concert with matter, because this is the type prescribed by the unified field theories that describe the universe's earliest instants. We start our calculations just after the 300,000th year. Radiation and matter have just become decoupled, and the electrons are bound in atoms. Their removal has a decisive influence on the shape of the structures that gravity sculpts by attracting matter toward the density fluctuations, making them more and more massive, until they collapse on themselves.

Before the 300,000th year, the free electrons exerted an equal pressure on all sides of the density fluctuations. There was no preferential direction, and gravity, if it had been able to act, would have created spherical bodies. After the imprisonment of the electrons in atoms and in the absence of their isotropic pressure, gravity forms flattened pancakelike objects or filaments. We can understand this if we recall that all motions in space may be referred to a set of three perpendicular axes: up and down, left and right, backward and forward. A spherical shape would require the collapse along each of these axes to occur at precisely the same rate, and begin at exactly the same time—which is highly improbable. Try to arrange a meeting with three other persons with extremely busy and different schedules, and you will soon realize that it is well-nigh impossible. It is much easier to arrange to meet two of them, or even better, to see one at a time. Similarly, it is far more likely that the density fluctuations will collapse preferentially in one direction, with a much slower motion of collapse or expansion in the other two directions. This will give rise to structures that are flattened like pancakes and with very high densities (Fig. 37).

Let us follow our model universe further in its evolution. The flattened structures continue to grow by accumulating matter, and eventually join up with one another to weave an immense, three-dimensional, cosmic spider's web. The flattened regions that contain all of the luminous matter occupy only 10% of the total volume: The remaining 90% is empty. If the model is evolved beyond the present age of the universe, 15 billion years, we discover that the cellular structure is but temporary and that it will be replaced in the future, by larger and larger, irregular concentrations of matter. As far as the formation of these structures is concerned, the current age of the universe is neither too young (because the structures do exist), nor too old (because the cellular structure still persists). Galaxies are thought to form from these large flattened pancakes through a process of fragmentation, the physics of which is still poorly understood. Very rough calculations suggest that the

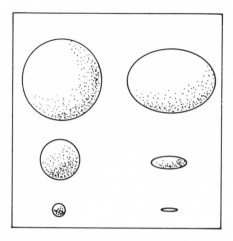

Fig. 37. *Pancakelike superclusters.* The largest structures of matter known in the universe are superclusters of galaxies. These are not spherical, but have the form of flattened pancakes. How could such structures have arisen from the initial density fluctuations (*see* Fig. 36)? A density fluctuation that was originally perfectly spherical would retain its spherical shape as it collapsed under the influence of its own gravity (left-hand sequence). In general, the fluctuations are not perfectly spherical. The right-hand sequence shows how a fluctuation that is initially slightly smaller in its vertical dimension will collapse. The initial disparity is amplified by gravity during the collapse; so the vertical dimension shrinks much faster than the other, horizontal ones. The final result is a highly flattened object, having a shape not unlike that of superclusters.

typical mass of a single fragment is about 100 million solar masses, that is, comparable with that of a dwarf galaxy*.

In an open universe with adiabatic fluctuations, therefore, the largest structures (the superclusters*) come first, and the smaller objects appear later by fragmentation. Gravity will make dwarf galaxies merge with one another to form normal galaxies, and the latter will in turn assemble into groups and clusters of galaxies. All these structures are remarkably like those we observe in the universe today, and we might think that the problem of galaxy formation has been solved. Alas, nothing could be further from the truth! A closer examination reveals aspects of the model that directly contradict observations. To begin with, the mass of the first structures to appear is only 100,000 billion (10^{14}) solar masses, which is 10 to 100 times less than the observed masses of superclusters. The pancakes are too small. Even more serious is the fact that, for the initial seeds to grow in 15 billion years into galaxies of the size we observe today, they would have to be quite large. But no such irregularities are observed in the background radiation, which dates from the same epoch. The COBE satellite has only found very small fluctuations (0.001%). Our triumph has proved to be very short-lived.

A Universe with Massive Neutrinos

But all hope is not lost. We can ask the computer to calculate another model universe. The previous model needed large seeds, because it had a low mean density of matter (just one-fifth of the critical density), and the force of gravity was not strong enough for small seeds to grow to the size of galaxies during the time that has elapsed since the Big Bang. What would happen if the universe had exactly the critical density? Gravity would be stronger, the seeds could be smaller, and the size of the density fluctuations would no longer be in conflict with observations of the background radiation. In addition, the critical density gives a universe that is flat, which is the most likely geometry according to the inflation theory. We have seen how the dramatic expansion of the universe in the inflationary era resulted in space becoming perfectly flat.

But hold on! If we increase the density or, equivalently, the total amount of matter, we must still take into account all of the other observational constraints. Any additional matter must not be associated with galaxies. The latter may appear to contain 10 times as much dark matter as luminous matter, but not 50 times as much. The additional matter cannot be in the form of baryons*, because the presence of additional protons and neutrons during the first 3 minutes of the universe's history would produce amounts of deuterium and lithium which would be inconsistent with those observed in stars and galaxies. There is no lack of candidates for this new, invisible, nonbaryonic matter. The various theories that attempt to unify the four existing forces into a single force predict the existence, in the first few fractions of a second after the Big Bang, of a vast multitude of particles, called by the strangest names, but that do not lack a certain poetry: neutrinos, axions, photinos, higgsinos, gravitinos, magnetic monopoles, pyrgons, maximons, newtorites, etc. Some physicists even suggest exotic objects such as quark nuggets or primordial black holes.

"But," you ask, "why, apart from the neutrinos and primordial black holes, has none of this weird fauna been mentioned in the course of our discussions of the first moments of the universe?" The reason is simple: With the exception of the neutrinos, not one of these objects or particles has yet been discovered, either in the universe or in the laboratory. For the time being, they are merely figments of physicists' fertile imaginations. Nevertheless, like good fairy godmothers, the scientists have endowed their inventions with two specific properties that would allow these exotic objects to play an important role in the development of structures in the universe. First, they all have mass. Second, unlike protons, electrons, or photons, they interact very weakly with other particles, which means that they can move about freely. This freedom of movement will largely determine the scale of any structure that appears.

The recipe for our new model universe is as follows: First a pinch (one-fiftieth) of baryonic, luminous matter (protons and neutrons). Second, an appreciable quantity of also baryonic, but dark, matter (about one-fifth). So far the recipe is the same as for the previous model. But what do we take for the remaining four-fifths? We need matter that is both dark and nonbaryonic. With such a vast pool of potential candidates, we do not lack choice. But to keep at least partially within the bounds of reality, let us settle on the neutrino. Unlike all the other candidates, we do at least know from our laboratory experiments that it exists. Unfortunately, we are still not sure if it has a mass. According to the predictions made by grand unification theories, it should have about one ten-thousandth of the mass of an electron, but this has still to be confirmed. For the moment, let us assume that the neutrino does have a mass. Let us introduce adiabatic seeds into our model, and study the evolution of a universe where the dominant mass is made up of neutrinos.

At first sight, massive neutrinos appear to be a wonder drug that will cure our previous model of all its ills. As the universe expands, the neutrinos, like the photons, lose energy and slow down. While at the very beginning of the universe's existence they travel at nearly the speed of light, they are already moving much more slowly long before the epoch of the 300,000th year, when matter and radiation separated. Gravity takes advantage of this lack of speed, and attracts them toward the density fluctuations. The neutrinos, unlike the photons, can move toward the fluctuations without impediment, thanks to their lack of interactions with other particles. They are captured by the fluctuations, which become larger and larger. The initial seeds of galaxies are thus able to begin to grow immediately after the Big Bang, unlike the situation in our previous model, where gravity had to wait for 300,000 years, until matter and radiation were decoupled, before it can act. Not only is gravity stronger (the density of matter is greater), but the fluctuations also have more time in which to grow. The initial seeds may thus be smaller. The resulting temperature variations in the fossil background radiation are much reduced and are in good agreement with the limits established by observation. One of the problems of the previous model has been solved.

As before, there is a wholesale destruction of the smallest fluctuations. Neutrinos trapped within them are just as reluctant to accept their imprisonment as photons were in our previous model. They escape, causing their prisons to disintegrate. Because they are able to travel much farther than photons, they will "erase" much larger fluctuations. This destruction occurs well before the age of 300,000 years, by which time the neutrinos have lost most of their energy because of the expansion of the universe. The fluctuations that do escape elimination will grow, and the smallest structure that survives will have a mass of between 1 and 10 million billion (10^{15} and 10^{16}) solar masses, precisely the mass that we find in superclusters. As in the previous model, these structures take the form of flattened pancakes weaving an immense cosmic tapestry. This time, however, the pancakes are not too small, but are the right size. The remaining problem has also been solved.

Does this mean that a universe containing massive neutrinos is the answer? Unfortunately, we still cannot claim success. A careful examination of the present model, after an evolution of 15 billion years, reveals that it does not look anything like what we observe. The superclusters may be the right size, but the clusters are too massive, and the voids too large. This discrepancy comes about because in the early universe the neutrinos traveled too fast, "erasing" the smallest structures (Figs. 38a, b).

"Cold" Dark Matter

What can we then do to lift the mystery surrounding the formation of the structures in the universe? How can we get out of this dead end? There is no lack of suggestions. Physicists are rarely at a loss for imagination. If massive neutrinos will not work, other particles can be used instead. To overcome the lack of small structures in the massive-neutrino universe, we can include, instead of neutrinos, other particles thought to be even more massive, which move more slowly, and therefore allow smaller structures to survive. Because a particle's energy of motion may be described in terms of temperature—"hot" particles move rapidly, whereas "cold" particles are slow moving—these slow-paced particles constitute what is known in the jargon as "cold dark matter*." (Neutrinos are "hot" dark matter.)

There are plenty of possible candidates. We need only to make a judicious choice out of the whole zoo of particles with exotic names. At present, axions and photinos are thought to be the most likely components of this cold dark matter. Unfortunately, we are now venturing into completely unknown territory. At present, these particles exist only in physicists' imaginations, and as a product of their highly complex equations. No one has actually seen them. If we are not even sure that they exist, how can we say anything about their masses? Is this real progress when, to solve a mystery, we have to invent objects that are just as mysterious themselves? Science, and astrophysics in particular, must be based on experiment and observation. If it fails to heed these principles, it loses all credibility. Physicists are well aware of this. The search for the particles that occupy the top of cold dark matter's current "hit parade" has begun, using the most advanced technologies. So far, however, such efforts have proved fruitless, and cold particles have failed to appear.

In the meantime, models where four-fifths of the critical density is made up of cold dark matter do not appear to contradict observations (Fig. 38c). The first structures to appear are still in the form of flattened pancakes, but their mass, as expected, is much smaller. Instead of having masses comparable with those of superclusters (10^{16} solar masses), the objects are more like dwarf galaxies of about 1 billion solar masses. Normal galaxies form subsequently from the coalescence of dwarf galaxies, and groups and clusters are formed from normal galaxies assembling together. These various structures form an immense cosmic network, resembling what is observed.

a

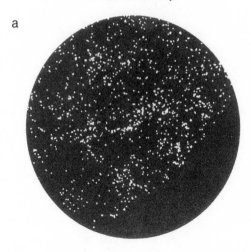

Fig. 38. *Model universes.* To compensate for the fact that they cannot carry out laboratory experiments, cosmologists perform numerical experiments on powerful computers, building models to simulate the formation and evolution of the structures present in the universe. These models are constructed by giving the computer certain initial conditions: the overall density of matter; the various components of this matter—protons, neutrons, electrons, neutrinos, etc.—and suitable seeds (which may be adiabatic or isothermal, *see* Fig. 36). The computer then calculates the evolution of the matter according to the laws of physics. The motion of some 30,000 to 1 million particles of matter are followed. At the end of 15 billion years of evolution (the age of the universe), that is, when the distance between any two points in the model has increased 1000 times, the model is compared with what is observed. (a) The universe as it is observed (in fact, this is the projected distribution of the 1000 brightest galaxies in the northern hemisphere). (b) A model where the matter density is equal to the critical density (its expansion will only cease infinitely far in the future), with a mass dominated by massive neutrinos. It is obvious that the distribution of galaxies in this model does not resemble what we actually observe. Instead of long, thin structures, we find large, irregular concentrations of galaxies. (c) A model where the mass density is one-fifth of the critical density, and where the mass is predominantly in the form of cold dark matter (photinos, axions, etc.). The distribution of galaxies shown in this model resembles more closely what we actually observe. However, this model also has a lot of flaws: First, cold dark matter only exists, at present, in physicists' unrestrained imaginations. In addition, the predicted temperature fluctuations are too large, when compared with those observed in the cosmic background radiation. Finally, a mass density less than the critical density is in conflict with grand unification theories.

Unlike the previous model, however, the order of appearance is reversed: Small structures form first, and larger ones later.

Can we finally claim victory? Have we really succeeded in solving the mystery of the formation of the structures in the universe? Once again, we have to restrain our enthusiasm, because a serious problem arises. In this model, all the cold dark matter is associated with galaxies, clusters, or sub

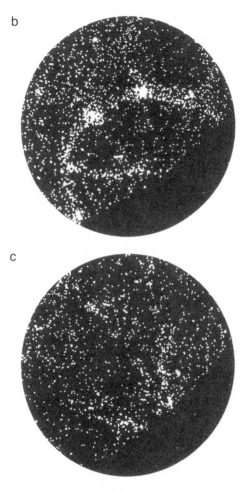

c

Fig. 38 (continued)

perclusters. It has the critical density of three hydrogen atoms per cubic meter. Yet we know from observation of the motion of galaxies (which we will describe in the next chapter) that the density of the matter (visible and invisible) associated with galaxies amounts to only one-fifth of the critical density. This is a quite intolerable contradiction! Cosmologists, who are never short of ideas, affirm that we are being deceived by the galaxies, and that these give a very distorted view of the distribution of mass in the universe. Let's assume, they say, that the universe is uniformly filled with dark matter, and that this matter is only luminous at a very few places (the densest), where it appears as galaxies. The matter within galaxies would thus represent only a small fraction of the total mass. It is as if we were flying at night over a vast ocean, and wanted to measure its extent. Here and there we can see the lights of a few ships. If we go by the distances between the ships, we are bound to underestimate the size of the ocean. In the universe, instead of obtaining the critical density from galaxies, we find merely one-fifth of its true value.

Were the First Building Blocks Extremely Tiny?

Perhaps galaxies are not telling us the whole truth. This is certainly an attractive idea, but one that is not firmly grounded in reality. Not only do we have no proof that cold dark matter actually exists, but we have even less idea of whether it exists throughout the whole universe. This last model, which is based on such uncertain assumptions, is nevertheless favored by many cosmologists, and is currently the most fashionable model because it is able to reproduce many of the observed features of the cosmic tapestry. Personally, I feel that hiding the major fraction of the dark matter well outside galaxies is not very satisfactory. Without galaxies to act as "cosmic lighthouses," it will be extremely difficult for us to detect or make any measurements of such dark matter. In that case, what is the use of a theory that cannot be verified experimentally? Before the jury renders its final verdict, however, let us wait and see if the highly impressive efforts being deployed to detect or capture particles of invisible cold dark matter coming from space are successful. In the meantime let us consider other solutions.

So far, our models have been seeded with adiabatic fluctuations containing both matter and radiation. Such seeds are favored because grand unification theories predict their existence. What would happen if the seeds were isothermal, that is, if they contained only matter? This would lead to a fundamental difference: With no radiation compressed in the isothermal fluctuations, none of the smallest structures would be destroyed, since there are no photons trying to escape. The first structures to emerge in the 300,000th year would be tiny: globular clusters (the same ones that Shapley used to determine the Sun's location in the Galaxy), with masses of about 100,000 solar masses. Larger structures would form later through gravitational attraction: The globular clusters gather together to form galaxies; galaxies assemble to form clusters; and clusters to form superclusters. But again, there is a significant problem with the present model: It is extremely difficult to create the flattened or filamentary shapes of superclusters from small, spherical, globular clusters.

Thus our current understanding of the formation of the various structures in the universe—those oases in the cosmic landscape that will later harbor life—is at best very sketchy. Our knowledge of the period after 300,000 years is still very uncertain. We do not even know in which order the various events occurred. Were the first structures to form the very largest (the superclusters), and did these later fragments give birth to the smaller objects (galaxies), or were the first structures the smallest (globular clusters or dwarf galaxies) that subsequently gathered into galaxies, clusters, and superclusters under the influence of gravity (see Fig. 36)? Most of our models were unable to satisfy two key observations simultaneously: the extreme uniformity of the early universe and the cosmic landscape that we observe today, a network of flattened, pancakelike superclusters, filaments, and voids. The only model that has been able to reproduce the cosmic tapestry, the one filled with cold dark matter, does not have a firm grounding in reality.

Cosmologists continue probing, feverishly devising new models with the help of their computers. Until now, gravity has played the starring role in our models. We have assumed that the immense voids, some tens of millions of light-years across, completely empty of galaxies, were all its work. Gravity has been so successful in gathering all the galaxies into clusters and superclusters that the rest of space is completely empty! The recent discovery that these voids are almost spherical (Fig. 33) suggests that perhaps gravity was not the only factor. As we shall see later, the appearance of the first structures—whether globular clusters, dwarf galaxies, or superclusters—may have been accompanied by the formation of vast numbers of massive stars, which last only a few million years (a mere blinking of an eye on the cosmic time scale), and come to a violent end in tremendous stellar explosions known as supernovae. The combined effect of these cosmic fireworks may have emptied the voids of galaxies, pushing them out to the edges of gigantic, spherical bubbles. Gravity is perhaps not the only cause of the filamentary structure of some superclusters. Unification theories suggest that during the phase changes that took place when it was very young, the universe—like a crystal that develops defects when it undergoes too rapid a phase change from liquid to solid—may have developed flaws in the fabric of space in the form of filaments that stretch for billions of light-years. These flaws are known as "cosmic strings*." They are incredibly massive and dense: A single centimeter might weigh a million billion tons, as much as a whole mountain range such as the Alps, and they are one million billion (10^{15}) times thinner than an atomic nucleus. These cosmic strings might be the origin of the gigantic, almost linear, structures that galaxies trace out across the universe (Fig. 39). After the discovery of cosmic seeds in the fossil radiation by the satellite COBE in 1992, cosmic strings have fallen somewhat out of fashion with astrophysicists because they would have caused temperature variations in the fossil radiation larger than those observed by COBE.

In any case, no model will draw unanimity until some light is shed on the question of the exact amount of matter in the universe, and on the nature of the invisible mass. The range of models would already be severely reduced if the mass of the neutrino were measured, or if the existence of certain exotic particles were to be confirmed. In the meantime, we have to doff our hats to the universe. Beginning with an almost perfectly homogeneous, initial soup, with essentially no organization, it has, somehow, been able to create a veritable patchwork where galaxies form an immense spongelike structure, honeycombed with gigantic voids.

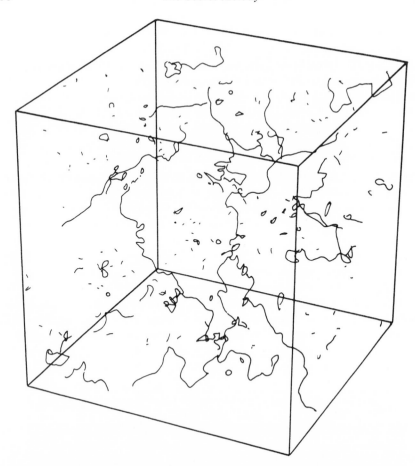

Fig. 39. *Cosmic strings.* Grand unification theories that attempt to unify fundamental forces in the very early universe predict that, as it cooled in the first few fractions of a second after the Big Bang, the universe developed flaws in the fabric of space, in the shape of long, thin "strings" running through space. These are shown here in a computer-generated model. Although cosmic strings have yet to be seen and their existence would be in contradiction with certain observations, some cosmologists believe that they may be at the origin of the filamentary structures that galaxies trace in the sky.

The Universe Through a Magnifying Glass: Galaxies and Stars

We have been admiring the overall pattern of the universe as revealed by its largest individual structures. Now let us examine the smaller objects in detail. Once again, we are forced to marvel at the beauty and richness of the motifs we find. The basic building blocks of the universe are the "nebulae," which Kant called "island universes," and which we now call galaxies. Within our cosmic horizon, we are able to see some hundreds of billions, and the number grows by about ten every year, as the horizon recedes with the passage of time.

Galaxies were born when the universe was between 2 and 3 billion years old. Gravity, which had been working relentlessly since the period of 300,000 years after the Big Bang—or perhaps even longer—valiantly opposing the expansion of the universe, which tended to undo its work, has finally succeeded in turning the density fluctuations into embryonic galaxies. Diffuse clouds of hydrogen* and helium*, which resulted either from the coalescence of small clouds the size of globular clusters or dwarf galaxies, or else from the fragmentation of enormous masses the size of superclusters, collapsed as a result of their own gravitational attraction. This collapse compressed and heated the gas, transforming it into hundreds of billions of small spheres of extremely hot gas (where the central temperature reached tens of millions of degrees), which we call "stars."

The final outcome for these embryonic galaxies depends on how effectively they convert the gaseous material into stars. Some are so efficient that practically all the gas is turned into stars within a billion years. They then become "elliptical galaxies*," so called because they appear as oval patches of light in the sky (Fig. 40). Three galaxies out of every ten are of this type. The others, less efficient, succeed in turning only four-fifths of the mass of gas into stars. The remaining one-fifth flattens out to form a thin, rotating

Fig. 40. *A giant cannibal elliptical galaxy.* As its name indicates, the shape of this galaxy, as seen projected against the sky, is an ellipse. The photograph shows Messier 87 (M87), which is the largest galaxy in the Virgo Cluster, at a distance of about 45 million light-years. Astronomers believe that M87 became so large by consuming some of its companions inside the cluster (photograph: Hale Observatories).

disk, which completes one rotation in some two hundred million years. This gaseous disk continues to give birth to stars, albeit at a much slower rate, and preferentially, along spiral-shaped arms that appear quite early on (some hundreds of millions of years after the formation of the disk). These are "spiral galaxies*'" (see Fig. 20). They dominate the galactic population, with six out of ten belonging to this general type. Others, which are frankly lazy, take a very long time to convert their gas into stars. These are generally dwarf galaxies, which contain just a few billion stars, 100 times less than the spiral galaxies. The mass of gas within them is comparable with that in the stars. They have no particular shape, being neither elliptical nor spiral, so they are known as "irregular galaxies*'" (Fig. 41). One galaxy in ten is of this type. Why do galaxies vary so much in their capacity to convert gaseous material into stars? The reasons are still obscure, but it would seem that this is related to where they originated. In a region where the density fluctuations were very great, the gas was more compressed and hotter, and therefore had a greater tendency to "turn on" and become stars. Clusters of galaxies, which are exceptionally dense regions, contain a predominance of elliptical galaxies. Galaxies that are isolated, or occur in small groups, which are considerably less dense regions, are generally spirals.

Cosmic Collisions

Does this mean that once galaxies have been formed, they preserve the identities and characteristics that they "inherited" at birth forever? Certainly not, because, as we have seen, galaxies do not live in splendid isolation, but among other galaxies. They interact with their environment, and their inborn properties may change. Some galaxies may even be consumed by others. Such effects are particularly important in the heart of clusters of galaxies, where the density of galaxies is very high (between 1000 and 10,000 galaxies in a cube some millions of light-years on a side), and about 100 to 1000 times greater than in the Local Group to which the Milky Way belongs. There is relatively more empty space between the stars in a globular cluster—even though they are very dense—than there is between galaxies in a cluster. The latter are separated, on average, by about five times their diameter, whereas the stars in a cluster are some 100,000 stellar diameters apart.

Inside a cluster, galaxies move at a velocity of approximately 1000 kilometers per second. The traffic inside a cluster is so congested that galactic "accidents" are frequent. A galaxy risks colliding with a companion every 100 million to one billion years. The average age of a nearby cluster of galaxies is four billion years; so any galaxy that we see now in a cluster must have undergone between four and forty collisions in the past. Most collisions do not involve direct impact, and only slight damage results, with a loss of some of the stars in the periphery of the colliding galaxies, which were torn loose and hurled into space by intense gravitational forces. These stars no longer belong to individual galaxies, but instead form a sea of intergalactic

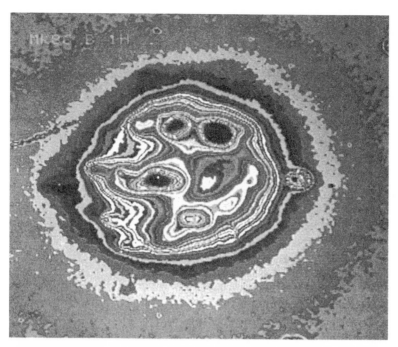

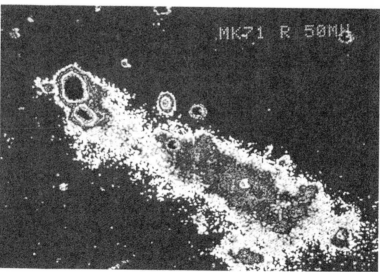

Fig. 41. *Dwarf irregular galaxies.* These images show two examples of irregular galaxies (those that are neither spiral nor elliptical; see also Fig. 17, showing the Magellanic Clouds). These galaxies are approximately 15,000 light-years across and have masses of about one billion solar masses. They contain a lot of gas, which they are actively converting into stars. Within them, the darkest regions (the most luminous areas) represent enormous stellar nurseries (photograph: T. Thuan and H. Loose).

stars, surrounding the galaxies in the cluster. They emit a faint, diffuse light, distinct from that of the galaxies.

From time to time a head-on collision occurs. The two galaxies then lose their individual identities, and merge to form a new, more massive, more luminous, galaxy. If both of the galaxies happen to be spirals, the violence of the impact ejects their gaseous disks into intergalactic space. The new galaxy, for lack of gaseous material, changes into an elliptical. The transformation is as radical as someone changing sex (Fig. 42).

Cannibal Galaxies

In addition, horrible dramas are taking place deep within the heart of clusters. Galaxies disappear, devoured by a giant elliptical galaxy. The latter is the brightest in the cluster, and may be around ten times as large and as luminous as the other galaxies. (Such a galaxy may be seen in the cluster shown in Fig. 30.) Its enormous mass exerts gravitational forces that brake the motions of any galaxy passing nearby. The galaxies that stray within range find themselves drawn into a gradual spiral toward the giant elliptical, which eventually "swallows" them. The largest galaxy becomes larger and larger, devouring its smaller companions, and thus becoming more and more massive and luminous. It is the law of the jungle, but applied to galaxies. The "strongest" becomes "stronger and stronger," at the expense of the "weakest" galaxies, which are threatened with extinction. On average, about one billion years occur between the cannibal galaxy's successive "meals." So at least four galaxies should have been victims since the cluster was formed. The intergalactic stars that are produced in galactic collisions are also attracted by the giant galaxy's strong gravitational field. They are drawn toward it and form a halo of diffuse light around the galactic cannibal.

We all have a combination of "inherited" and "acquired" characteristics. The inherited traits are transmitted by our parents, through the genetic code inscribed in the DNA double helix. The acquired traits are the result of our interaction with our particular environment: relations, friends, loves, work, school, etc. In just the same way, galaxies possess a combination of inborn and acquired characteristics. Their "genetic" properties, determined at the time of their birth, have been inevitably modified by their environment. During their lifetime, galaxies may lose stars or their disks of gas, or grow by devouring their companions. Just as sociologists have difficulty in separating inborn from acquired traits in people, it is not easy for astrophysicists to disentangle the inherent properties of galaxies from those molded by their environment. In this respect, the Space Telescope (see Fig. 12) should help considerably. It should allow us to see galaxies at far greater distances, and thus at a much younger age, which have not been affected by environmental factors for so long. By comparing the characteristics of distant galaxies, which are primarily inherent, with those of closer ones, which have more acquired traits, it should be possible to decide on the relative significance of inborn and environmental factors.

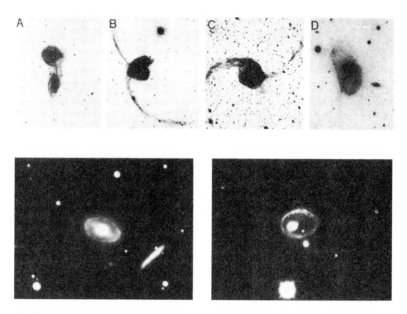

Fig. 42. *Nature vs. nurture in the world of galaxies.* Galaxies do not live in splendid isolation, but interact with their environment. As a result, their inborn properties are modified. They may collide with one another, lose their identity and merge to form a single more massive and luminous galaxy. In the course of these cosmic collisions, many stars are torn off from their parent galaxies by gravitational forces and thrown into intergalactic space, where they form diffuse luminous structures. (a) Four examples of collisions of galaxies, with luminous arms spurting out from the shapeless pile of stars resulting from the colllision. In time (less than 1 billion years), when the effects of the collision will have disappeared, the stellar heap will become an elliptical galaxy. (b) If the colliding galaxies are both spiral galaxies with galactic disks, the collision may bore a hole through one of the disks, creating a ring galaxy. The hole will not last forever: The stars at the borders will fill it up in less than 1 billion years and the ring galaxy will also become an elliptical galaxy (photograph: F. Schweizer).

The First Stars

The universe came to an abrupt halt on its ascent toward complexity, at an age of about 3 minutes, when the primordial helium did not combine into more complex atoms. The universe had to find a way out of this impasse, and give itself a chance to continue its evolution. It did so brilliantly, with the help of gravity, by giving birth to galaxies and stars. Galaxies are needed to escape the general cooling and dilution produced by the expansion of the universe. The contents of galaxies, bound by gravity, are no longer part of the universal expansion, and can therefore retain their heat and energy. But galaxies alone would not do. They are not dense enough to allow the collision and combination of hydrogen and helium atoms into others, sine qua non condition for resuming the march toward increasing complexity. On average, an embryo galaxy contains just one atom of hydrogen (10^{-24} gram) per cubic centimeter, which is millions of billions times less dense than the air we breathe. The "vacuum" inside galaxies is more extreme than anything we can create in the laboratory.

The universe needed denser regions to attain greater complexity; so it invented stars. During its gravitational collapse, an embryo galaxy breaks up into hundreds of billions of tiny clouds of hydrogen and helium gas. Again under the influence of gravity, which gives them their spherical shape, these clouds themselves collapse. The density in their interiors gradually increases. It soon becomes more than 160 times that of water. The temperature rises and reaches tens of millions of degrees. In their cores, atoms of hydrogen and helium born in the first few minutes of the universe collide vigorously, releasing electrons, hydrogen nuclei (protons), and helium nuclei. The scene is somewhat reminiscent of the one when the universe was 3 minutes old. The only difference is the absence of free neutrons. Never mind! The spheres of gas are about to initiate nuclear reactions that use just protons. Four of these combine to form a nucleus of helium-4. A nucleus of helium-4 actually consists of two protons and two neutrons. Two of the original protons change into neutrons, emitting two antielectrons (positrons) and two neutrinos (Fig. 43). The agents responsible for these reactions are, once again, the high temperature and quantum uncertainty. These overcome the electromagnetic force, which causes the protons to repel one another. The fusion of protons into helium nuclei releases energy, which appears in the form of radiation. The spheres of gas light up. The first generation of stars is born. Their date of birth is not very well known, but is thought to be around two to three billion years after the universe was formed.

There is another consequence of this liberation of nuclear energy: The contraction of the spheres of gas comes to a halt. An equilibrium is established between radiation pressure, which tries to disperse the star, and gravity, which tends to compress it. But what is the mysterious source of energy of the stars? Einstein has the answer: If we compare the mass of four free protons with that of the helium-4 nucleus that is produced when they fuse, we find, to our surprise, that the mass of the helium-4 nucleus is not equal to that of four protons, but is slightly less. What has happened to the mass difference? It has been converted into energy, which feeds the fires of stars. The amount of energy may be calculated by multiplying the difference in mass by the square of the velocity of light.

A Second Chance for the Universe

So far, the invention of stars has added nothing new. Helium nuclei already existed after the first three minutes of the universe. Will stars prove capable of overcoming the helium obstacle and give the universe a second chance? To know the answer to this question, we have to wait until the core of a star exhausts its supply of protons—until it runs out of its hydrogen fuel. The length of time this will take depends on the mass of the star. Like people,

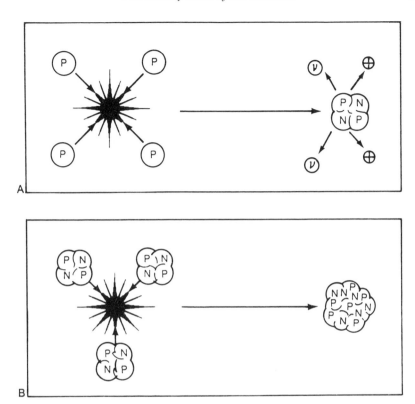

Fig. 43. *Nuclear alchemy within cosmic crucibles.* A star like the Sun produces energy and shines thanks to the numerous nuclear reactions that take place in its core, which has a temperature of some 10 million K. (a) Four protons (or hydrogen nuclei) fuse to form one helium nucleus with the liberation of two positrons and two neutrinos. When the star has consumed most of its hydrogen supply, it no longer produces enough radiation to counteract its gravitational attraction. The core collapses slightly, and the central temperature rises to 200 million K. (b) Three helium nuclei then begin to fuse to produce a nucleus of carbon-12, and this provides the star with an additional source of energy.

stars come in various sizes and masses. The smallest and least massive have about one tenth of the mass of the Sun. The largest and most massive, on the other hand, contain about 100 solar masses of material, and naturally have larger cores. Naïvely, we might think that they may last longer with their larger hydrogen reserves. Not at all! The richest people are often the most spendthrift. A star with a mass 60 times that of the Sun burns the candle at both ends. It exhausts its hydrogen supply after a few million years, which is nothing when compared with the universe's 15 billion years. A star with a mass like that of the Sun is far more moderate. It will exhaust its hydrogen supply after some nine billion years. The most stingy is a star with a mass just one tenth of that of the Sun. It will continue to burn hydrogen for some 20 billion years, more than the present age of the universe.

Let us follow the fate of a star that has the same mass as the Sun. Events will approximately be the same for other stars, but the speed at which they occur will vary. In more massive stars, everything will happen much faster, and in less massive ones, much slower. Also, because of their higher central temperatures, heavier stars will be able to make more heavy elements. Let us rejoin the star when it has just exhausted its hydrogen supply. The radiation diminishes, and is no longer able to counteract the force of gravity. The star contracts. The density increases and the temperature rises to about 100 million K. The helium nuclei, which were created by the fusion of protons, still refuse to combine in pairs. But a miracle occurs: They fuse in threes to form a carbon-12 nucleus (Fig. 43b). This carbon-12 is the carbon found in trees, in the paper of this book, or in the paintings of Van Gogh or Matisse. This miracle occurs because the mass of the carbon-12 nucleus is very nearly that of three helium nuclei. In fact, it is slightly less, and the mass difference is converted into radiation, as before. With a new source of energy, radiation is again able to balance the star's gravitational compression. The contraction of the stellar core, which has become smaller and denser, stops. At the same time as the core contracts, the star's atmosphere swells incredibly, driven by the enormous surge of energy produced by the helium fusion within its core. The star's diameter increases a hundredfold. The radiation, which now escapes over a surface that is tens of thousands of times larger, is correspondingly weakened. The surface of the star cools, and its color shifts to the red. The star becomes a red giant*.

Why has the star been able to overcome the helium barrier, whereas the primordial universe failed so miserably? Because it is very difficult for random processes to bring three helium atoms together. Time is needed, a commodity the expanding universe did not have. The material within it became inexorably less and less dense with every tick of the cosmic clock. The probability of a three-way encounter dwindled from second to second, and became practically nil at the end of the first 3 minutes. The red giant does not have to worry about the universal expansion diluting its material. It has millions of years—a whole eternity by comparison with just 3 minutes—in which suitable collisions can occur. This is why it succeeds where the universe failed.

The march towards increasing complexity can begin once again. Inventing stars has saved the universe on more than one count. It now has at its disposal cosmic furnaces in which it can forge, at its leisure, the chemical elements that are essential for life. It is no longer condemned to sterility. In addition, as we have seen, stars create disorder. They allow the universe to create complexity without violating the laws of thermodynamics, which say that disorder must always increase.

Stars with Layers Like an Onion

The fusion of helium into carbon will last only some 2 billion years, about 5 times less than the period over which hydrogen was turned into helium. At the end of that time, the core of the red giant, with insufficient radiation to support it against gravity, will collapse yet again. For a star like the Sun, this would be the end of the road. Because of its relatively small mass, its central temperature cannot rise enough to trigger burning of heavier elements. But for stars with the mass of 4 Suns or more, life goes on. Gravitational contraction heats their core to a temperature of 600 million K, and carbon burning begins. Stars more massive than about 9 Suns take the nuclear alchemy even further. In the searing heat of their central cores, more complex and familiar elements such as neon, oxygen, sodium, magnesium, aluminum, and silicon, or even phosphorus and sulfur, make their appearance. These massive stars go on as far as they can on the path towards increasing complexity. The same sequence of events is repeated again and again: When a fuel runs out, the core collapses and becomes denser and hotter. A new source of heavier fuel is tapped, producing new heavier and heavier elements. The pace of events quickens, and the cycles take less and less time. More than 20 new elements come into existence in a few million years.

The action is not confined to the core of the red giant. The radiation released by the nuclear reactions at the center heats all the outer layers and allows them to burn fuels as well. The temperature is not the same throughout the star, however; it decreases outward from a few billion K in the core to a few thousand K at the surface. Because hydrogen is converted into helium only at a temperature of 10 million K, helium to carbon at 100 million K, etc., the fuel and the endproducts vary according to the depth of the layer in the star. The star therefore acquires a structure that is like an onion's, with the outer layers made of progressively lighter elements. Towards the end of its life, the star's core consists of iron, cobalt, and nickel, which are the products of silicon burning. In the layer above, carbon fuses into silicon, phosphorus, and sulfur. Yet farther above, helium fuses into carbon, oxygen, and neon. Finally, hydrogen is converted into helium in the layer above. Some 60% of the star's mass is thus involved in some form of burning. The remaining 40% is too cool to join in and retains the hydrogen born in the first 3 minutes.

Iron, the Recalcitrant Element

The star pursues its evolution until iron-56 appears. It has gone a long way on the road towards complexity. Beginning with the simple building blocks of protons and neutrons, the star has managed to build elements as impressive as iron-56, whose nucleus contains 26 protons and 30 neutrons. It already contains the hydrogen, carbon, nitrogen, and oxygen that will make up more than 90% of the atoms in our bodies, as well as the other chemical elements that are responsible for the wide variety of shapes and colors in the world around us. But events take a different turn with the arrival of iron. The star is stopped dead in its tracks. It is powerless to progress any further. Iron-56 cannot be used as fuel. It is powerless to provide the star with energy and help it in its unending battle against gravity. What is the reason for this impasse? In the earlier fusion processes, the mass of the final product is always less than the sum of the masses of the nuclei that are fused. (The mass of a helium nucleus is less than that of four protons; the mass of a carbon nucleus is less than that of three helium nuclei, etc.) The conversion of this mass difference into energy allows the star to shine, and prevents it from collapsing under the influence of its own mass. The conversion of hydrogen into helium provides the most energy. (This is, unfortunately, the source of the devastating power of the hydrogen bomb. It constitutes also an energy source we are trying to master. Herculean efforts are still required to control thermonuclear fusion, but significant advances are being made, and the price is worth the effort: a source of clean power—there are no radioactive waste products—that is inexhaustible since hydrogen nuclei can be obtained from seawater.) Fuels heavier than oxygen provide the star with the least energy, but they do enable it to linger on. Everything changes when iron-56 enters the scene. It does not combine with other nuclei unless it is provided with energy. Starting with iron, the mass of the final fusion product is greater than the sum of the masses of the parent nuclei. Iron-56 demands a heavy price for its participation in nuclear reactions. With no energy to draw on, the star cannot satisfy that demand. Short of fuel, it stops shining. Gravity, no longer held in check by radiation, takes control, and causes the star to collapse and die. Depending on its mass, a star's death may be peaceful or violent.

Three Ways to Die

Let us follow the fate of a star that has a mass less than 1.4 times that of the Sun. It has a serene death. When it runs out of fuel, the star shrinks from the size of a red giant (with a radius of some 50 million km) to that of the Earth (with a radius of about 6000 km). The star becomes a dwarf (Fig. 44). It is quite hot, because the energy generated by the collapse motion has been converted into heat. Its surface temperature is about 6000 K. This heat is radiated away into space, and the color of this radiation is white, similar to that of the Sun; so these stars have come to be known as "white dwarfs*." The density is enormous: A cubic centimeter weighs about one metric ton. But what stops the white dwarf from collapsing even further? What is there to resist gravity? It certainly cannot be the radiation, which has become far too weak. The answer was provided by the German physicist, Wolfgang Pauli, one of the founders of quantum mechanics. In 1925, he discovered that two electrons cannot be squeezed together: They are mutually exclusive. (Pauli's discovery is known as the "exclusion principle."[12]) As the star collapses, it forces all the electrons within it into a smaller and smaller volume. The tighter they are packed together, the greater their resistance, and

12. The "exclusion principle" was first applied to white dwarfs by the Indian-American astrophysicist S. Chandrasekhar.

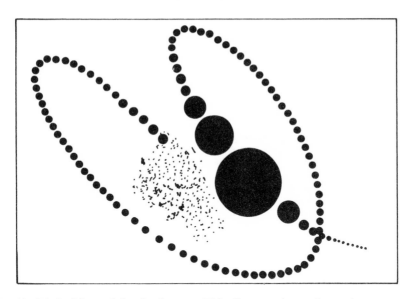

Fig. 44. *Birth, life, and death of a star.* This diagram shows the various stages, at 100 million-year intervals, in the life of a star similar to the Sun. It is born following the collapse of an interstellar cloud; burns hydrogen for 9 billion years as a star; then turns into a red giant, burning helium, for 2 billion years, before finally collapsing into a white dwarf. At the end of its life it becomes a black dwarf, a stellar corpse lost in the darkness of the immensity of space.

the more they try to escape. This resistance creates a pressure, which opposes the force of gravity, and stops the white dwarf from collapsing. This mutual repulsion is not a result of the electromagnetic force, which causes like electrical charges to repel one another, but is instead one of the consequences of quantum mechanics.

As the core collapses, the outer layers are expelled from the star. They are illuminated by light from the white dwarf and appear as a yellowish or reddish ring of gas, which is called a "planetary nebula*." [This term is a misnomer, because planetary nebulae (Fig. 45) and planets have nothing in common.] Most stars (including our Sun) will die peacefully in this way, because most stars have masses less than 1.4 solar masses. A large telescope is required to detect white dwarfs, because they are so faint. Sirius, the brightest star in the sky, has a white-dwarf companion. A white dwarf takes billions of years to cool. Finally, transformed into an invisible "black dwarf," it becomes just one among the innumerable stellar corpses that litter the Galaxy. As for the planetary nebula, it will disperse into space, enriching it with the heavy elements made in the stellar crucibles.

What happens to a star with more than 1.4 solar masses? Its end is far more violent. Once again, the final outcome bifurcates depending on the mass of the star, and whether it has more or less than five solar masses. The additional mass compresses the star more. The collapse occurs so fast (in

Fig. 45. *A planetary nebula.* This photograph shows the Ring Nebula in Lyra. It is the shell of gas ejected by a dying star with a mass less than 1.4 solar masses. When it ran out of fuel, the star collapsed into a white dwarf (which can be seen in the center of the nebula). The radiation from the white dwarf illuminates the planetary nebula. The name is misleading: There is no connection with planets (photograph: Hale Observatories).

just a fraction of a second) that the electrons are taken by surprise and do not have time to organize their resistance to gravity. The limit of 6000 km for the radius of the white dwarf is fast broken. The radius of the star shrinks to just 10 kilometers. The final density is extremely high. It may attain one billion metric tons per cubic centimeter. This is rather like compressing the mass of 100 Eiffel Towers into a space the size of the ball at the tip of your pen. The nuclei cannot resist this extreme compression, and break apart into protons and neutrons. The electrons are pressed so hard against the protons that they are forced to combine with them, producing neutrons and neutrinos. The neutrinos, which we met in the early moments of the universe's existence, live up to their reputation. They do not interact with matter; so they escape immediately. The core of the star becomes a gigantic "nucleus" of neutrons. The latter, which survive for just 15 minutes when they are free, lose their tendency to decay when they are trapped. They now resist further gravitational contraction, and stop the collapse of the neutron star*. As with the electrons, neutrons obey an exclusion principle, and cannot be compressed too closely together.

   At the end of the core collapse, an immense explosion takes place. The onion layers of the star, enriched with heavy elements, are propelled into space at velocities of thousands of kilometers per second. The explosion reaches a brightness equivalent to that of 100 million Suns. A brilliant point of light appears in the sky. This is a "supernova." The sudden halt to the core collapse caused by the neutrons' aversion to being too close together is at the origin of this cataclysmic explosion. A shock wave is created, which propagates toward the surface, and blows the star apart.

These explosive stellar deaths occur about once every hundred years in galaxies. Since humans began to record observations of the sky, about ten of these events have been seen in the Milky Way. In 1572, the young Tycho Brahe saw a "new star" in the constellation of Cassiopeia, and this discovery sowed doubt in his mind about the immutability of the skies as proclaimed by Aristotle. The remnants of the supernova now bear his name. On February 23, 1987, a supernova in the Large Magellanic Cloud, which is one of the Milky Way's dwarf satellite galaxies and lies at a distance of some 150,000 light-years, shook the astronomical world. The whole range of modern instrumentation (large ground-based telescopes, space-borne observatories, and other equipment that Tycho Brahe could not even have faintly imagined) were pressed into service to study this extraordinary event. Even the neutrinos that escaped from the collapsing core of the dead star were captured by detectors located several kilometers beneath the surface of the Earth, in abandoned gold mines. But one of the most famous supernovae in the history of astronomy is without question the one whose remnant is now known as the "Crab Nebula." This "guest star"—as it was called poetically by Chinese astronomers—appeared on the morning of July 4, 1054. As bright as Venus, it was visible during the day for several weeks. Yet no mention of it can be found in Western astronomical writings of the period. Their authors must have had greater confidence in Aristotle's teaching of the immutable, unchangeable universe than in the evidence of their own eyes. . . .

It has been a long time since the guest star was visible with the naked eye. Now the faint supernova remnant may be seen with a telescope as having the crablike shape that inspired its name (Fig. 46). But what makes it famous is the discovery, in 1967, that its central regions harbored a neutron star. This discovery confirmed that such an object, which had been proposed on theoretical grounds as early as 1934 by the American astronomers Walter Baade and Fritz Zwicky, could indeed be the result of the death throes of a star. It was found to flash on and off 30 times a second, which gave rise to the name of "pulsar." This strange behavior comes first from the fact that the neutron star does not radiate over all its surface. Its radiation (which consists largely of radio waves) emerges in twin narrow beams, similar to that of a lighthouse. Second, the neutron star is rotating very rapidly, giving the impression, as one of the beams of radiation sweeps across the Earth, that it is flashing on and off (Fig. 47). The pulsar will serve as a celestial lighthouse for several million years. Its energy reserves, generated during the collapse, will finally be exhausted. It will rotate slower and slower, and eventually cease radiating. It will finally lapse into a deadly silence, and this stellar corpse will be neither seen nor heard from again. In our Galaxy, one star in every thousand ends its life as a pulsar.

We finally come to the most extreme form of stellar death. This fate awaits any star that has a mass of more than about five solar masses. Such a large mass provides an extremely violent collapse. This time it is not just

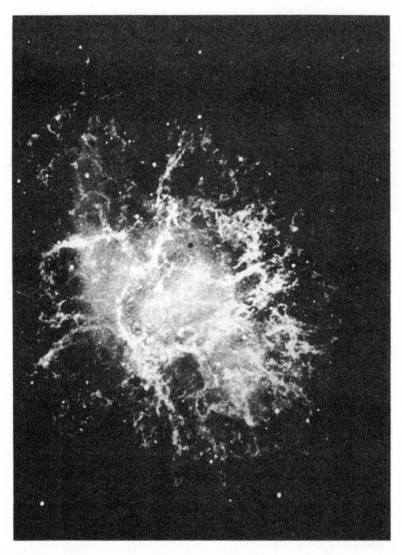

Fig. 46. *The Crab Nebula.* The photograph shows the remnants of a star in our Galaxy that exploded into view on the morning of July 4, 1054. The core of the star collapsed into a neutron star about 10 kilometers in radius. It lies in the center of the nebula and emits radio signals, which are received at regular intervals on Earth, causing it to be also known as a pulsar (*see also* Fig. 47). The tattered fragments of the upper layers of the star continue to expand, driven by the original explosion, and they are now scattered across hundreds of billions of kilometers in space, enriching the interstellar medium with heavy elements that were produced during the star's lifetime and during the explosion.

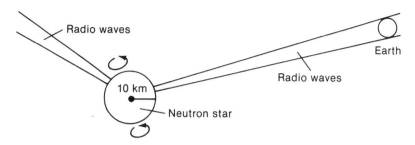

Fig. 47. *A pulsar.* A pulsar is a neutron star, with a radius of about 10 km, which is rotating very rapidly, a kind of cosmic lighthouse. The pulsar does not radiate over all its surface, but emits radiation primarily in two beams. (The radiation is mainly at radio wavelengths.) An observer on Earth will receive a radio signal each time one of the beams sweeps across the Earth. Successive signals are separated by the time it takes for the star to complete one rotation. The fastest pulsar detected to date has signals coming every 1.6 millisecond, which means that it rotates 600 times every second. It is a fantastic celestial spinning top!

the electrons, but the neutrons as well that are taken by surprise. They cannot resist the extreme gravitational forces, which overwhelm them. These compress the matter at the core of the star into such a small volume that the resulting gravitational field is enormous. The core becomes a black hole.

As in the previous case, the violence of the collapse triggers a gigantic explosion, which expels the outer layers of the star into space. The birth of every black hole is greeted by a supernova explosion. This time, the dead star does not even leave a stellar corpse behind. It will make its presence felt only by the gravitational effects it exerts on surrounding matter. It will slow down time. It will turn foolhardy astronauts into spaghetti, break and swallow them. For a terrestrial observer, a black hole is extremely difficult to detect, unless, as we have seen, it is part of a binary, together with another star that is still shining. The black hole will attract some of the gas from the other star's atmosphere, and the atoms of gas will emit x-rays as they fall toward the black hole, thus revealing the latter's presence. It is believed that a black hole exists in the constellation of Cygnus, where there is a very intense x-ray source (see Fig. 26). In our Galaxy, black holes are much rarer than white dwarfs and pulsars; massive stars make up a very small fraction of the galactic population.

Supernovae Do Us a World of Good

We have seen how the universe "invented" galaxies and stars to circumvent the helium barrier. But all the products of stellar alchemy, all these wonderful nuclei of heavy elements, would have been of no use at all if they had remained eternally locked within the cores of stars, because the next stage in the road towards complexity is the construction of atoms from these nuclei. The electromagnetic force has to combine nuclei and electrons and build atoms. Such combination is impossible within stars. They are too hot for atoms to survive. Cooler and calmer environments are needed. What better than the vast expanses of space that lie between the stars in galaxies, the interstellar medium*? The temperatures found there range from a glacial cold around 100 K (− 173°C) to some ten thousand K. The interstellar medium is heated to temperatures higher than the 3 K (− 270°C) of intergalactic space by the radiation from hot and massive stars, and by the supernovae that accompany their violent death.

How are the fusion products to be extracted from the interiors of the stellar furnaces? One method is by means of planetary nebulae, but this is very inefficient, because the nebulae have a very low mass, little more than a few tenths of a solar mass. In addition, they are produced by low-mass stars (less than 1.4 solar masses), which have not carried heavy element production very far. A star with a mass less than half that of the Sun, for example, does not make anything more complex than helium. The temperature of its core is not large enough to overcome the electromagnetic forces and allow for more complicated combinations of protons and neutrons. There is yet another way. Massive stars (those with several solar masses or more) slowly lose their outer layers. Their intense radiation overcomes the gravitational attraction holding the envelope to the star, and pushes it into space. Like autumn winds snatching the leaves from the trees, the stellar wind, driven by the strong stellar radiation, blows away the envelope of the star, and disperses it into interstellar space. But this method is again very inefficient. A stellar wind does not transport much mass, and the products of stellar burning leave their birthplace at a snail's pace. So the universe resorts to drastic measures, and blows up the stars. This method works the best. Several solar masses of enriched material are ejected into space. This is the first good bestowed by a supernova: It enriches the space between the stars with heavy elements.

But that is not all a supernova achieves. It makes use of its immense energy resources to pursue the alchemy interrupted in the core of the star. You will recall that the climb towards complexity by constructing heavier elements stopped with iron. The iron nucleus is the most stable of all elements, and would not combine with other particles to form still heavier, more complex elements, unless additional energy is provided. A supernova has energy to spare: Superheating sets off iron burning, and a whole chain of fusion reactions. About sixty elements see the light of day during the explosion. This time, nature has found the proper recipe. Stellar alchemy can now be brought to the ultimate. This is how all elements heavier than iron come into being. We celebrate in passing the birth of gold and silver that will later adorn beautiful women, of the far more mundane lead, and of the uranium responsible for the bomb that destroyed Hiroshima. Henceforth the gamut of all 92 naturally occurring, stable elements—that is, those that do not spontaneously disintegrate after some time—is complete. They range from the simplest and oldest, hydrogen (one proton) and helium (two protons), to uranium (92 protons), with iron (26 protons), the most stable, on the way.

A supernova can boast of one final achievement. Its incredible energy propels into interstellar space streams of electrons, protons, and other nuclei manufactured by the creative alchemy of the star. Moving at velocities near the speed of light, these particles embark on a long interstellar travel and cross the Galaxy from end to end. Some of these particles will eventually encounter the Earth. They are captured by detectors of terrestrial physicists, who call them "cosmic rays*." Biologists believe that bombardment by these cosmic rays may cause major changes in the molecular structure of the genes within our bodies. Supernovae, through cosmic rays, are thus probably at the origin of the genetic mutations that have punctuated Darwinian evolution, which has led from primitive cells to us.

Quasars

Let us return to our story. The first stars have been born. They have admirably fulfilled their role in creating heavy elements. Having begun with gas that was poor in metals (the collective term used by astronomers for all elements heavier than hydrogen and helium), they have returned it considerably enriched. The lives of massive stars have lasted a mere twinkling of an eye. A few million years of brilliant light and extreme temperature, and they have gone. All that remains are dense and compact stellar remnants, neutron stars and black holes. As for the less massive stars, they are more thrifty and parsimonious and burn up their energy more sparingly. They will last for billions of years. Some of them are still alive today.

There are good reasons for believing that, in certain galaxies, this first generation of stars was particularly rich in massive stars, leading to a particularly bountiful production of metals. After several million years, about one billion stellar corpses in the form of black holes can be found littering the vast expanse of these galaxies. A billion years pass, and those billion black holes, under the influence of their mutual gravitational attraction, have coalesced to form a gigantic black hole with the mass of a billion suns at the center of the galaxy. This giant black hole preys upon its host galaxy. It ensnares any star passing in its vicinity, shapes them into spaghettilike filaments, tears them apart, and engulfs them. Gas from the disintegrating stars streams at high velocities toward the black hole, forming a flattened disk around it. It heats up and radiates at extreme energies just before it crosses the radius of no return of the black hole, beyond which it will be invisible forever. The surroundings of the black hole shine brilliantly. The luminosity is 1000 times that of the host galaxy, as much as one hundred thousand billion Suns. A quasar* is born. The incredible energy produced by a quasar comes nevertheless from a region that is scarcely 100 times the size of our Solar System. A quasar occupies a region a few light-months across, less than one hundred-thousandth of the average size of a galaxy. The monster at its heart, which provides all the energy by greedily devouring unfortunate stars, is even smaller. Its radius of no return is just a few billion kilometers, roughly the size of the Solar System. To terrestrial observers, the quasar appears as a tiny, brilliant point of light, like a star (Fig. 48), hence its name,

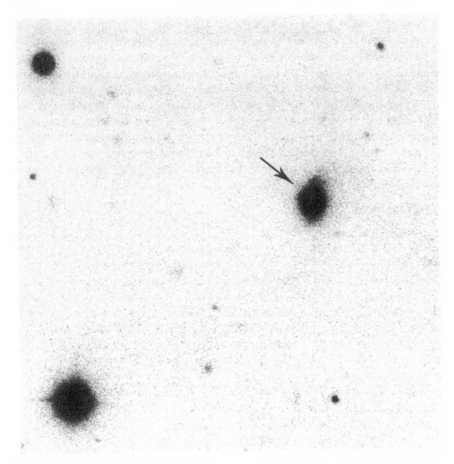

Fig. 48. *A quasar.* If it were not for the extreme red shift of its light and its slightly elongated shape, the quasar indicated here by an arrow, and known as 3C 48, would be easily mistaken for a star. (Compare its image with the two bright stars that are also visible.) The name *quasar* comes from the term "quasistellar object." The majority of astronomers believe that quasars are the most distant and brightest objects in the universe, and that their incredible energy derives from a gigantic black hole of one billion solar masses at the center, feeding on torn-up stars from the host galaxy. The latter is just visible in this photograph, and is responsible for the slightly elongated shape of the quasar (photograph: Hale Observatories).

which derives from "quasistellar object." It was this stellar appearance that completely astounded astronomers in the early 1960s. When the light from quasars was analyzed, it was nothing like the light from any known type of star. The puzzle was bandied around for several years, and only solved in 1963. The light from quasars did resemble that from a large collection of stars, but it was shifted so far towards the red that it was not immediately recognizable. Hubble taught us the greater the red shift, the farther away the galaxy. The enormous red shifts of the light from quasars implied that they were at the edge of the observable universe, much farther away than most galaxies. When we look out to great distances we are also looking far back in time; so quasars appear to us today as they were when the universe was still young, when it was no more than a few billion years old (see Appendix A).

Quasars constitute the only benchmarks that we currently have to locate the birth date of the first stars and galaxies. The light from quasars indicates that they already contain a lot of metals. They were therefore born after the first generation of stars had done their alchemy. The latter probably came into being between the ages of 2 and 3 billion years. Might we one day happen on a galaxy actually being born—what we might call a primordial galaxy? The explosive death throes of the massive, first-generation stars in primordial galaxies should provide a marvellous, celestial fireworks display. Their light should be detectable across a vast expanse of space, and should shine like beacons in the darkness. We may well witness such a wonderful sight thanks to the Hubble Space Telescope because it can see seven times farther than ground-based telescopes. The heavy mists that veil our knowledge of that mysterious epoch will then perhaps be finally dispelled.

Galaxies containing quasars are not the only objects to nurture a voracious cannibal at their hearts. Other galaxies, known as "galaxies with active nuclei," also have black holes at their centers. These are generally 10 to 100 times smaller than the black holes in quasars, and cause less damage. But they also rip stars apart and devour them, producing intense radiation that covers a wide range from gamma rays and x rays to radio waves. Even our Galaxy may harbor a black hole with a mass of somewhat less than 1 million solar masses at its center. Aristotle's unchanging universe is most definitely dead. Our telescopes, which are sensitive to the whole gamut of electromagnetic radiation, reveal events of extraordinary violence in the centers of certain galaxies.

Molecules in Space

We have now reached the end of our story and the last page of our book on the history of the universe. Billions of years go by. The universe continues its implacable expansion (distances increase with time to the power $\frac{2}{3}$; see Appendix E), thinning out and cooling down. Superclusters, clusters, and groups of galaxies are slowly woven together into a cosmic tapestry. Galaxies live out their lives, and the formation of metals, which began with the first generation of stars, continues within them. Several generations of stars go past. Like living things, stars are born, live out their lives, and die. With each generation that passes, two parallel sequences of events occur. Massive stars live their short lives quickly, and in their explosive agony, hurl gaseous material enriched in metals into interstellar space. This material condenses into interstellar clouds, which contract in turn under the effect of gravity, to give birth to new stars. As for the lighter stars, they last considerably longer. They join those from the earlier generation, and the generations of stars overlap. Bit by bit, as they transform gaseous material into stars, galaxies become richer in metals. By now the latter form 2% of their mass.

What happens now in the gradual ascent towards greater complexity? At this stage, a galaxy contains the nuclei of heavy elements supplied by the creative alchemy inside massive stars, and that are now scattered into interstellar space. With this material, atoms and molecules can be constructed. But how may the interactions between these nuclei—an absolutely essential condition for the formation of more complex structures—be encouraged? Interstellar clouds have too low a density to favor such interactions. Their density is some 10^{22} times less than that of water. Interstellar dust grains, born in the atmospheres of red giant stars and pushed out into interstellar space by stellar winds, are far more suitable for such a role. Tiny particles, some thousandths of a millimeter in size, they possess a solid nucleus, which consists mainly of silicon, oxygen, magnesium, and iron (like the Earth's crust), and which is covered by a thin layer of ice. The surface of these grains proves to be a fertile breeding site, where nuclei of heavy elements repeatedly encounter each other. They tried out every possible combination, resulting in a veritable orgy of unions. Molecules containing two, three, four, and up to thirteen atoms see the light of day. About 100 of them have been discovered by radiotelescopes. Each emits radio waves of a particular nature, which allows it to be identified. The molecules of hydrogen (H_2) and carbon monoxide (CO) are among the most abundant. Molecules of water (H_2O), methane (CH_4), and ammonia (NH_3) are less abundant.

All molecules seem to have a marked preference for a quartet of atoms: carbon (C), hydrogen (H), oxygen (O), and nitrogen (N). More than 99% of the bodies of all living things are made of these four elements. The basic building blocks of life have already entered the scene. "But," you will say, "we are still a long way from the real basis of life, the proteins, enzymes, and nucleic acids, which contain thousands of atoms, or even millions in the case of DNA." That is true, of course; Nature has not taken complexity to the ultimate in interstellar space. But the lesson to be learned from this vast proliferation of interstellar molecules is not that Nature failed, but that she was, on the contrary, triumphantly victorious. She has proved to be inventive in the extreme, and was able to create a whole zoo of molecules in an extremely inhospitable environment: in the icy cold and almost perfect vacuum of interstellar space. Astronomers were the first to be caught by surprise. They did not have the slightest expectation of finding such a diverse and various fauna. In any case, given Nature's wonderful dexterity in producing molecules, it would be presumptuous of us to think that she had given us the exclusive rights to life.

The Invention of the Planet

Time passes. The cosmic clock registers 10.4 billion years. Among the hundreds of billions of galaxies that populate the observable universe, we will concentrate our attention on one with rather pretty spiral arms. It is

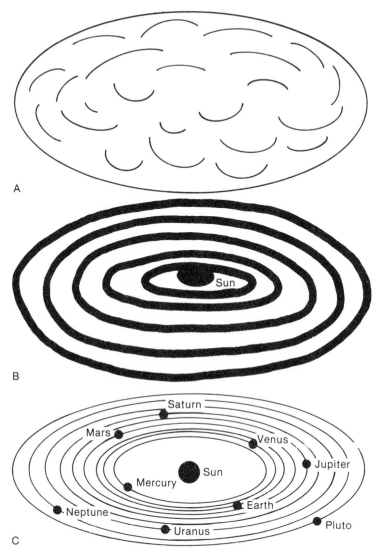

Fig. 49. *The formation of the Solar System.* The Solar System resulted from the collapse of a cloud of interstellar gas, originally several thousands of billions of kilometers across. As it collapsed, the cloud flattened (a), and its center became so hot and dense that nuclear reactions began (*see also* Fig. 43): The Sun was born (b). The surrounding less dense gas condensed into planets and asteroids. The Solar System has come into existence (c).

known as the Milky Way. In a small, remote corner of this galaxy, about two-thirds of the way from the center to the edge, an interstellar cloud begins to contract. Its collapse may have been triggered by a nearby supernova. The temperature at its center begins to mount. From a few tens of degrees, it soars, over a few million years, to some 10 million K. Nuclear reactions begin, and the gas cloud ignites and becomes a star. The Sun, a third-generation star, is born.

Nature continues on her long ascent towards complexity. Molecules consisting of just ten or so atoms are not enough. The conditions in interstellar space, with its temperatures far below the most Arctic conditions, and its vast, nearly empty spaces, are far too extreme for still more complex structures to be formed. Life needs a more benign environment. Nature thus invents planets, using the interstellar dust grains scattered throughout the cloud as building material. As the cloud contracts, some dust grains escape from it. They begin to orbit the Sun, settling into pretty rings rather like the ones encircling Saturn. Within these rings, some of the slightly larger dust particles begin to capture others and grow. The masses gradually increased: 1 gram, 1 kilogram, 1 metric ton, then finally billions of tons. Eventually almost all the material in the rings has become concentrated into nine, solid, spherical (gravity likes spherical objects) bodies. The family of planets in the Solar System has been born, with Jupiter as the major body (Fig. 49). Smaller clumps condensed in turn to form satellites around every planet except Mercury and Venus. Earth has its Moon, and Jupiter and Saturn are each surrounded by twenty-odd satellites. The remaining material became meteoroids or asteroids*. Many of these crashed into the newly wrought planets. The craters that cover the surface of the Moon and Mercury bear mute witness to this period of intense bombardment.

The period of major upheaval lasted several hundred million years. At the end of it, the Earth, with a mass of 6000 billion billion tons, began to evolve the conditions necessary for life. It was rotating very fast. The Sun raced across the sky above a landscape dotted with numerous volcanos, which disgorged vast flows of red-hot lava over the surface. It was just two and a half hours from sunrise to sunset, because initially the Earth rotated once in every 5 hours. Since then the gravitational forces exerted on it by

Fig. 50. *The history of the universe.* (a) Detailed illustration of the first billion years of the universe's existence and its gradual cooling. Recent findings in elementary particles physics have allowed cosmologists to outline the history of the universe all the way back to 10^{-43} second after the Big Bang. Present knowledge does not allow us to go any farther back. This would require a quantum theory of gravity. At that time of 10^{-43} second, the four fundamental forces (electromagnetic, gravitational, and strong and weak nuclear forces) were perhaps unified into a single superforce. As time passed, the universe expanded and cooled. The inflationary phase, that is, the period when the universe expanded exponentially, began at 10^{-35} second and ended at 10^{-32} second. During this period, the strong nuclear force became distinct from the electronuclear force, the universe acquired its homogeneity and flatness, and the seeds of future galaxies were sown. Quarks, electrons, neutrinos, and their antiparticles came into being from the vacuum. Nature showed an infinitesimally greater preference for quarks over antiquarks (one-billionth more). This meant a future universe dominated by matter rather than antimatter, and that there would be one billion photons for every particle of matter. At 10^{-12} second, the weak nuclear force split from the electromagnetic force. Subsequently there were four forces. The

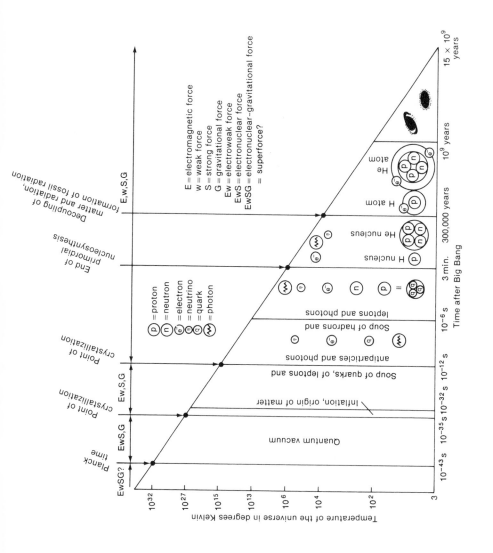

181

the Moon have continuously slowed it down. It will continue to rotate more and more slowly. The 24-hour day will stretch to 48 hours, weeks, months, years. . . . But those who hate to change their habits need not worry. Even if they reach the age of 100, they will not have to add more than about 30 seconds to the length of a day. The Earth's slowing down is almost imperceptible over the course of a human lifetime.

The Ascent of Life

One billion years after the birth of the Sun, the Earth had cooled considerably. On the sea of lava, a mass of grey rock—a proto continent—begins to take shape. The solidified lava kept growing, eventually covering 30% of the Earth's surface. As it solidified, the lava gave off the huge quantities of gas contained within it, blanketing the Earth with an atmosphere 100 times thicker than today's. The primordial atmosphere, composed of hydrogen, ammonia (NH_3), and methane (CH_4), plus water vapor (H_2O) and carbon dioxide (CO_2), was noxious and unsuitable for life. As the planet continued to cool, water condensed out of the early atmosphere, and rain began to fall. Torrential rains flooded the Earth, until finally oceans covered three-quarters of its surface.

Subject to the incessant bombardment by energetic ultraviolet light from the young Sun [ultraviolet radiation was able to penetrate the primitive atmosphere very easily, because the ozone (O_3) layer did not yet exist; all the oxygen was trapped in the water], and to the powerful electrical discharges of lightning from the storms permanently raging, the simple molecules in the primitive atmosphere went on a synthesis binge. The terrestrial atmosphere was millions of billions of times denser than the interstellar medium; so collisions were far more frequent. "Organic" compounds made their appear-

Fig. 50. (continued)
soup of quarks, electrons, neutrinos, photons, and their antiparticles existed until 10^{-6} second, when most of the protons and neutrons—which formed from the quarks—were annihilated along with their antiparticles. From then on, the universe was populated by photons and neutrinos, with a very small number of protons and neutrons (one billion times less numerous than the photons) and electrons. By the third minute, the synthesis of hydrogen and helium nuclei had taken place. Then nothing important happened until about 300,000 years. By this time, the universe had cooled sufficiently for electrons to combine with the nuclei to form hydrogen and helium atoms. This was accompanied by the emission of the fossil background radiation we observe today at a temperature of 3 K. The electrons bound within atoms no longer prevented radiation from propagating through space. The universe became transparent (*see also* Fig. 28). Around one billion years, the first quasars and proto-galaxies were formed. (b) Detail of the most significant events that took place in the last 14 billion years: the formation of the Sun and the Solar System, 4.6 billion years ago; the formation of the Earth's atmosphere and the continents, 500 million years ago; the appearance of life on Earth, and the evolution of species until the arrival of the first *Homo sapiens*, just 2 million years ago.

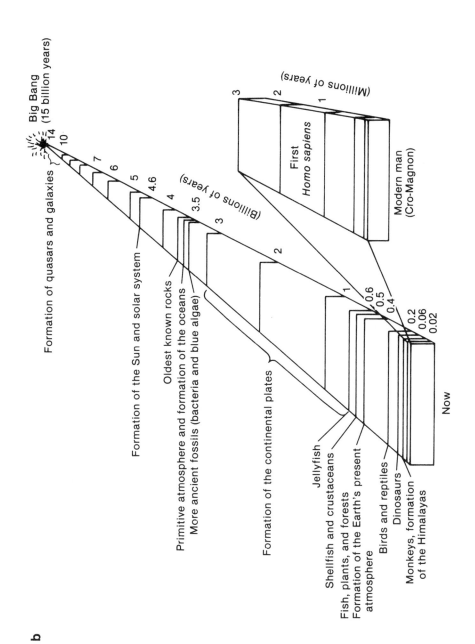

Big Bang
(15 billion years)

Formation of quasars and galaxies

(Billions of years)

Formation of the Sun and solar system

Oldest known rocks

Primitive atmosphere and formation of the oceans

More ancient fossils (bacteria and blue algae)

Formation of the continental plates

Jellyfish

Shellfish and crustaceans

Fish, plants, and forests

Formation of the Earth's present atmosphere

Birds and reptiles

Dinosaurs

Monkeys, formation of the Himalayas

Now

(Millions of years)

First
Homo sapiens

Modern man
(Cro-Magnon)

b

183

ance. Amino acids, clusters of some 30 atoms, began to proliferate. Carried by the rains, these organic materials fell to the oceans. Water now began to fulfill its great mission as the catalyst of life. Because of its considerable dissolving power, a wide range of alien molecules can readily adhere to it. Millions of times denser than the terrestrial atmosphere, it is the ideal breeding site where all sorts of encounters and combinations may occur. In addition, as a perfect host, it could offer guest molecules protection from the harmful effects of the Sun and raging storms.

In this ideal medium, amino acids assembled into long chains to form, some hundreds of millions of years later, proteins, which in turn linked up to form the double helices of DNA molecules. These strands held the secret of immortality: self-replication. They would later transmit the genetic code of all living things. Then, when the universe was 11.5 billion years old, cells, each containing millions of DNA molecules, entered the scene. They in turn served as building blocks of life. The primeval ocean teemed with single-celled bacteria and blue-green algae, then the most advanced forms of life. The oldest fossils date from this epoch. Three billion years went by. During this long gestation period, the continental plate formed. When the cosmic clock struck 14.4 billion years, 600 million years ago, the first multicellular organisms, jellyfish, appeared. Another 100 million years, and the first mollusks and crustaceans made their entrance, followed some 100 million years later by the first fishes. At about the same time, some 450 million years ago, the land became carpeted with flora. These plants and trees were to have a major impact on the Earth's atmosphere. Water in the ground was absorbed through their roots, and carbon dioxide in the air through their leaves. Through photosynthesis they tapped the energy of sunlight and converted these substances into sugars, releasing in the process oxygen into the atmosphere. The oxygen atoms combined in threes to form ozone. This resulted in the ozone layer, which blocks ultraviolet rays from the Sun and protects us from contracting skin cancers. Of all planets in the Solar System, the Earth is the only one that possesses free oxygen in its atmosphere. It is precisely this same ozone layer that makes newspaper headlines nowadays. By dint of polluting his ecological environment, humans have with aerosol propellants and refrigerants created a large hole in the ozone layer, which appears to continue to grow bigger.

The stage is now set for life, protected by the atmosphere from the harmful effects of the Sun, to emerge from the water and colonize the land. Some 300 million years ago, birds and reptiles (such as snakes, lizards, and tortoises) made their entrance. Another 75 million years later, and dinosaurs came on the scene. They ruled the Earth for some 160 million years. During this period, about 230 million years ago, the original single continental plate broke up, and the continents began drifting across the globe. Then, suddenly, 65 million years ago, the dinosaurs disappeared. Some astronomers believe that their demise was caused by an asteroid, some ten kilometers across, that collided with the Earth. The shock created an enormous crater in the Yucatan peninsula, in Mexico (which is no longer clearly visible, having been erased by movements of the Earth's crust) and sent a vast and opaque cloud of dust into the atmosphere, which prevented sunlight from reaching the surface. Plants, unable to carry out photosynthesis any longer, died off. The Earth cooled down, entering a long winter. The dinosaurs (and other species that fed on plants) could not resist the cold and hunger, and became extinct. This theory, however, is not universally accepted, and the sudden death of the dinosaurs remains a mystery. In any case, misfortune for one is good luck for another. The mammals, which first appeared about 100 million years ago, were able to evolve freely and safely, in the green prairies that covered the landscape. (Everything had returned to normal after the dust settled to the ground, and sunlight could again warm the surface of the planet.) Primates entered the scene some 20 million years ago, at about the same epoch that the Indian continental plate collided with the Asian plate, lifting the Himalayas. The first member of *Homo sapiens* trod the Earth only about 2 million years ago.

The long ascent towards increasing complexity, begun some 15 billion years ago, had finally led to Man (Fig. 50). Beginning with a vacuum, and then with the initial soup of elementary particles, the universe has come up with human beings made up of 30 billion billion billion (3×10^{28}) particles, capable of reproducing and multiplying. Towards the end, cosmic evolution had greatly accelerated. Mankind is but a twinkling of an eye in the history of the universe. If the whole of that history were reduced to a single day, the Sun and Earth would form only around 17:00. Most of the ascent towards Man would take place in the last hour. Jellyfishes would appear at 23:02, fishes at 23:22, birds and reptiles at 23:31. Dinosaurs would enter the scene at 23:38 and leave it 15 minutes later. Primates took over at 23:58, and humans did not make their entrance until 11.5 seconds before midnight. The whole of human civilization and technology over the past 4000 years would occupy no more than the last 200ths of a second, hardly as long as a photographic flash. Will Mankind be wise enough to escape from nuclear and ecological suicide, continue to live on this beautiful planet, and accompany the universe as it continues to evolve, even if only for a few billion years to come? Life has probably not finished its long ascent towards complexity. What surprises does it hold in store for us?

| 6 |

The Invisible and
the Fate of the Universe

We have spent a long time scrutinizing the history of the universe. We have been involved with the various events that have occurred: getting worried when it reached the helium impasse, then sighing with relief when stars were invented. We have applauded when supernovae fertilized interstellar space with new elements, and cheered when the Sun and the Solar System were born. We have admired the irresistible ascent towards life on Earth, which eventually led to ourselves. The history of the universe has proved fascinating, and we naturally want to know what happens next. We are curious about the universe's future (Fig. 51). Will the expansion continue indefinitely, and the universe become infinite in extent; that is, is it "open*"? Or will its expansion come to a halt one day, and will it collapse back on itself; that is, is it "closed*"?

The future of the universe is not read from cards or seen in a crystal ball, but is contained in three numbers, known as cosmological parameters*, according to the Russian mathematician Alexander Friedmann. In 1922, he devised a mathematical model based on Einstein's theory of relativity, published in 1915. This model, which also has the advantage of being the simplest, best describes the behavior of the observed universe. The first parameter is linked with the age of the universe. (It is often called the "Hubble parameter*"). It provides information about the pace at which events occur, the time it takes for each episode to unfold. Does the evolution of the universe proceed at a breakneck speed, like in an old, speeded-up, Charlie Chaplin movie, or does it go at a snail's pace? We know that the evolution

186

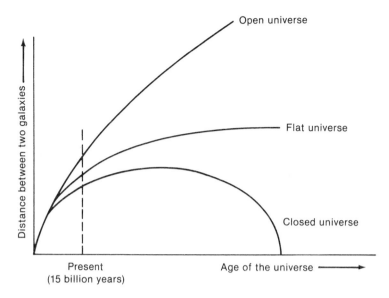

Fig. 51. *The future of the universe.* The gravitational attraction exerted by matter (both visible and invisible) slows down the expansion of the universe. However, because we are not yet able to measure accurately the total amount of matter, we cannot say anything definite about the eventual fate of the universe. If the density of matter is less than the critical density of three hydrogen atoms per cubic meter, gravity will not be able to halt the expansion, which will go on forever. The distances between the galaxies will increase continually, and the universe is "open." Current observations appear to favor this kind of universe. If the density is precisely equal to the critical density, the expansion of the universe will come to a halt at an infinite time in the future. This universe, which is described as "flat" (*see* Fig. 23), will have a similar fate to one that is open. If the density is higher than the critical density, the universe will reach a maximum size, and then collapse back on itself. In such a universe, which is described as "closed," the galaxies will get closer and closer to one another, and will eventually disintegrate into particles of radiation and matter at an extremely high density and temperature. Will the universe be reborn from its ashes (*see* Fig. 56)?

of the universe has taken between 10 and 20 billion years. We cannot pinpoint its age with more precision, because we do not know, as yet, how to measure cosmic distances very accurately.

The second parameter describes the deceleration* of the universe's rate of expansion (the deceleration parameter*). Every galaxy feels the gravitational attraction of all the other matter (both visible and invisible) in the universe. The galaxy's motion is slowed down; it is decelerated.

The third parameter is connected to the second. But instead of describing the rate at which the universe is decelerating, it addresses the cause of the deceleration. It characterizes the mass or, more precisely, the density (the mass divided by the volume) of the universe. (It is known as the density parameter*.) If the universe contains less than three hydrogen atoms per cubic meter, the expansion will never cease. On the other hand, if there are more than three atoms per cubic meter, events take a different turn: The universe is condemned to collapse back on itself at some time in the future. This critical density, which determines the fate of the universe, is an extraordinarily small number, particularly when one realizes that one gram of water contains one million billion billion hydrogen atoms. On average, the universe is exceptionally empty.

An enormous amount of observational effort has been exerted in the past 6 decades to determine these cosmological parameters. Considering what is at stake, it is worth it. These parameters will not only reveal the fate of the universe, but (as we have seen in the discussion of the origin of the structures within it) will provide a key to the past. Despite prodigious efforts, however, and definite progress, it has not yet been possible to establish the cosmological parameters with the required accuracy. As for anything fundamental in nature, the secrets are not so easily revealed. The sphinxlike universe has yet to answer.

Let us admire from closer up the unrelenting effort of astronomers, and their boundless ingenuity and imagination. Having witnessed the arduous struggle to make the universe reveal its true age, let us examine now the methods used to measure its slowing down.

The Deceleration of the Universe

If you want to measure the deceleration of your car, all you need to do is to note the speed at two different instants: when you start to brake, and when the automobile comes to a stop. If you were traveling initially at 100 km per hour (or 27.8 meters per second), and you take 2 seconds to come to a standstill (i.e., for your velocity to become zero), the deceleration, which is simply the decrease in velocity per unit time (here 1 second), is $(27.8 - 0)$ meters per second, divided by 2 seconds, that is, 13.9 meters per $(second)^2$. In principle, the deceleration of the universe could be measured in just the same way. One could measure the velocity of any one galaxy (using the Doppler effect) at two different times, and divide the change in velocity by the time difference to obtain the deceleration of the universe. In practice, however, such a program cannot be carried out. The rate at which the galaxies are slowing down is so minute that it is imperceptible during a human lifetime of 100 years, or even over the 2 million years that human beings have existed. We would have to wait for billions of years, which would not do at all.

Luckily, generous Nature comes to our rescue and offers us an invaluable way of uncovering her secrets. Because we cannot wait for time to pass, she allows us to reach back into the past. Light, which carries information across the universe, permits us to travel back in time, because its propagation is not instantaneous. Although moving faster than anything else in the universe, it still travels only 300,000 km per second, and takes time to reach us. The universe is always seen with a certain delay. By capturing light from more and more distant galaxies, we are going back up the river of time: Telescopes are truly time machines. Thus, in principle, to measure the deceleration of the universe, all we have to do is to measure the velocity of two galaxies at different distances. This is equivalent to measuring the expansion velocity at two different times and should therefore give us the deceleration.

Once again, the problem of cosmic distances rears its ugly head. To obtain the deceleration parameter, we must determine the age of galaxies at the same time as their velocities. To date them, we need to know their distances. At night, we use the apparent brightness of the headlights of an oncoming car (the intrinsic brightness of which we know) to estimate the distance between the car and us. Similarly, as we have seen, astronomers use cosmic lighthouses with known intrinsic brightness (or "standard candles") to measure distances.

Among the zoo of objects that populate the universe, the giant elliptical galaxies in the heart of clusters of galaxies seem to be ideal standard candles. They are extremely bright: Given that they are about five times as bright as their more normal-sized companions, they can be seen at very great distances; they allow us to journey back into the past over at least half the age of the universe. In addition, their actual brightness is well known, and varies very little from galaxy to galaxy, at least as far as nearby giant galaxies are concerned (the variation is less than 40%). A constant intrinsic brightness is absolutely essential in any lighthouse; a sailor needs to be certain that if a lighthouse appears faint, his ship is well away from the coast. He needs to know that there is no possibility that he is actually very close to crashing his ship on the rocks, because the lighthouse is fainter than normal. To decipher the mystery of the deceleration of the universe, astronomers measure the apparent brightnesses (which, when taken in conjunction with the true luminosities, give the distances and thus the ages) and the red shifts (the Doppler shift gives the velocities) of hundreds of giant elliptical galaxies. (In practice, it is essential to observe as many giant ellipticals as possible to compensate for the slight variations in intrinsic brightness from galaxy to galaxy, and for the errors in individual measurements.)

To accomplish this task, astronomers push telescopes and electronic detectors to their limits. They track down the most distant galaxies—galaxies so far away that the light from an entire galaxy is reduced to just a few photons. They try to reach back as far as possible into the past: The greater the difference in time, the larger the difference in the universe's expansion velocity, and the easier it is to measure. As for us, we stand on the sidelines rubbing our hands with glee: We shall finally discover the deceleration parameter. At long last, the universe will reveal its ultimate fate.

Giant Elliptical Galaxies Are Not Good Standard Candles

Success was only possible if the true brightness of giant elliptical galaxies did not vary in time or space. We now know, however, that they do vary in at least two ways. First, there are changes in brightness caused by stellar evolution*. Galaxies are made of stars, which are born, live out their lives, and die, and which, consequently, vary in brightness. The brightness of the galaxy, which is simply the integrated brightness of all the stars within it, must therefore change with time. It declines by several percent every billion years. Distant galaxies observed during their exuberant youth are therefore intrinsically brighter than nearer galaxies seen in their more peaceful maturity. If the changes in brightness due to stellar evolution are not taken into account, the distances of remote galaxies will be underestimated—we will think that they are closer because they are brighter. The deceleration parameter will be overestimated—remote galaxies will appear closer to nearby ones, and the deceleration will therefore appear greater. A decline in brightness by a few percent per billion years may seem an insignificant amount at first sight. The error it introduces into the determination of the deceleration of the universe, however, is such that it is no longer possible to distinguish between an open universe, with perpetual expansion, and a closed one that will one day collapse back on itself.

We have failed because giant elliptical galaxies are not good distance indicators, because their brightness varies with time. "What does that matter?" you may well say, "We know the reason for the variation; so all we have to do is to calculate how much there is, and correct the brightnesses of the giant ellipticals to make them truly standard candles." Unfortunately, this is easier said than done. To calculate the evolution of galaxies through stellar evolution, we need to know the rate at which stars are formed in galaxies as a function of time, and also the distribution of stars by mass. Both of these are very poorly known at present. The problem of star formation remains one of the most fundamental problems of astrophysics, and as long as the fog that hangs over it is not dispelled, the evolution of galaxies cannot be calculated with the desired accuracy.

Galactic Cannibalism

The other cause of evolution among giant elliptical galaxies is their insatiable appetite for devouring their smaller companions. We have seen that in the very dense cores of clusters, galaxies interact gravitationally. These interactions slow down the smaller galaxies and make them spiral gradually toward the giant elliptical in the center of the cluster, which eventually engulfs them. As a result of this "galactic cannibalism," the largest galaxy becomes larger and larger and brighter and brighter.

This "dynamical" evolution has an effect opposite to that produced by stellar evolution. Nearby mature giant ellipticals are intrinsically brighter than young, remote galaxies; they have had longer to cannibalize their smaller neighbors. Once again, the change in brightness is a few percent per billion years, which is more than enough to upset completely our attempts to determine the deceleration parameter. As before, our ignorance of galactic cannibalism is such that we cannot allow for it with the desired accuracy.

To sum up, the prospects for determining the deceleration parameter are not rosy. Evolutionary effects confuse the issue and prevent us from distinguishing between an open and closed universe. Can we bypass this problem by using other standard candles rather than giant elliptical galaxies? Quasars, discovered in 1963, raised everyone's hopes. They seemed to be ideal lighthouses: extremely bright and at the very edge of the universe. Unfortunately, disillusionment soon set in. The intrinsic brightness was found to vary, from one quasar to another, by a factor of more than 1000, which made them quite unsuitable as standard candles. Other candidates have been suggested, but in the last analysis, the fundamental but poorly understood problem of the evolution of galaxies has always proved to be an insurmountable barrier. Might the Space Telescope (Fig. 12) one day provide the answer? It should be able to peer much deeper into space and to detect galaxies at a far younger age, and therefore at an earlier stage of evolution. Such observations, when compared with those of nearby galaxies, should allow one to refine the corrections for galactic evolution. Perhaps only then will the universe reveal the rate at which it is slowing down.

The Invisible Mass in the Universe

In the meantime, are we to give up any hope of knowing the ultimate fate of the universe? Not in the least, because the third parameter—the mean density of the universe—can help us solve this problem. The deceleration of the universe depends on the amount of matter it contains. If we can make a tally of all the mass within it, we shall have another method of predicting its future.

Flashback to Pasadena, California, in the year 1933. In an office of the California Institute of Technology, the Swiss astronomer Fritz Zwicky checks and rechecks his calculations. He has just finished a series of observations of the motions of galaxies in the Virgo Cluster. These allow him to determine the total mass of the cluster. He draws up a list of all the possible uncertainties: those that result from observational errors or a lack of data; oversimplified assumptions, etc. There is no escaping the conclusion. He has to believe the evidence: The total mass of the cluster is considerably greater than the sum of the masses of the individual galaxies.

For the first time, observation has suggested the presence of enormous quantities of invisible matter*. Sixty years later, the problem of the "missing mass" continues to haunt and obsess astrophysicists. All the observations that have been accumulated since have only confirmed Zwicky's conclusion. The invisible mass appears to be everywhere. It pervades all the known structures of the universe, from tiny dwarf galaxies to the largest superclusters. Even particle physicists have recently become involved. They argue vociferously for large quantities of invisible matter. It has become their pet subject. By hiding everything behind a veil of invisibility, they hope to give an air of respectability to all the exotic, massive particles they maintain were created in the very early universe and that so far exist only in their unrestrained imaginations.

Very reluctantly, astronomers have come to accept the idea that we live in an "iceberg universe" where nearly all the mass escapes direct detection by their instruments. But there is a fundamental difference between an iceberg and the universe: We know what the submerged part of an iceberg is made of, whereas the nature of the invisible mass remains a powerful challenge to human understanding. In our inventory of the matter content of the universe, we shall have to be very careful to include both the visible and the invisible mass.

Weighing the Universe

How can we measure the invisible? At first sight, the task seems impossible. Luckily, Newton gave us the key, and the solution is simpler than it first appears. When matter is present in the form of several distinct bodies, whether they be stars or galaxies, Newton's universal gravitation runs the show. To obtain the total mass of a collection of stars or galaxies that are bound together by gravity, all we need to do is to study the velocities of the individual stars or galaxies, by making use of the Doppler effect. (We assume that the whole system is in equilibrium, i.e., that it is neither expanding or contracting.) High velocities imply large masses. They must be high to compensate for the important gravitational attraction of the large mass. Conversely, low velocities indicate less mass.

The essential point is that the motions measure *all* the mass that is present, whether luminous or nonluminous, visible or invisible. To illustrate this point, consider the Solar System, where the planets move gracefully under the gravitational influence of the Sun. Imagine that some giant hand compresses the Sun beyond its radius of no return, turning it into a black hole. The planets would continue to orbit around the black hole as if nothing had happened. The only catastrophic result would be the extinction of all life on Earth, for lack of energy. Imagine that extraterrestrials visited the Solar System. By observing the motions of the planets, they would have no difficulty in deducing that there was an invisible body, with the mass of the Sun, in the center. Similarly, the French astronomer Le Verrier and the English astronomer John Couch Adams discovered the planet Neptune in 1846 by examining the perturbations caused by some unknown mass on the orbit of Uranus. Detailed study of the orbits of the American Lunar Orbiter probes revealed that the Moon was not homogeneous and that various "mascons" (mass concentrations) existed beneath the surface. The motion of various bodies thus enables astronomers to "map" gravitational fields. Thanks to Newton, we are now ready to take an inventory of the mass in the universe. Let's go in search of the invisible mass.

The Cosmological Principle

It is quite obvious that, in practice, we cannot draw up a list of *all* the contents of the universe. Several lifetimes would not suffice. We need to adopt

a simplifying assumption to make our task more manageable. At this point
the phantom of Copernicus makes himself heard again. He reminds us that
neither the Earth, the Sun, the Milky Way, the Local Group, nor the Local
Supercluster is special or privileged in any way. These structures are reproduced billions of times throughout the cosmos. Our tiny corner of the universe has no special significance. Why then not postulate that the universe
is perfectly similar everywhere (a quality called "homogeneous*") and in
every direction ("isotropic*")? Cosmologists have answered positively. The
"cosmological principle*" that assumes that the universe is homogeneous
and isotropic is one of the fundamental tenets of Einstein's theory of general
relativity*. It has been confirmed in a spectacular fashion by the discovery
of the 3 K background radiation that bathes the whole universe. The temperature of this radiation does not vary by more than 0.001% from one side
of the sky to the other. We can heave a big sigh of relief. All we need to do
is to draw up a list of the mass in our neighborhood and measure its mean
density. According to the cosmological principle, the mean density of matter
anywhere else in the universe is the same. We must, however, make sure
that our mass inventory is carried out within a large enough volume that it
is truly representative of the universe as a whole. This volume will have to
include at least the Local Supercluster.

Something Dark Around Galaxies

Let us begin our search for the invisible. We start in our own backyard, the
solar neighborhood. Let us follow the movements of the stars in a region
around the Sun some 300 light-years across. Like riders on some gigantic
galactic merry-go-round, the stars move ceaselessly around the center of the
Milky Way. Tearing through space at a speed of about 230 kilometers per
second, the Sun completes one orbit every 250 million years. Like the
wooden horses that go up and down on a merry-go-round, stars wander
above and below the galactic plane* (at a speed of about 10 km/s) in their
travel around the galactic center. This up and down motion reveals the total
density of matter in the disk. Rapid motion would imply a high density,
whereas a slow movement would correspond to a low density. Let us add
up all the mass that can be seen with telescopes in the Solar neighborhood:
stars, white dwarfs, clouds of atomic and molecular hydrogen, etc. This
leads to a first surprise: The density of visible matter is only about half the
total density. The invisible is beginning to make its presence felt. Even in
our own backyard there is as much invisible matter as visible.

Let us continue our exploration. Because of the absorption of visible
light by interstellar dust* grains in the galactic plane, we cannot see much
farther than our immediate surroundings with optical telescopes. (To observe the distant extragalactic worlds, we have to point our optical telescopes at right angles to the galactic plane, to minimize absorption by dust.)
The Galaxy is, however, transparent to the radio waves emitted by the
atomic hydrogen gas, and those waves can cross the Milky Way from one

end to the other without being absorbed. In addition, the hydrogen gas in the plane of spiral galaxies generally extends about twice to three times as far as the stars, up to some 100,000 light-years from the center (Fig. 52). The study of the motions of the hydrogen atoms allows us to explore a region 300 times as large as the solar vicinity. Instead of just our backyard, it is the whole surrounding neighborhood that opens up to our scrutiny.

Just like the stars, the atoms of hydrogen move in quasicircular orbits around the center of the Galaxy, at a few hundreds of kilometers per second. These motions hold the secret of the mass distribution in the Milky Way (or

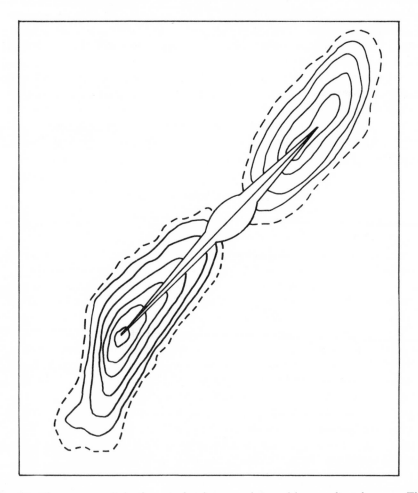

Fig. 52. *The gaseous disk of a spiral galaxy as observed by a radio telescope.* The disk of gaseous hydrogen in a spiral galaxy is invisible, but emits radio waves. With a radio telescope it may be seen to extend two or three times as far as the visible disk of stars. Studying the motion of this hydrogen gas allows us to deduce the existence of an enormous massive invisible halo, whose nature is still unknown, which surrounds the visible part of the spiral galaxy (*see* Fig. 53).

in any other spiral galaxy). The greater the mass within the orbit of one of these hydrogen atoms, the higher its velocity will be, and vice versa. All we need do to map the mass distribution is to examine the rotation of the gas as a function of its distance from the galactic center (to establish the "rotation curve" of the galaxy, in astronomers' jargon).

Radio astronomers threw themselves into this work with great enthusiasm. And, once again, a considerable surprise awaited them at the end of the road. The hydrogen atoms revealed an astonishingly constant rotation velocity. Wherever their location (except very close to the galactic nucleus), they obstinately moved at the same velocity, 230 kilometers per second. The astronomers were flabbergasted. They had imagined that all the Galaxy's mass lay within the visible disk. But if this were the case, the hydrogen atoms beyond the edge of the disk should have been moving more slowly. (Their rotation velocity should have decreased inversely as the square root of their distance from the center.) The very fact that they showed no sign of slowing down implied the presence of invisible dark matter well outside the visible disk.

In the case of the Galaxy, the gaseous disk ceases some 70,000 light-years from the galactic center. How can our exploration of the invisible be carried further? We have to find more distant objects that are still bound gravitationally to the Galaxy. We make use of the motions of the globular clusters—the same ones that contributed to dislodge the Sun from its central place in the Galaxy—and the dwarf satellite galaxies (such as the Magellanic Clouds) that orbit the Milky Way. This permits us to probe the total mass out to 200,000 light-years from the galactic center. The motions are analyzed, and we are faced with the same conclusion: There is dark matter surrounding the Galaxy. Where is this mysterious mass? Is it distributed in a spherical region around the Galaxy, or is it confined to an invisible disk that is merely an extension of the visible disk? Although the various motions can reveal the existence of this dark matter, they do not provide any information on how it is distributed. We have to fall back on more indirect, theoretical arguments. We have seen how the stars in the galactic disk behave as if they were on some fantastic merry-go-round, revolving around the galactic center at the great speed of 230 km/s, but moving perpendicular to the galactic plane at the leisurely speed of 10 km/s. Calculations show that because these transverse motions are very small in comparison with the rotation, the stars ought to be organized into a gigantic bar, from the ends of which the spiral arms would spring, after only one complete rotation, that is, some 250 million years later. But the Galaxy does not show any such structure (although there is the suggestion of a small bar at the center). The lack of such a large bar may be explained if a massive invisible spherical halo* surrounds the visible galaxy. It would prevent any large central bar from forming. By using the motions of stars, hydrogen atoms, globular clusters, and dwarf galaxies as probes for the invisible matter, the mass of the halo is found to increase proportionally with the distance from the center of the visible galaxy. The mass of the halo within the sphere defined by the Sun's orbit (i.e., with a radius of 30,000 light-years) is 100 billion (10^{11}) solar masses, equal to the mass of the disk. If we move out six times as far (to a radius of 180,000 light-years), to the distance of the most far-away dwarf galaxy which can be used as a probe of the invisible matter, the mass of the invisible halo becomes six times as great, to reach a mass of 600 billion Suns.

The Extent of the Invisible Mass

But where do these invisible galactic halos stop? How can we determine their limits? We need to find even more distant probes. We recall that galaxies are often found in pairs (known as "binaries*"), bound together gravitationally. In principle, the motion of one galaxy relative to the other should enable us to determine the total mass, both visible and invisible, contained in the pair. This allows us to probe the invisible mass out to distances about seven times the visible radius of a galaxy, that is, out to about 300,000 light-years. We are now exploring not just our immediate neighborhood, but the whole "city." Once again, the invisible matter manifests its presence. Studies of the motion of paired galaxies reveals that there is dark matter surrounding each of the galaxies in the pair out to at least 300,000 light-years. The invisible mass inside that radius (which remains proportional to the radius) is now ten times the mass of the visible disk, amounting to 1000 billion (10^{12}) solar masses.

It appears that we still have not reached the boundaries of the invisible halos. We have to enlarge tenfold the domain of our exploration, carrying our search for the invisible from binary galaxies to groups of galaxies. We need to explore the county surrounding the city. Let us first examine the Local Group*. Apart from the Milky Way, it contains our nearest large galactic neighbor, the Andromeda Galaxy*. As far as gravity is concerned, the Local Group may be considered as a binary that consists of just the Milky Way and the Andromeda Galaxy, because the other galaxies are smaller galaxies with less mass. How can we determine the mass of the Local Group? We apply our tried and tested method, and scrutinize the motion of the Andromeda Galaxy.

We are in for a surprise. Instead of receding from the Milky Way, like most of the other galaxies, the Andromeda Galaxy is approaching, at a velocity of 90 km/s! Instead of its light being shifted toward the red, it is blue shifted! Yet originally, at the time of the formation of galaxies, some 13 billion years ago, the Andromeda Galaxy must have been moving away from the Milky Way, just like the other galaxies. The expansion of space caused by the original explosion made this inevitable. The motion must have been reversed some time during cosmic evolution. The total mass of the Local Group must be large enough for its gravity to have made Andromeda turn around. A simple calculation indicates that this mass must be approximately 2000 billion solar masses. If our Galaxy and the Andromeda Galaxy, which are otherwise very similar, have about the same mass, each would then contain about 1000 billion solar masses, which is ten times as much as the mass in visible matter. This mass is familiar: It is exactly the same as found in binary galaxies. Yet the separation between the Milky Way and the Andromeda Galaxy (2.3 million light-years) is seven times as great as that between galaxies in binaries. Only one conclusion is possible: The limits of the invisible haloes were already reached with binary galaxies. Exploring larger systems did not reveal any further invisible matter.

To be absolutely certain of that result, let us examine the motions of galaxies in other groups. Their average size, 6 million light-years across, allows us to explore a region twenty times as large as that sampled with binary galaxies. We reach the same conclusion: There is no more invisible matter in groups of galaxies than we have discovered in binary galaxies. The invisible matter is distributed in giant halos 300,000 light-years in radius,

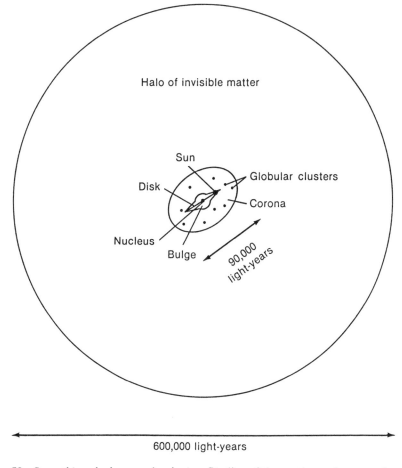

600,000 light-years

Fig. 53. *Something dark around galaxies.* Studies of the motions of stars and gas in spiral galaxies show that the latter are surrounded by enormous invisible halos, five to ten times as large and massive as the visible part. This invisible matter appears to be present everywhere in the universe, but its true nature remains a mystery.

containing 1000 billion solar masses, and amounting for 90% of the total mass in galaxies (Fig. 53).

Invisible Intergalactic Matter

Let us now take yet another step in the hierarchy of structures in the universe and extend our search for invisible matter to clusters of galaxies. One galaxy in ten lies within a cluster (most of the others are in groups). Our zone of exploration is now some 15 million light-years across. Instead of a single county, we are now examining a large part of the country. It was in one of these clusters of galaxies, the Virgo Cluster, that Fritz Zwicky first stumbled across the problem of invisible matter. Let us again scrutinize the motions of galaxies inside clusters. This time the result is different. Within clusters, the quantity of invisible matter is two to three times larger than it is in groups or binary galaxies. The dark matter, which accounts for more than 90% of the total mass in clusters, is composed therefore of two components in comparable quantities: one bound to individual galaxies in the cluster, and the other made up of invisible intergalactic matter lying between the galaxies.

We have to take one final step, and extend our exploration to the largest structures in the universe, the superclusters. Let us focus in particular on the Local Supercluster*. Our field of investigation is now the whole country. Extending over some 60 million light-years, the Local Supercluster contains approximately 10,000 galaxies, gathered into some ten groups and clusters, all bound together gravitationally. These galaxies are organized into a vast flattened pancakelike structure. The Local Group is at the edge of the pancake, while the Virgo Cluster occupies the center (see Fig. 31). The Local Group is not stationary; it is tearing through space—that space that is bathed in the fossil background radiation—like a ship through water, at a velocity of 600 km/s. This motion is betrayed by an extremely slight increase (by about 0.2%) in the temperature of the fossil background radiation ahead of the Local Group and an equivalent decrease in its wake. (As the Local Group cuts through space, the fossil photons that are ahead rush toward it, and the background fossil radiation, as observed from the Earth, is blue shifted as a consequence of the Doppler effect. The apparent temperature is higher. Conversely, fossil photons behind the Local Group are receding away from it; they are red shifted and appear cooler.) The Local Group is moving because it is being pulled by two gravitational forces, one exerted by the Virgo Cluster within the Local Supercluster, and the other by the nearby Hydra-Centaurus supercluster (see Fig. 27). Once again, we can use the infalling motion of the Local Group toward the Virgo Cluster to probe the invisible matter in the Local Supercluster. (The infall velocity is about 250 km/s towards the Virgo cluster, while it is higher toward the Hydra-Centaurus Supercluster, about 550 km/s.) No surprise this time. The amount of invisible matter is the same on the scale of superclusters as it is for clusters.

Can we now call a halt to our search for dark matter? Is the volume of the Local Supercluster sufficiently large to contain the whole range of structures and motions in the universe, and thus to give us an accurate determination of the total mass? Are there no other motions besides the expansion of the universe on scales larger than those of superclusters? A few years ago, this appeared to be the case, but recently some dark clouds have appeared on the horizon. The Local Supercluster and the Hydra-Centaurus Supercluster, instead of simply following the expansion of the universe, seems to be falling in the direction of the Southern Cross constellation at a velocity of several hundred kilometers per second, possibly because of the gravitational attraction of an enormous concentration of matter whose nature is unknown. The latter has been called the "Great Attractor*," for lack of any further information (see Fig. 27). Its mass might be 30 times that of the Local Supercluster and equal to that of tens of thousands of galaxies. Astronomers are busily working to confirm the existence of the Great Attractor. If they succeed, we shall have to revise upward the mass and density of the universe. Stay tuned . . .

Failed Stars and Planets

Our search for the invisible matter must stop here, at least for the time being. Meanwhile, it makes its presence felt, through its gravitational effects, on every scale that we have investigated. All-pervasive and omnipresent, nearby or far away, it exerts a dominant influence on all the visible systems that we have considered, while making up about 90% of the total mass of the universe. Fritz Zwicky had uncovered one heck of a problem!

Once the vertigo is over, astronomers—and the reader—ought to assemble their thoughts, and try to find out more. What is the nature of this invisible matter? What does it consist of? Let's take a plunge into the deep and icy waters of the cosmic ocean to try to determine the nature of this newly discovered iceberg universe, whose tip is all we can see.

We might as well say it straight away: The nature of the dark matter remains utterly unknown. It is shrouded in deep mystery. After 60 years of research, we have made very little progress. Some observations allow us to state what the dark matter cannot be, but there still remains a host of possibilities. Deprived of the photons to carry information, astronomers are quite literally in the dark. They are left with very indirect, and thus very uncertain, methods for probing the nature of the dark matter. The state of our ignorance is such that, if it were not for its utter absurdity, I might as well suggest that the dark matter consists of innumerable copies of this book, floating in space!

Let us now review some of the candidates that have been proposed for this dark matter. None has received universal acclaim. First, we have failed stars, those with a mass less than one one-hundredth that of the Sun, and whose central temperatures are not high enough to trigger thermonuclear reactions. Because they are small relative to true stars, and because they do not shine, these failed stars are known as "brown dwarfs*." Jupiter, the colossus planet in the Solar System, with a mass of about one one-thousandth that of the Sun, is an example of a failed star. It is not capable of emitting light on its own: The pretty light it sheds on clear crisp nights is but the reflected light of the Sun. Pierre Corneille may have been think-ing of such failed stars when he evoked "that dim light that falls from the stars . . ." Then come the failed planets, the asteroids*, which the Little Prince in Antoine de Saint-Exupéry's fable loved so dearly. These are lumps of rock with rough, irregular landscapes, ranging in size from a few kilome-ters to a few hundred kilometers. Their masses are so low that gravity has not been able to sculpt them into spheres. Until we learn more, these failed stars and planets remain potential contenders for the dark matter. So far, the jury has not found sufficient evidence to dismiss them.

Is God Playing with Cosminos?

The same cannot be said for comets, those large balls of snow and ice that light up when they enter the inner part of the Solar System by reflecting sunlight. If they really made up 90% of the mass of the universe, we would have seen far more of them visiting the Solar System. And how about dead stars? After all, we know that stars die, leaving behind white dwarfs*, neu-tron stars*, or black holes*. Such compact, nonluminous, stellar remnants would seem to be ideal candidates for the dark matter, provided they exist in sufficient numbers. But we can soon discard that idea, because, as we have seen, when stars die they expel heavy elements that have been manu-factured in their interiors. If 90% of the mass of the universe consisted of such stellar remnants, the quantity of metals would far exceed the amount observed today in stars and galaxies. There remain primordial mini black holes*, born perhaps in the earliest moments of the universe's existence. They have nothing in common with stars; so the constraint on the abundance of heavy elements does not apply to them. But they should be now exploding with the brightness of millions of billions of galaxies. Such explosions re-main desperately absent.

Clouds of atomic hydrogen then? They would emit far more strongly at radio wavelengths than what is observed. Clouds of hot ionized hydrogen? Atoms of hydrogen in a cloud of heated gas collide violently. Under such circumstances, the electromagnetic force is not strong enough to continue binding the proton and electron into an atom of hydrogen. The proton and electron are freed, and the atom is ionized. In the central regions of clusters of galaxies there is admittedly hydrogen gas heated to millions of degrees, whose presence is revealed by a strong x-ray emission, detected by space-borne instruments. But the mass of this hot gas amounts only to about 10% of the mass of the clusters.

Nothing very convincing so far. Our plunge into the icy water has yielded precious little. Faced with this situation, we have to turn to other far less conventional possibilities. Speculations about the nature of the dark matter have therefore taken a very different turn, thanks to recent developments in particle physics, in particular the unification theories, which attempt to weld the four fundamental forces of nature into a unique force during the very earliest moments of the universe. As we have seen, these theories predict the existence of a host of particles, all possessing mass, and each having a name that is more exotic and poetic than the last: neutrinos, gravitinos, photinos—we shall call all those with names ending in "-ino", "cosminos"—axions, etc. Until now, apart from the neutrinos, all these particles have existed only in the physicists' imaginations. As for measuring the mass of the neutrino, that is yet another thorny question. American, Russian, German, and French physicists are working night and day trying to determine the mass of the neutrino. But, so far, it has remained elusive.

In the meantime, we should note that if neutrinos had a mass equal to just one one-millionth of that of the electron, they would, by virtue of their enormous numbers, dominate the mass of the universe. They would even be able to stop the universe's expansion if their mass were one ten-thousandth that of the electron. However, as we have seen, a universe with massive neutrinos cannot produce the observed cosmic tapestry of galaxies. Cold dark matter (photinos, higgsinos, etc.) would do better, but its existence remains to be confirmed. Einstein said, "God does not play dice." Would he play with cosminos? Nature has more than one trick up her sleeve, and the final solution to the dark matter problem will no doubt surprise us all. In any case, it now seems certain that 90 to 98% of the mass of the universe is invisible. We can but agree with the wise words of Saint-Exupéry's fox to the Little Prince: "What is essential is invisible to the eye." The trouble is: What constitutes the essential?

Cosmic Mirages

The nature of the dark matter still remains shrouded in mystery. There is, however, a glimmer of hope on the horizon. In recent years, astronomers have discovered a new phenomenon that promises to cast some light on many questions about the total amount of matter (visible and invisible) in the universe, and its distribution in space. This is the phenomenon of gravitational lenses.

Gravitational lensing occurs when two (or more) celestial bodies at different distances from the Earth appear perfectly (or almost perfectly) aligned along the line of sight. To reach us, the light from the background object, a quasar, for example, has to go through the gravitational field of the foreground object that bends its path (see Fig. 25, which shows the deviation of light by the Sun). This light deflection changes the image of the quasar, which appears distorted or multiplied. In 1936, Einstein showed that general relativity* predicted that if two stars were aligned along the same line of sight from the Earth, the more distant star would show an additional image, in the form of a ring, surrounding the usual point of light (Fig. 54a). The ring of light is a sort of cosmic mirage, an optical illusion. The case is very similar

a

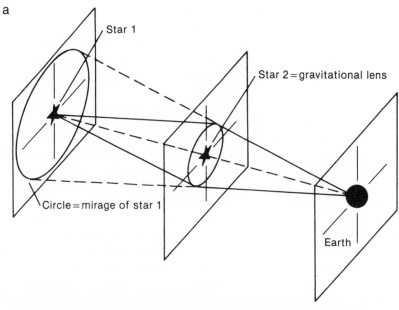

Star 1

Star 2 = gravitational lens

Circle = mirage of star 1

Earth

b

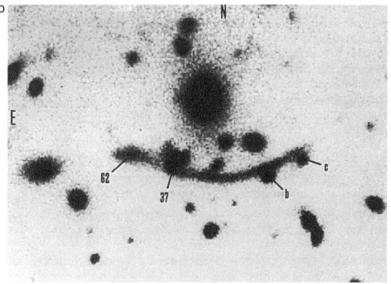

N

E

62

37

b

c

to that of a desert mirage, where the parched and weary travelers find, to their despair, that the inviting oasis where they had hoped to quench their thirst is not actually there. A mirage in the desert is also the result of the deviation of light, due not to the gravitational field of a celestial object, but to the exceptionally hot air over the desert. The foreground object, whose gravitational field bends the light from the background object, is known as a "gravitational lens*." Just as the lenses in your glasses bend light to correct for your myopia, gravitational lenses bend the light from celestial objects to create gravitational mirages.

Einstein thought that the exact alignment of two stars with the Earth was highly unlikely and that the phenomenon of gravitational lenses would remain a purely theoretical idea. The Swiss astrophysicist Fritz Zwicky (the very one who discovered dark matter) took up Einstein's ideas a year later, in 1937, but suggested that galaxies and clusters of galaxies, and not stars, might serve as gravitational lenses. This was how things stood for the next 42 years, because no one had ever seen a gravitational mirage in the sky. In 1979, however, a pair of quasars were discovered. They were very close to one another on the sky and their properties were absolutely identical: same red shift, same color, etc. This amazing similarity started astronomers thinking: The resemblance could not be accidental. Could one of the quasars be

Fig. 54. *Gravitational lenses and the dark matter.* (a) In 1936, Einstein used general relativity to show that if two stars were aligned along the line of sight of a terrestrial observer, the foreground star (2) would act as a gravitational lens to produce a deformed image of the background star (1). The gravitational field of star 2 would bend the light from star 1 (*see* Fig. 25), and the terrestrial observer would not only see a point of light, but also a ring of light centered on it. This ring is therefore like a cosmic mirage: It does not really exist, and is merely the distorted image of the star. Einstein thought that the notion of a gravitational lens was a purely theoretical concept, because he felt that the perfect alignment of two stars with the Earth was too improbable. Einstein was wrong. In 1987, was discovered not a large ring of light, but a very considerable portion of a ring, a giant luminous arc lying in the direction of a cluster of galaxies, known as Abell 370, some 4 billion light-years away. (b) Photograph (after Soucail *et al.*). This arc is explained in a way similar to that shown in (a), although here, instead of star 2, it is the cluster of galaxies Abell 370—and particularly its central region, which is the densest—that acts as a gravitational lens. As for star 1, we have instead a galaxy. The image of the galaxy is distorted by the central region of the cluster into a gigantic luminous arc. The image is not a perfect circle, because the galaxy and the center of the cluster are not exactly aligned. The above scenario has been confirmed by measurements of the distances involved, using the red shifts of the galaxies in the cluster and of the arc. The latter is found to lie at a much greater distance than the cluster, at about 6 billion light-years. More than a dozen giant luminous arcs are now known, all in the direction of clusters of galaxies. The properties of these arcs (their shape, length, orientation, etc.) allow us to deduce the total mass, both visible and invisible, in the central regions of the clusters causing them. Gravitational lenses may therefore be considered as new tools helping astronomers in their search of the dark matter in the universe.

the gravitational mirage of the other? This would explain in a natural way why they so resembled each other. But a mirage required a gravitational lens. When the area of sky around the quasars was scrutinized, a galaxy was discovered superimposed on one of the quasars. This confirmed that the first cosmic mirage had been discovered. There was, in fact, only one quasar. The galaxy lies on the line of sight between Earth and the quasar; so a terrestrial observer sees the galaxy and quasar images superimposed on the sky. The galaxy acts as a gravitational lens, producing the mirage of a second quasar. Since then, other cosmic mirages have been discovered. Even luminous rings (or rather parts of rings), which Einstein predicted, have been found in the direction of clusters of galaxies (Fig. 54b). The existence of gravitational lenses is no longer in any doubt.

What is the connection between these gravitational lenses and the dark matter? A cosmic mirage is the result of complex interaction between light from a celestial object (a star, quasar, galaxy, or cluster of galaxies) and the gravitational field of another object, which acts as a lens. The gravitational field depends on the total mass (both visible and invisible) and its spatial distribution within the lensing object. In addition, the light's path is also influenced by the gravitational field of all the intergalactic material (both visible and invisible) between the celestial object and the lens, and between the lens and the Earth. Cosmic mirages can therefore inform us not only of the dark matter in the lenses themselves, but also in intergalactic space.

As I write these lines, cosmic mirages have not yet helped us very greatly in unlocking the mystery of the dark matter. But they have confirmed some known facts about it. For example, the existence of luminous arcs in some clusters of galaxies (Fig. 54b) implies that 10 times as much nonluminous as luminous matter exists in the centers of these clusters. This conclusion was already reached from our study of the motion of galaxies within clusters. After more than 10 years' research, we now know of only about two dozen cosmic mirages. Their relative rarity tells us that intergalactic space is not full of massive black holes (with masses of about one billion suns), because these would be excellent gravitational lenses. Let us hope that, as measurements continue and theoretical understanding improves, cosmic mirages may one day reveal the secret of the dark matter in the Universe.

According to the Latest News, the Universe Is Open

Let us return to our main concern. We want to obtain the density of matter in the universe to predict its fate. For that, there is no necessity to know the exact nature of the dark matter. All we need to know is that it exists. Let us first calculate the density of luminous matter in stars and galaxies. This amounts to scarcely one-fiftieth of the critical density. If we now add the dark matter, whose mass is ten times that of the visible matter, the total density of the universe amounts to no more than one-fifth of the critical density.

We have chosen to use galaxies to deduce the density of matter. We could have used a completely different method, based on the abundance of deuterium. As you will remember, this element was created during the first 3 minutes after the Big Bang. Its existence is very sensitive to the density of matter in the universe. The amount of deuterium in stars and galaxies also implies that the density of matter is one-fifth of the critical density. If the universe were denser, all traces of deuterium would have been destroyed by thermonuclear reactions in the very early stages of the universe.

Two completely independent methods have given exactly the same answer, which is reassuring. The average density of matter in the universe is less than one atom of hydrogen per cubic meter. At least five times as much matter is required to halt the expansion of the universe. (If the universe were to have precisely the critical density, it would take an infinitely long time for its expansion to cease.) For the time being, the universe is open. Unfortunately, the answer is not as firm as one would like. New facts may emerge that would throw everything back into question. In the course of our travels in the universe, we have already come across massive, menacing monsters, such as the "Great Attractor," lurking in the shadows! If the existence of such monsters is confirmed, the density of the universe will have to be revised upwards. Then again, using galaxies to measure the density of the universe may be misleading. The method assumes that all the mass in the universe is bound up in galaxies and clusters of galaxies. But might galaxies be fooling us? What if there were a massive, invisible component, uniformly distributed throughout the universe that did not follow the distribution of the galaxies? It would not be revealed by any method that relies on the motion of galaxies to measure the mass. We would then be like the man who drops his door key somewhere in the street at night, but who persists in searching only under the lamp posts, because those are the only areas that are fully lit.

You may say that we can still rely on the deuterium method, which gives the same answer. But here again, there are flaws in the argument. Deuterium can only tell us about dark matter that consists of protons and neutrons, of what is called baryonic matter*. It says nothing about neutrinos, photinos, gravitinos, and other cosminos*, in short about any nonbaryonic matter, because the latter is not involved in the creation or destruction of deuterium. Yet physicists suspect that such exotic particles may exist abundantly to permit the unification of the four fundamental forces in the very early universe. They also tell us that if the universe really did go through an early stage of exceptionally rapid expansion (in the inflationary era), it cannot have *anything but* the critical density.

Let us bring our long quest for the three cosmological parameters to a close. What conclusions can we draw from all this? A universe 15 billion years old, containing matter amounting to one-fifth of the critical density, and thus continually expanding, does agree well with all we observe. Its age is in accord with those of the oldest stars in globular clusters. It is also capable of producing the observed quantities of helium and deuterium.

But there remain many problems. The age of the universe, which depends on the Hubble parameter, is uncertain by a factor of two. Its value is based on a rather shaky ladder of distance indicators. The situation, as far as the deceleration parameter is concerned, is hopeless. The effects of the evolution of galaxies are so large that they mask all deceleration effects. As for the density parameter*, there is the distinct possibility that a significant fraction of the mass (five times the mass currently known to exist, or more) still completely escapes detection. Although it amounts to at least 90% of the mass of the universe, we do not have the slightest idea of what the dark matter might be.

The definitive determination of the cosmological parameters still remains to be carried out. Over more than 50 years, we have certainly made progress, but we still have a long way to go. As often happens in science, as soon as some questions are settled, new, even more complicated problems arise. The farther we get, the longer the track seems to become, and the farther we appear to be away from the winning post. The Hubble Space Telescope (Fig. 12) and the new large ground-based telescopes (Fig. 11) offer a gleam of hope. But we have a Herculean task ahead of us if, at the end of the road, the universe is to reveal its true destiny.

The Sun Goes Out

Our attempt to predict the future of the universe has only been partially successful. We believe that its expansion will continue forever, but we are not absolutely sure. Nevertheless, we cannot resist playing the role of prophet. Because the universe can follow only two possible lifelines, why not follow both of them and see where they lead? We have extrapolated from the present back into the past, apparently with success. Why not do the same for the future? The equations do not know the direction of time. Of course, we can look back in time with our telescopes and check whether the past agrees with our equations. We cannot do the same for the future. Light brings us news from the past, but it has nothing to say about what is to come. Without any method of verifying our results experimentally, at least on a human time scale, predicting the future of the universe may seem a rather futile exercise. In science, however, it is imprudent to refuse to explore all possible avenues. Who can tell what marvels await us?

So let us embark on a voyage into the future. Before starting out, however, we need to lay down the rules of the game. Our predictions will be based on the extrapolation of current physical laws. They are to remain unchanged during our odyssey. Gravity, for example, should not become weaker in the future. The past 15 billion years appear to confirm that the laws of physics do not change with time. Then we need to assume that all physics is known, which is a much more doubtful hypothesis. The first half of the twentieth century alone saw the appearance of two new forces (the strong and the weak nuclear forces) and the birth of quantum mechanics and relativity. Why should we know it all now? But we must make do with what we have. Finally, we have to bet that human intelligence, which has proved to be so harmful for the Earth's ecological balance, will never be able to alter the universe's evolutionary course.

Let us first consider what the future holds in store for an open universe[13] since that is the model favored by observations. The universe will continue to thin out and cool down as it expands for the next few tens of billion years. There will be no striking events on a cosmic scale, and the universe will go on its peaceful way. But several events, although insignificant as far as the universe is concerned, will nevertheless disturb the humdrum daily routine in our own Galaxy. In 3 billion years, the Large Magellanic Cloud, a dwarf galaxy currently orbiting the Milky Way at a distance of some 150,000 light-years, will collide with the Galaxy. This dwarf galaxy, whose motion is being slowed down by the Galaxy's gravity, has already started to spiral inwards, a course that will eventually take it into our Galaxy's hungry mouth. The cannibal's brightness will increase through gaining another billion stars. Some 700,000 million years later (in 3.7 billion years), the Andromeda Galaxy, our close neighbor (at a distance of 2.3 million light-years), which is now rushing toward us at a velocity of 90 km/s, will crash into the Galaxy. The collision will not cause too much damage, because there is a vast amount of space (about 3 light-years on average) between stars. The two galaxies will pass through one another, but will lose their disks of hydrogen gas. The Sun's orbit will change a little, the orbits of the planets in the Solar System will be slightly perturbed, and there may be an increased risk of earthquakes. But all in all, nothing to make a fuss about.

Things become more serious when the next event occurs, because the human race's very survival depends on it. In about 4.5 billion years, the Sun will exhaust its supply of hydrogen. It will begin on its reserves of helium. This new source of energy will cause its atmosphere to expand to about 100 times its current size, and it will become a red giant*. Mercury will be engulfed in its searing atmosphere. Earth's inhabitants will see the Sun's red disk invade a considerable part of their sky (about one-fifteenth). The red giant will heat our planet to a temperature of about 1200°C. The atmosphere will be lost. The oceans will evaporate. Rocks will melt. Jungles will burn. Life will no longer be possible. Our great-great-great- . . . grandchildren will embark on their spacecraft and go settle at the edge of the Solar System, on the distant worlds of Neptune and Pluto, far from the burning clutches of the red giant. But the respite will be (relatively) short. Another 2 billion years and the helium reserves will also be exhausted. The Sun will go out, leaving a white dwarf* behind, which will inexorably cool into a black dwarf (see Fig. 44). It will be time to go in search of another source of energy, of another Sun. Perhaps will begin then the colonization of the galaxy, of which science fiction authors are so fond.

---

13. Detailed calculations of the future of an open universe and life within it can be found in F. J. Dyson, *Reviews of Modern Physics* **51**, 1979, p. 447.

A Long Night

In the long term, all the stars in the universe will go out. The galaxies will stop giving off light. They will have exhausted all their hydrogen supplies and will no longer be able to create new stars. The wonderful creative alchemy that goes on inside stars will cease forever, and never illuminate the universe again. The galactic landscape will be littered with stellar corpses. An abundance of black holes, neutron stars, and black dwarfs will lie scattered around among innumerable planets, asteroids, and meteoroids. An endless, dark night will descend on the universe (at least as far as human eyes are concerned: the universe will continue to be bathed in an ever cooler background radiation, and would be visible to radio eyes). The universe will then be 100 times its current age. The average distance between galaxies will have increased from 1 to 20 million light-years. The era of stars will have lasted 1000 billion years (10^{12}) in all.

At the end of that time, the 100 billion stellar corpses in each dead galaxy will continue to be bound by gravity, and will ceaselessly follow their old orbits. But gravity, with unlimited time at its disposal, will tend to alter things. It will make the ex-stars interact and exchange energy. Some will gain while others will lose (energy is always conserved). Those that gain will convert their additional energy into velocity. They will move more rapidly, enlarge their orbits, and drift out toward the edge of the galaxy. Carried along by their impetus, they will escape from the gravitational hold of their parent galaxy and become lost in the immensity of intergalactic space. Those that lose energy will slow down and fall toward the galactic center, making the nucleus denser and denser. When the cosmic clock strikes one billion billion (10^{18}) years, the galaxy will have literally evaporated. It will have lost 99% of its dead stars. The tiny percentage of stars that are not expelled (one billion of them) will lie within the nucleus of the galaxy. The latter, which will become more and more massive and dense, will continue to contract until the radius of no return, equal to about half the distance between the Sun and Pluto, the most distant planet is reached. It will become a black hole. As the nucleus contracts, there is less and less space to move about, and innumerable head-on collisions occur between the stellar remnants, causing them to shatter. The martyred matter will light up. Brilliant fireworks will erupt, illuminating the inky blackness of the night which envelops the dead galaxies. The whole range of the electromagnetic spectrum will be present, from gamma rays to radio waves, and including visible light. The spectacle will continue after the formation of the black hole, which will capture dead stars and tear them apart, heating up the matter and causing it to shine. The galaxy will once again exhibit the brilliance that it showed long ago when it harbored a quasar in its heart, a few billion years after the Big Bang. This period of splendor will last only 1 billion years and then, once again, absolute darkness and glacial cold will return. Nothing will remain of the galaxies of years past other than galactic black holes with masses of about one billion solar masses.

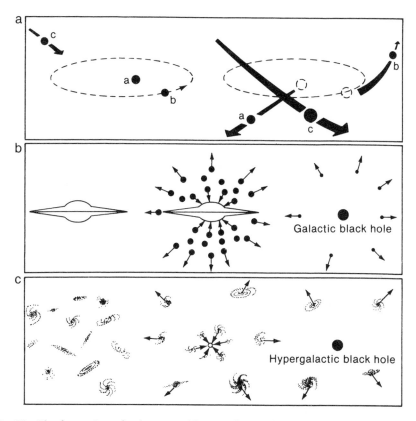

Fig. 55. *The formation of galactic and hypergalactic black holes.* (a) Stars exchange energy through gravitational interaction when they pass close to one another. Star *c* loses energy to stars *a* and *b*, which are bound by gravity and orbit one another in a binary system. Thanks to the energy that they gain, both stars *a* and *b* fly off in different directions, disrupting the binary. (b) In this game of energy exchange, 90 to 99% of the stars in a galaxy (the least massive) gain energy. They acquire enough energy to escape from the parent galaxy, which literally evaporates at the end of 10^{18} years. At the same time, the most massive stars (between 1 and 10% of the total) lose energy and fall toward the galactic center. Gravity causes them to collapse into a galactic black hole with a mass of one billion solar masses, sole remnant of the original galaxy, which once possessed 100 billion suns. (c) The same phenomena of evaporation and black-hole formation take place in clusters of galaxies. There, the players in the energy-exchange game are not stars in a galaxy, but entire galaxies, each consisting of 100 billion stars. After 10^{27} years, 90 to 99% of the galaxies (the least massive) have evaporated from their parent cluster. All that remains of the cluster is a hypergalactic black hole with a mass of 1000 billion solar masses, the result of the gravitational collapse of the galaxies that lost energy (the most massive ones).

Clusters of galaxies also participate in the action. They also turn into black holes after 10^{27} years. Like the stars in the previous case, each of the galaxies in a cluster will swap energy through gravitational interactions. Once again, those that gain (99%) will leave the cluster and become galactic black holes, while those that lose will gather together in the center of the cluster to form a hypergalactic black hole, with a mass of 1000 billion (10^{12}) solar masses and a radius of no return of 3000 billion kilometers (Fig. 55).

So, when the cosmic clock rings one billion billion billion (10^{27}) years, when the galaxies will be separated by 1 billion billion light-years, the wonderful cosmic tapestry of galaxies and clusters of galaxies will have literally evaporated into space. Lost in the darkness of the universe will be innumerable galactic and hypergalactic black holes, carried along by the universal expansion. In the midst of these gigantic monsters, there will be swarms of debris of all sorts (asteroids, comets, planets, black dwarfs, neutron stars, and small black holes of a few solar masses), winners in the energy game that were ejected from their parent galaxies or clusters of galaxies.

The Evaporation of Black Holes

Black holes are not eternal. We have seen how quantum mechanics, in a total defiance of classical mechanics, causes them to evaporate and dissolve into radiation. It achieves this tour de force by temporarily drawing on the gravitational energy of the black hole to bring swarms of virtual particles into existence just beyond the radius of no return. These particles, once in the real world, turn into radiation as they collide with their antiparticles. The borrowed energy causes a loss of mass for the black hole. (Einstein showed us that mass and energy are equivalent.) As the black hole shrinks in mass, it heats up (its temperature is inversely proportional to its mass), and it radiates more strongly. The process accelerates until the black hole evaporates completely in a blaze of radiation. Naturally, it takes a long time for quantum mechanics to accomplish its task. The more massive the black hole, the longer it takes. (The lifetime of a black hole is proportional to the cube of its mass.) But quantum mechanics is in no hurry. It has all eternity in which to act . . .

We also saw that quantum mechanics cannot begin the evaporation process at any time. It can start its action only when the universe has cooled sufficiently. Just as boiling water can evaporate only when it is in contact with cooler air—heat travels from hot to cold, not the other way around—black holes can evaporate only when they are hotter than the background radiation surrounding them. The temperature of a black hole with a mass of the Sun is one ten-millionth (10^{-7}) of a degree; that of a galactic black hole is one ten-million-billionth (10^{-16}) of a degree, while that of a hypergalactic black hole is one ten-billion-billionth (10^{-19}) of a degree. By the year 10^{20}, the universe will be cool enough for black holes with the mass of the Sun to evaporate. Galactic black holes can begin to evaporate when the universe is 10^{34} years old. Hypergalactic black holes will have to wait until 10^{39} years before they begin to turn into radiation. Time will pass, during which quantum mechanics will patiently continue its work. Solar-mass black holes will be fully converted into radiation by the year 10^{65}. By the year 10^{92} it will be the turn of galactic black holes to disappear. Hypergalactic black holes will suffer the same fate around the year 10^{100} (see Appendix C). All of them will briefly illuminate the overwhelming darkness in their final, explosive death throes. In that extremely distant time, all that will remain will be neutron stars, black dwarfs, planets, meteoroids, comets, asteroids, and other tiny bodies, once the gainers in the energy-exchange game and lost by their parent galaxies, and now immersed in a vast sea of radiation, which will grow evermore colder. (Its temperature will be only 10^{-60} K above absolute zero by the year 10^{100}.)

Diamonds Are Not Forever

In our society, diamonds are symbols of both luxury and durability. They not only adorn beautiful women, but they are considered as indestructible and eternal, being the hardest of all natural minerals. This immortality is certainly borne out on the scale of a human lifetime of a 100 years or so, on that of the 2 million years during which humans have existed, or even over the 15 billion years of the universe's existence. But it does not hold over a period of 10^{65} years. At the end of this very, very long time, the temperature of the diamond will be so close to absolute zero that the atoms of which it consists ought to be locked into a permanent crystalline structure by electromagnetic forces. This would be the case if classical mechanics had the last word, but, once again, quantum mechanics will defy it. It will slacken the electromagnetic bonds between the atoms and give them back their freedom. From time to time, the atoms will be able to move and twirl around inside the diamond and rearrange themselves. The previously solid diamond will behave like a liquid free to change its shape. The precious stone that was once so exquisitely shaped by the skilled hands of a diamond cutter will then assume a perfectly ordinary shape, that of a sphere, under the influence of gravity.

But the task of quantum mechanics will not be over. It will then try to change the very nature of the atoms in the diamond. Nature is lazy. It follows the policy of least effort, and does not like to expend any more energy than necessary. Every system tends to evolve towards a state of minimum energy. In the world of atoms, iron represents the lowest energy state. It is the favorite element of Nature, which has endowed it with the most stable atomic structure (an iron nucleus has 26 protons and 30 neutrons), and wants to turn all other atoms into iron. Quantum mechanics will force atoms heavier than iron to split and to release all superfluous protons and neutrons. Lighter elements will be constrained to fuse and acquire the missing protons and neutrons. Everything will turn into iron. You will recall that it is just these fusion reactions that cause stars to shine. They occur in large numbers in stellar cores heated to temperatures of tens of millions of degrees. In the glacial cold of the future universe the reactions will be much rarer. It will require time, a lot of time, for them to have any effect. Quantum mechanics will take 10^{1500} years to convert all matter into iron. (The lucky owners of diamonds can sleep peacefully: They will not wake up tomorrow and find their beloved jewels turned into balls of iron!) A new Iron Age will begin. Apart from neutron stars, everything, from the smallest grains of dust to black dwarfs with masses of several solar masses, including planets and asteroids in between, will become spheres of iron. The nuclear reactions leading to iron will release energy, but will be quite incapable of slowing down the inexorable cooling of the universe toward absolute zero. (The temperature at that stage will be 10^{-1000} degree.)

The iron spheres are not at the very lowest rung of the energy ladder. Spheres of neutrons are lower, and black holes lower still. Quantum mechanics will go to work again. It will convert the nuclei of iron into neutrons. Protons inside the nuclei will combine with electrons to produce neutrons and neutrinos. The neutrinos will be lost to space, while the balls of iron change into spheres of neutrons. The universe will enter the neutron era. Quantum mechanics will patiently continue its labor. It will then try to turn nearly all the matter into black holes. Inside the spheres, the neutrons are perfectly lined up in a crystalline structure, like rows of highly disciplined soldiers. Quantum mechanics will allow some of the neutrons to slip out of this structure. Some of the neutrons close to the surface will be lost to space. Others will fall toward the center and, as they accumulate there, will cause the neutron spheres to collapse into black holes. The transformation of the iron spheres, first into spheres of neutrons, and then into black holes, will be marked by innumerable fireworks that release not only neutrinos and x rays, but also visible light, which will briefly dispel the utter darkness pervading the entire universe.

All these events will unfold extremely slowly. By the time this episode comes to an end, the age of the universe will be about $10^{10^{76}}$ years (the exact age depends on the smallest mass that can collapse into a black hole). Such an age is quite unimaginable. To write out such a number, I would have to follow the figure 1 with as many zeros as there are hydrogen atoms in all the hundreds of billions of galaxies in the observable universe. All the books in the world since the dawn of humanity would not suffice to hold such a number. Even if I had the necessary paper, I would never try to write it down. Writing one figure per second, it would take me until the year 10^{68} to finish, by which time all the galaxies and clusters of galaxies would have been transformed into black holes, and all black holes of one solar mass would have evaporated.

In this extremely distant future (see Table 2), black holes will evaporate in a comparatively short time. Thus nearly all matter (including our beloved diamonds) will end as radiation. The universe will then be no more than a vast sea of radiation (photons) and neutrinos (as well as some much rarer particles, such as protons and electrons), which will become evermore

Table 2　The distant future of an open universe

Time (years)	Event
10^{12}	Stars go out
10^{18}	Galaxies become galactic black holes
10^{27}	Clusters of galaxies become hypergalactic black holes
$[10^{31}$–$10^{36}]$	[Protons decay]
10^{100}	Black holes evaporate
10^{1500}	All matter (except neutron stars) turns into spheres of iron
$10^{10^{76}}$	Neutron stars and spheres of iron collapse into black holes; the latter evaporate into radiation

colder. The temperature will continue to approach absolute zero, without ever quite reaching it. Here and there, scattered in the frigid darkness, and carried along by the expansion of the universe, a few microscopic dust grains (those whose mass was less than 20 micrograms, and that did not collapse into black holes) will be found. The universe, as it continues to expand, will become less and less dense: The matter and radiation content in a given volume of space will keep on decreasing. But the "vacuum" will never become total. Virtual particles, which will appear and disappear in accordance with the withdrawals and repayments made through Nature's energy bank, and which exist thanks to quantum uncertainty, will be there to populate it for all eternity.

Is the Proton Mortal?

Diamonds are not forever. They will eventually collapse into black holes and evaporate into radiation. But that will take nearly all eternity. The number describing the length of time required is so vast that it would be impossible to write it all down, even if one spent one's whole life doing so. Diamonds will decay far sooner, and we will be able to write down their lifetime far more easily, if one of the predictions made by the grand unification theories is proved correct. This prediction says that the proton may decay. Until now it had been considered immortal along with the photon, the neutrino, and the electron. All the other particles disintegrate into lighter particles after a short period of time, under the action of the electromagnetic force or the weak nuclear force. As we have seen, a free neutron has an average lifetime of just 15 minutes.

According to grand unification theories, a proton, although no longer immortal, still has a very respectable lifetime. It will not decay for some 10^{32} years or more. So far, no one has seen a proton decay. Although our body contains some 3×10^{28} protons, we would have to wait at least 3000 years (30 times a human lifetime) for one of these protons to disappear. To make up for the brevity of a human lifetime, physicists, who are in a hurry, go searching for the death of a proton by using large masses of material. In 1000 metric tons of matter there are about 5×10^{32} protons, and about five of those should decay every year. These experiments are generally undertaken several kilometers underground to shield them from cosmic rays*, which contain protons and electrons ejected into space by supernovae*. (The interaction of these cosmic rays* with the detector material could mimic the decay of a proton.) An old gold mine in South Dakota, another at Bangalore in India, and two tunnels under Mont Blanc on the border between France and Italy have been used for this purpose. Gold miners have been replaced by physicists intensely watching for the decay of a proton. So far, not a single proton has been caught in the act of disintegrating. Is the proton closer to being immortal than we thought? Stay tuned . . .

In any case, if the proton were mortal, the future course of the universe would take a dramatically different path after the formation of the galactic and hypergalactic black holes. The various debris littering space would suffer quite a different fate. Black dwarfs and neutron stars would see 10^{25} out of their 10^{57} protons decay every year into positrons (antiparticles of the electrons), neutrinos, and photons. The neutrinos, which remain averse to any form of interaction, would escape from the stellar corpses and get lost in the immensity of space. The other particles would be reabsorbed by the stars, which would heat up slightly—to a temperature of 3 K for the least massive black dwarfs, and to 100 K for the neutron stars. The dead stars would appear to be rejuvenated. They would radiate very feebly for some 10^{30} years. When the year 10^{33} arrives, all their protons will have decayed, and they will die. Our beloved diamonds and other objects in the universe suffer the same fate. The universe will then be just a vast, thinned out sea of electrons, positrons, neutrinos, and photons. The galactic and hypergalactic black holes will form a few invisible islands scattered here and there. These will drag out their existence until they completely evaporate into radiation at the end of 10^{100} years, as before.

What will happen to the electrons and their antiparticles, the positrons? If they come into contact, they will also disappear in a flash of light. In an open universe, however, where the density of matter is less than the critical density, the density is so low that there is little likelihood of electrons and positrons encountering one another. Out of 10^{42} electron–positron pairs, perhaps one may vanish in a burst of radiation. Electrons and positrons will survive independently for a very long time.

Even in the case of a denser universe—one that has exactly the critical density, for example—the particles will still be extremely far apart. Their average separation will be one billion times the size of the universe today. But they will still be close enough for the electromagnetic force to act and bind the electron–positron pairs into gigantic atoms of positronium. Quantum mechanics will allow the electrons to spin and twirl about in the vast ballrooms of the positronium atoms, billions of billions of light-years across, for some 10^{120} years, until they encounter the positrons and disappear in a flash of radiation.

As in the case where the proton is immortal, the open universe will become an ocean of radiation, neutrinos, and electrons, with the addition of positrons, but without the protons and microscopic dust grains; only the universe will reach this state much more quickly than if it had to wait for the spheres of iron to collapse into neutron spheres and black holes. At the rate of one zero per second, I need only 2 minutes to write down the complete number 10^{120}.

A Blazing Inferno

Because there may be monstrous, massive objects like the Great Attractor* lurking somewhere out in the darkness of space, and because the universe may contain a considerable amount of mass in the form of vast numbers of particles that were born during the first few fractions of a second of the universe and that have been conjured up by physicists' fertile imaginations, the possibility that the density of matter is greater than the critical density, and thus that the universe is closed, cannot be dismissed. In this case, the expansion will come to a halt, and the universe will collapse back on itself. The freezing darkness will be replaced by a fantastic apotheosis of light and heat, and the fires of hell will replace the desolate, frigid cold of space.

The future development of a closed universe will take place at a rate determined by the density of matter. The latter brakes the expansion through its gravitational influence. The lower the density, and thus the gravitational attraction, the longer it will be before the expansion is brought to a halt. A universe with exactly the critical density would continue to expand indefinitely, but more and more slowly. Its destiny would be the same as that of an open universe. A universe with twice the critical density (i.e., with six hydrogen atoms per cubic meter) would cease expanding after 40 billion years. (Its age would then be 50 billion years.)

Let us follow the changes of this last universe. Initially, the events that occur in a closed universe resemble those in an open one. First, the Large Magellanic Cloud will be destroyed by the cannibal action of our own Galaxy, after 3 billion years. This will be followed by a collision with the Andromeda Galaxy after 3.7 billion years. Then, at 4.5 billion years, the Sun will swell into a red giant, and the torrid heat will drive humankind away from Earth. The death of the Sun will come next, leaving behind a pitiful stellar corpse, a white dwarf, which will become, after another 10 billion years, a black dwarf. The universe will continue to expand, slowing down all the time as it does so. Galaxies will have time to harbor several more generations of stars, and the galactic graveyards will be filled with many more stellar corpses.

When the cosmic clock strikes 50 billion years, the universe will have tripled its size. There will be, on average, 3 million (rather than 1 million) light-years between the galaxies. The temperature of the background radiation will have dropped by a factor of three, to 1 K above absolute zero. The force of gravity will finally overcome the expansion imposed by the initial explosion. From now on, gravity will call the tune. The universe will begin to contract under its own gravitational attraction. The galaxies will cease to recede, reverse their motion, and begin to fall back together. Our descendants will see the light of nearby galaxies shifted toward the blue, through the Doppler effect. The light from more distant galaxies will continue to be shifted toward the red because news of the collapse of the universe in faraway regions (shown by a blue shift in the light from the galaxies) will not reach our great-great- . . . grandchildren until tens of billions of years later.

The universe will keep on contracting for the next 100 billion years, without anything extraordinary happening. The galaxies will continue to get closer and closer together. The background radiation, compressed into a smaller and smaller universe, will gradually get hotter and hotter. When the universe has shrunk to about one-fifth of its present size, clusters of galaxies crammed together like sardines in a tin will begin to merge and lose their identities. Events will begin to accelerate. Another 900 million years and the universe will be one one-hundredth of its present size. It will be the turn of galaxies to merge and combine. The universe will then be a vast expanse of stars, bathed in a background radiation at 300 K (27°C), as hot as on Earth. The collapse of space will have given considerable energy to the stars, which will be rushing through space at some 3000 kilometers per second.

The contraction will continue: 100 million years will pass and the universe will be one one-thousandth of its present size. The background radiation will have risen to 3000 K, close to the temperature of the surface of stars. Night will be no more. The light from the sky will be as blinding as the light from the Sun. The stars will be hurrying through space at velocities so close to that of light (300,000 kilometers per second) that they will be flattened like pancakes. Another 900,000 years and the temperature will reach 10,000 K. The surfaces of the stars will begin to evaporate. Molecules and then atoms in the stellar atmospheres, no longer able to withstand the extreme heat, will begin to break down, liberating nuclei and electrons, which will fill the whole of space. Light will have more and more difficulty in making its way through the jungle of electrons. The universe will become opaque, just as it was during its first 300,000 years. The pace of evolution will quicken. The temperature will keep on rising. Stars will evaporate more and more into nuclei and electrons. Some 100,000 years later, the temperature of the universe will be 10 million K, as hot as the center of the Sun. Such an extreme temperature will initiate nuclear reactions inside stars, causing them to explode and die. In a chaotic turmoil of nuclear explosions, the stars will be vaporized into showers of nuclei, electrons, and photons. Free nuclei, unable to survive in such a furnace, will rapidly disintegrate into protons and neutrons.

The universe will almost revert to its youthful appearance: It will be a sea of radiation and particles as before, but with, this time, an additional

Table 3 The countdown to the Big Crunch in a closed universe

Time before the final collapse (years)	Event
-10^9	Clusters of galaxies merge
-10^8	Galaxies merge
-10^6	Stars, flattened like pancakes, zip across the sky at velocities near that of light
-10^5	Stellar atmospheres evaporate into elementary particles; radiation can no longer propagate; the universe becomes opaque again
-10^3	Stars explode; black holes, corpses of bygone stars, devour the surrounding matter and rapidly grow larger
-1	The universe becomes a soup of quarks, electrons, neutrons, and their antiparticles

scattering of black holes, the remnants of bygone stars. These gluttonous black holes will swiftly engulf anything that comes within their grasp, be it particles or radiation, becoming even larger and more voracious. Their rate of growth will accelerate. Some will merge, and an increasing part of the universe will be swallowed by them. Events will happen even faster. One more thousand years, and the 10,000 billion K mark is passed. The photons will have so much energy that they will be able to transform into particle–antiparticle pairs. Swarms of antielectrons and antiprotons will appear. Protons, neutrons, and their antiparticles will disintegrate, freeing quarks and antiquarks. The universal soup will become a mixture of quarks, electrons, neutrons, and their antiparticles. The three forces (electromagnetic, weak and strong nuclear forces) will become a single force once more, but gravity will remain obstinately aloof. Everything will occur just as if a film of the Big Bang were run in reverse, except that this time the universe will be far less homogeneous, because of the presence of numerous black holes. The universe will continue to contract (down to 10^{-28} cm, 1 million billion times smaller than the nucleus of a hydrogen atom), and heat up, until the fateful temperature of 10^{32} K is reached. We shall not be able to follow the universe any further as it contracts. Its ultimate collapse (the Big Crunch) will evade us, like the very earliest moments of the Big Bang, when it was born. We shall be prevented from going further back by the Planck barrier*, the ultimate boundary to our knowledge. Known physical laws stop there and cannot be extrapolated beyond (Table 3).

A Cyclic Universe?

What will happen behind the Planck barrier? Will the universe die with an infinitely high density and temperature? Or will some still-unknown mechanism prevent its final collapse and allow it to "bounce" and start again in another Big Bang? Might it be eternally reborn, like the phoenix, from its own ashes? Will it undergo an infinite series of cycles of expansion and contraction? All these questions will remain unanswered as long as we are unable to describe how gravity acts in the world of the infinitely small. One thing is, however, certain: Cycles may succeed one another, but they will never be the same. Each subsequent cycle will have a net energy gain, because the heating during the contraction phase exceeds the cooling during the expansion. This increase in energy will allow the universe to expand more and more before it begins to collapse. The cycles will become longer and longer (Fig. 56). Each cycle will also add a new harvest of black holes and its contribution of disorder (or entropy*). Succeeding universes will become more and more heterogeneous and disordered. The fact that our universe was so extraordinarily homogeneous at its beginning, as shown by the background radiation, implies that if the universe is cyclic, we are in one of the very early cycles (which would not be to the liking of the phantom of Copernicus, because that would imply that we are living at a privileged time in the history of the universe), or that the universe, beyond the Planck barrier, knows how to erase all the "roughness" created by black holes, and convert disorder into order.

A final enigma concerns the arrow of time. We have seen that the direction of the cosmological arrow of time, which is the same as those of psychological and thermodynamic times, is linked to the expansion of the universe. Does that mean that the direction of cosmological time will reverse when the universe is contracting? But if the direction of psychological time

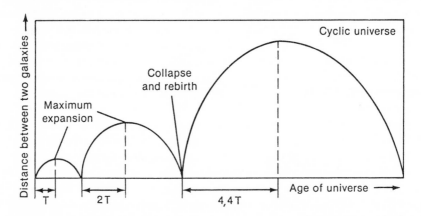

Fig. 56. *A cyclic universe.* Could a closed universe that collapses (*see* Fig. 51) be reborn from its ashes, like the phoenix, and undergo another cycle, perhaps with new laws of physics? No one knows, because current physics break down at extremely high temperatures and densities. In any case, even if the universe does start a new cycle, cycles may follow one another, but they will not be the same. The universe would gain energy from cycle to cycle, which would cause each successive cycle to last longer and the maximum size of the universe to become larger and larger.

did the same, would not our descendants have the impression that they were also in an expanding universe? Once again, there are no definite answers. At first glance, however, nothing very special happens when the universe begins to contract: Light from distant galaxies continues to be red shifted, and only becomes blue shifted after billions of years. Why then, should time change its direction?

Taming Black Holes

The future of the universe is not particularly rosy. Depending on its density of matter, it will end up either being crushed in hellfire or else stretched out in unending, glacial cold. Could life persist in such a distant future? Assuming that our descendants have enough sense to avoid nuclear and ecological suicide, and that the pace of technological and scientific progress does not slacken, what strategy could they adopt to survive and ensure that the universe remains inhabitable? Only the problems of survival facing our possible descendants in an open universe are of interest. The case of a closed universe is quickly taken care of. There is no way in which our distant grandchildren could avoid the inevitable: In a few tens of billions of years, they will all be cremated in a blazing inferno. Infinitely high temperatures will wipe out all forms of life.

In the case of an open universe, survival depends primarily on access to a source of energy. The death of the Sun will lead inevitably to an interstellar exodus, in search of another star. The latter will, in its turn, become extinct after some 10 billion years, and yet another star will have to be found. An so on, until the year 1000 billion (10^{12}), when all the stars will go out and the galaxies will cease to glow. That will be the end of the era of stellar thermonuclear energy. Immense swarms of stellar corpses will remain: white dwarfs, neutron stars, and black holes. Paradoxically, only black holes will prove useful as energy sources. Our descendants could first try to extract the rotational energy from these invisible objects. Black holes, like all stars, rotate. The Sun rotates once every 26 days, but collapsed stars rotate even faster, just as ice skaters spin faster as they bring their arms in toward their bodies. The fastest neutron stars may, for example, rotate as fast as about 1000 times per second. Rotation is one of the three parameters that describe a black hole. (The others are its mass and electrical charge. The latter is generally zero, because the progenitor stars are usually electrically neutral.) According to the British physicist Roger Penrose, it would be possible to extract energy from black holes by "feeding" these spinning monsters with radioactive waste. If, as a radioactive nucleus falls toward the gaping mouth of a black hole, it splits into two particles before it crosses the radius of no return*, one of the new particles will be devoured by the black hole, but the other, under certain circumstances, can escape with more energy than the nucleus originally had. It does this by extracting some of the rotational energy of the black hole, which, as a result, rotates a little more slowly. All our descendants will need to do then will be to capture these escaping par-

ticles. They will kill two birds with one stone: Get rid of their harmful radioactive wastes, and provide for their energy requirements (Fig. 57).

They will have to be constantly on the qui vive. The period between the
years 10^{12} and 10^{27} is one of great upheaval: The stellar corpses will be constantly swapping energy, with, as end result, the ejection of 90% of the black
holes from their parent galaxies. Our descendants will have to abandon the
black hole they are exploiting for energy as soon as there is a danger of its
being ejected, otherwise they will also be flung out into intergalactic space.
They will have to search for another black hole in the Galaxy and start extracting energy from it instead. After a long migration from black hole to
black hole, they will find themselves at the center of the Galaxy, exploiting
the large galactic black hole with a mass of one billion solar masses, which
will have resulted from the fusion of all the black holes that were losers in
the energy-exchange game. Other civilizations in the Galaxy (if they exist)
will also follow similar migrations and arrive at the same galactic black hole.

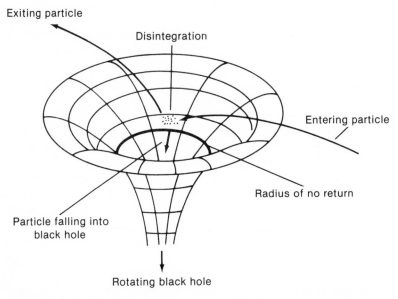

Fig. 57. *Exploiting black holes.* To extract energy from a black hole (represented
here by curved space), all that is required is to dump radioactive material into it. A
radioactive nucleus may split into two particles just before it crosses the radius of no
return. One of the two particles is devoured by the black hole, and the other escapes
with more energy than the original nucleus. This increased energy (which could be
exploited for our energy needs) occurs at the expense of the black hole's rotational
energy: It turns slightly slower. The same mechanism is at work when a black hole
evaporates. Here, the two particles do not come from the disintegration of a radioactive nucleus, but materialize by borrowing energy from the intense gravitational
field surrounding the black hole. As previously, one of the particles falls into the
black hole, while the other escapes, carrying with it some of the energy of the black
hole. The latter is thus slowly converted into radiation.

They will all gather around the black hole, whose radius of no return is about three light-hours (slightly less than the size of the Solar System). They will engage in long interstellar dialogues, and will exploit the black hole together. When the rotational energy of the galactic black holes is exhausted, then it will be the turn of the hypergalactic black holes, with masses of 1000 billion (10^{12}) solar masses. These will also gradually lose their rotational energy and slow down, until eventually they stop rotating.

Our descendants will then turn to the energy radiated by black holes as they evaporate. They will construct enormous reservoirs around black holes, beyond their radius of no return, in which to capture and store the precious energy. That energy will be far less than that provided by the rotation of black holes. Our descendants will become very frugal. They will be forced to hibernate for long periods to avoid consuming too much energy. The evaporation will gradually decrease the mass of the black holes. Their temperature will rise, and they will lose matter at an ever-increasing rate until they eventually explode. To slow down this mass loss, our descendants will hunt for space debris: planets, comets, asteroids, etc., that they will use to feed the black holes and slow down their evaporation. (The lifetime of a black hole is proportional to the cube of its mass, as explained in Appendix C.) Just as ancient Man domesticated and fed wild beasts to have food, our future descendants will tame and feed black holes to survive. In a last attempt to delay the eventual evaporation, our descendants may even cause black holes to merge. But any respite will be (relatively) short. After 10^{100} years, all the galactic and hypergalactic black holes will have turned into radiation.

Other Forms of Life

What will happen to life in the long term? Could it persist in a universe that has only a finite amount of energy? Could it survive when faced with an unsolvable energy crisis? Life as we know it, made of flesh and blood, of organic molecules and strands of DNA, could not survive in such a frigid universe. The metabolism of such forms of life requires a continuous source of energy, which would no longer be available. The Anglo-American physicist Freeman Dyson[14] believes that life could go on if it knows how to adapt its form and metabolism to an environment that is becoming ever more glacial. He adopts the optimistic view that the survival of consciousness and intelligence does not depend on the specific nature of the material that supports them, but on the level of complexity with which that material is organized. According to him, the intertwined double helix of organic molecules forming DNA is not required to make a brain. A cloud of microscopic grains of dust—for example, the ones that weigh less than 20 micrograms and cannot collapse into black holes—or, if the proton decays, a cloud of electrons and positrons, with a sophisticated organization, will do just as well. By

14. F. J. Dyson, *Reviews of Modern Physics* **51**, 1979, p. 447.

adapting their metabolism to a cold that is becoming ever more intense, and by hibernating during longer and longer periods to conserve energy, such life forms could possibly last for all eternity. Here, we are, of course, in the realms of pure science fiction. But we should not underestimate Nature. It took her just 15 billion years to create, from a void, a universe filled with stars and galaxies, as well as an intelligence sufficiently evolved to reflect on the universe that engendered it. Who knows what she is capable of, if she has all eternity before her?

| 7 |

An Accidental or
Necessary Universe?

The Phantom of Copernicus Is Called into Question

In the sixteenth century, Copernicus dislodged Mankind from its central place in the Solar System. Since then, his phantom has continued to haunt us and cause further havoc. First the Earth lost its central place, then the Sun was reduced to the rank of an ordinary star and relegated to a distant region in the Milky Way. The Milky Way itself was soon lost among the hundreds of billions of galaxies in the observable universe. Mankind was reduced to insignificance compared with the vastness of space. The emergence of intelligence and consciousness was simply the result of chance, an accidental event in the long history of the universe. Our existence did not matter in the least to the universe. It did not care a hoot whether we were here or not. This reduction of human consciousness to worthlessness plunged some thinkers into extreme despair. As early as the seventeenth century, Blaise Pascal uttered an anguished cry against "The eternal silence of infinite space," a cry that was echoed 3 centuries later by the similar cries of distress from the French biologist Jacques Monod: "Man knows at last that he is alone in the universe's unfeeling immensity, out of which he emerged only by chance,"[15] and by the American physicist Steven Weinberg: "The more the universe seems comprehensible, the more it also seems pointless."[16]

15. Jacques Monod, *Chance and Necessity,* New York, Knopf, 1971.
16. Steven Weinberg, *The First Three Minutes,* New York, Basic Books, 1977.

223

Faced with this bleak prospect, a resistance movement has arisen. The phantom of Copernicus has been called into question. In the past 20 years, a few physicists, who have become disenchanted with its devastating power, have tried to throw off its crippling yoke, and give humanity back its privileged position in the cosmos. According to them, Mankind did not emerge by chance in an indifferent universe. On the contrary, both are linked in a close symbiosis: The universe is the way it is because Mankind is there to observe and wonder about it. The existence of human beings is inherent in the properties of every atom, star, and galaxy in the universe, and in every physical law that governs the cosmos. Alter even slightly the properties and laws of the universe, we would no longer be here to discuss it. The nature of the universe and our existence are therefore inextricably linked. "The universe possesses very precisely just the properties needed to produce living creatures with consciousness and intelligence." The Australian astronomer Brandon Carter, one of the leaders of the concerted attack against the phantom of Copernicus and the author of the last statement, has coined the term "anthropic principle*" (from the Greek word *anthropos,* meaning "Man") to describe it. It remains to be seen whether this attack is justified. We need to examine the main lines of this argument.

Nature's Numbers

If you throw a ball in the air, it follows a graceful curve in space before falling back to the ground. This curve, far from being arbitrary, has a very precise mathematical shape. A physicist would tell you that it may be only an ellipse, a parabola, or a hyperbola. He will tell you how long the ball will stay in the air, and the exact point where it will land. For this, he will use two types of information: the laws of physics, and what he will call "the initial conditions." The physical law that controls the path of the ball is the law of gravitation: The gravitational force attracting the ball toward the Earth varies in proportion to the product of the masses of the ball and Earth, and inversely in proportion to the square of their distance. The qualitative statement "in proportion to" may be expressed quantitatively (i.e., in a way that allows one to make accurate calculations) if we introduce a number called the "gravitational constant," traditionally denoted as G, into the law. Newton's law may therefore also be expressed as follows: The gravitational force equals the product of G and the two masses, divided by the square of the distance between their centers. The value of G dictates the force of gravity. If it is large, gravity is strong. If it is small, gravity is weak. The value of G, which has been measured countless times in the laboratory, is very small: Gravity is the weakest of the four fundamental forces. But no physical theory—and here's the rub—can explain why G has that particular value and no other. We are "given" the value, and we have to live with it.

Newton's law is not sufficient to determine the graceful curve the ball follows through space. We also need to know the initial conditions, the exact point in space at which it leaves your hand, its initial velocity, and its initial direction. If it is hurled up in the air, it will have a high initial velocity, and will go a long way. Thrown less energetically, its initial velocity will be less, and it will not travel so far.

What applies to the ball also applies to the universe. Its evolution and its fate depend on the same two factors: the laws of physics, which are governed by a few numbers (these numerical parameters are known as the "fundamental constants of nature"), and the initial conditions—given to it by the fairies at its birth, at the time of the Big Bang.

How many "numbers" are required to describe the universe? Our current knowledge suggests that we need slightly more than a dozen. Just as G dictates the force of gravity, there are two other parameters that control the strength of the weak and strong nuclear forces. Then there is c, the velocity of light, and the highest velocity anything in the universe may attain. Then comes h, known as Planck's constant (named after the same physicist who discovered the Planck barrier), which dictates the size of atoms. Atoms are small because h is minute. Then there are the numbers that describe the masses of elementary particles. There is, of course, the mass of the electron, but also the mass of the proton. (Why not the mass of a quark, since the proton is supposed to be made of three quarks? The reason is that free quarks have never been observed, and thus remain theoretical entities at present.) Then comes e, the charge of the electron (the charge of the proton is equal and opposite). There exist a few more, adding up to about 15 in all. These numbers are, as their other name implies, constant. They do not appear to vary in space or time. Our remote descendants, or extraterrestrials living on the other side of the universe, would determine exactly the same values for these numbers. We have been able to confirm this by reaching out in space and back in time by observing distant galaxies. The properties of these galaxies do not appear very different from those of the Milky Way. This can be only understood if the constants have not varied in any appreciable way. (This constancy of the physical constants will be very useful the day we come into contact with extraterrestrial civilizations. We shall communicate with light, using the same physical parameters (the same mass of the proton, the same mass of the electron, the same velocity of light, etc.). If it were otherwise, the situation would be chaos, and any dialogue would be impossible.)

Have we cataloged all the "numbers" in nature? No one knows. If tomorrow other forces or particles are discovered, the list will lengthen. But it might also shorten. Physicists are actively working to reduce it. They want to unify the four fundamental forces into a single force. If they succeed, they will be able to describe the four forces by a single parameter rather than several. And then, might it not be possible to discover a more comprehensive theory that could predict the velocity of light, the constant governing the atomic world (h), and the mass of the electron? Or some grand unifying principle that would describe the whole universe, and would not depend on any "given" number? (In which case the number of parameters would be reduced to zero.) Today's physics is still a long long way from this all-embracing principle with no parameters. We have not yet succeeded in unifying gravity with the other three forces. And we are still less certain that it even makes sense to talk about an ultimate unifying principle. In the meantime, let us be content to accept these numbers as "given" values in physics. Let us accept that light travels at 300,000 kilometers per second rather than at 1 kilometer per second, and that the mass of the proton is 1826 times that of the electron and not the other way around, and let us see how these physical "constants" are responsible for the incredible range and richness of the structures found in the universe, and the existence of life within it.

The Things in Life

The numbers found in nature govern our everyday life. They determine the size and mass of everyday things. They cause the world to be the way it is, and not something completely different. What may appear as a statement of the obvious reflects in fact the infinite range of choices of mass and size at Nature's disposal when she was building the structures within the universe. The planets, for example, instead of being spheres some thousands of kilometers across, could have been the size of grains of dust. Human beings could have been the size of microbes. Why is the day, the length of the Earth's rotation period, only 24 hours? Why are all the things that make life worth living the way they are? Why is the highest mountain on Earth less than 10 km high? Why are flowers less than one meter high and trees less than a few tens of meters? Why is a raindrop only a few centimeters in diameter? The answer lies in those numbers, in the fundamental constants of nature.

Objects consist of atoms, and it is the latter that give them their solidity. An atom is the result of an equilibrium between two forces: the electromagnetic force, which attracts the electrons toward the protons in the nucleus, and depends on the mass of the proton m_p, that of the electron m_e, the charge e of the electron, and the force that opposes it, which results from the exclusion principle*. The latter prevents electrons from being too close to one another and involves Planck's constant (h). The gravitational constant G is not in the picture because gravity plays no role in the atomic world: The gravitational force between a proton and an electron is 10^{40} times weaker that the electromagnetic force between the two. The "numbers" m_p, m_e, e, and h control the two forces, with the result that the size of an atom is extremely small, approximately one hundred millionth of a centimeter. Nature could have built atoms the size of the Eiffel Tower by choosing other constants, but she did not do so.

Atoms are then organized into crystalline lattices to form the solid objects with varied shapes and colors that contributes to the pleasures of life: vases of flowers, Jefferson's Monticello, paintings by Degas and Monet, etc. Because of the small value of G, gravity does not play a role unless the mass of the object exceeds 100 billion billion metric tons (10^{26} grams), which excludes most of the things in life. Beyond that mass, the gravitational force is strong enough to have its say. It only likes spherical shapes, and it turns every mass greater than 10^{26} grams into the spherical objects we call planets. We had escaped narrowly! If G were far greater, gravity would have intervened at much smaller masses, and we would be living in a very dreary and gloomy world, where only spherical objects would be allowed! (We have seen how, in the very distant future, quantum mechanics will turn everything into a sphere. But we shall no longer be made of flesh and blood, and if we are in the form of clouds of dust or electrons, surrounded by frigid darkness, perhaps we would possess no more our aesthetic sense of beauty.)

For a planet to be habitable, it must have an atmosphere that will protect it against the injurious effects of the ultraviolet radiation from the Sun, and that will allow complexity to evolve, organic molecules to be formed, and life to arise. The planet must be massive enough to retain the atmosphere, but not so massive that the resulting atmosphere is so thick that it blocks off the light from the Sun, the source of its energy. The numbers tell us that such a planet will have a radius of approximately 6,400 kilometers and a mass of roughly 6×10^{27} grams, which is what we find for Earth. The "centrifugal" force that pushes a passenger in a car against the side when the driver takes a turn too rapidly also tries to dislocate the crystalline lattices within the Earth, break the atomic and molecular bonds, and split it apart. So the Earth should not rotate too rapidly. Once again, the physical constants tell us that a day 24 hours long will do quite well. What is the largest mountain that could exist on Earth? The one that the crystalline lattices within the Earth's crust can sustain without breaking. A simple calculation shows that the physical constants are compatible with a highest mountain not exceeding one one-hundredth of the planet's radius, that is, 64 kilometers. The highest mountains on Earth, the Himalayas, do not exceed seven kilometers.

The physical constants also tell us why human beings cannot grow taller than approximately 2 meters: A larger body would break after an accidental fall. (The same applies equally to horses, elephants, or giraffes. The physical constants do not distinguish them from human beings.) The universal numbers reveal to us that human beings occupy a very specific place between the atom and the planet. Their masses and sizes are the geometric mean of the masses and sizes of a planet and an atom. (The geometric mean is the square root of the product of two numbers; for example, 9 is the geometric mean of the numbers 3 and 27, while 15 is the more familiar arithmetic mean.)

Some 15 physical constants therefore determine the world that surrounds us. They control the magnificent hierarchy of structures and masses found in the universe, from the smallest atom to the largest supercluster of galaxies, including Mankind, planets, stars and galaxies (Fig. 58). They limit the height of mountains and ensure that we are not the size of microbes. Most extraordinary, however, they also, together with the universe's initial con-

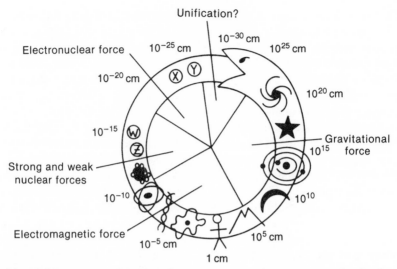

Fig. 58. *The fine tuning of the four forces in nature.* The serpent swallowing its own tail illustrates each force's realm of influence. The gravitational force controls the cosmos (galaxies, stars, planets, and the Moon). The electromagnetic force governs objects on scales ranging from human beings to atoms, including amoebas and the double helix of DNA. The strong and weak nuclear forces rule the world of atomic nuclei and elementary particles. At still smaller scales (like that prevailing when the universe began), the nuclear and electromagnetic forces combine to form the electronuclear force. Might gravity combine with the electronuclear force to form a single "superforce" at still smaller scales? The relative strengths of these four forces are finely tuned so as to allow the emergence of life and consciousness. Alter these forces by even the slightest amount, and we would no longer be here to talk about them. (After a diagram by S. Glashow.)

ditions, allowed life to develop and consciousness and intelligence to appear. The existence of life depends on a precarious balance and an extraordinary set of circumstances. Change the numerical values or the initial conditions by even the slightest amount, and the universe would be completely different and we would no longer exist.

Model Universes Are Sterile

By succumbing to their impulse to play God, physicists have realized how extremely improbable the emergence of life and intelligence is in a universe with arbitrary numerical parameters and initial conditions. To compensate for the fact that we have only a single universe (our own) to observe, they create a multitude of fictitious universes and follow their hypothetical evolution, using complex equations and powerful computers. Just as in their study of the formation of the wonderful cosmic tapestry woven by the galaxies, astrophysicists use models to scrutinize the conditions necessary for the emergence of life and consciousness. These models will all have different numerical parameters and initial conditions. In one, the weak nuclear force will be less intense, and in another it will be the electromagnetic force that is greater. The next may contain so much matter that it collapses after just a second, while yet another may be populated with so many black holes that it will be utterly inhomogeneous. The sixty-four thousand dollar question is: Will these model universes be able to work their way up the ladder of complexity and finally give rise to consciousness? Will they be able to form the galaxies, stars, planets, and primitive oceans that will give birth to life?

At first sight, the answer to these questions is surprising. Life as we know it would not have the slightest chance of emerging in any universe that was the slightest bit different from our own. All the model universes would be sterile and devoid of consciousness. The numerical parameters do not suffer any modification. If we slightly increase the value of the parameter that controls the intensity of the strong nuclear force (by just a few percent, for example), protons (the hydrogen nuclei) could no longer exist in a free state. They would combine with other protons and neutrons to form heavy nuclei. Without hydrogen, we can say goodbye to water, molecules of DNA, and life. Stars could form, but they would speedily burn themselves out in the absence of hydrogen fuel. Let us decrease slightly the strength of the strong nuclear force. We now see the opposite effect: No nucleus other than that of hydrogen can survive. Hydrogen atoms can no longer fuse to give helium. Nuclear reactions cannot occur, and stars, which are the vital source of energy, are no longer able to shine.

The strong nuclear force does not allow any change. Let us leave it aside and increase the "number" that controls the electromagnetic force. Negatively charged electrons will be more strongly bound to nuclei of opposite charge. The chemical reactions that rearrange electrons between the nuclei of various elements will require far more energy and will be much rarer, thus making the formation of the intertwined DNA strands extremely improbable. If we decrease the electromagnetic force slightly, neither chemical bonding nor complex organic molecules will be possible.

In desperation, let us reduce G. Gravity will be so weak that interstellar gas clouds will no longer be able to collapse under their own gravitational attraction and reach sufficiently high densities and temperatures to initiate nuclear reactions. Again, the stars will not light up. The various chemical elements and the energy required by life will never come to be. Increasing G does not help either. The cores of stars, highly compressed under the weight of overlying layers, will be so hot and dense that nuclear reactions will take place extremely rapidly. Their hydrogen fuel is rapidly exhausted, too fast for cosmic evolution to proceed through the steps necessary for the emergence of life.

Similar arguments apply to the other "numbers." The same conclusion is always reached: The numerical parameters are quite inflexible. Changing their values ever so slightly eliminates any possibility of life. What would happen if the universe's initial conditions were altered instead? Would we end up with models that are more hospitable to life? Once again, we shall come to the conclusion that our universe's initial conditions are very specific and that any modification leads to the suppression of life.

A Very Precisely Adjusted Universe

One of the most important initial conditions governing the universe is the amount of matter per unit volume (the density of matter). As we have seen, this quantity determines the fate of the universe. A universe that is very dense initially will have a very short life. After expanding for a year, a month, or even a second, it will collapse under the influence of its own gravity and will be crushed out of existence in a blazing inferno. Life will not have sufficient time to climb the rungs of the ladder leading to complexity. Living things are made of the nuclei of heavy elements (such as carbon) that are created within the cores of stars. For these elements to be available, at least one generation of stars should have lived out their lives and come to a violent explosive end, seeding the interstellar medium with the products of their combustion. Then it is necessary to wait for planets to be "invented" and be patient during the long progression from the amino acids to the human brain. In total, we must reckon on several billion years. "Easy," you may say, "Reduce the initial density slightly to produce a longer-lasting universe and give life time to develop." But then we must be careful not to fall into the opposite trap. A universe with a very low density would certainly last for a long time. But the matter within it would be spread so thin that stars and galaxies could no longer condense, and it would also be condemned to be lifeless.

In fact, only one universe with one extremely precise density can simultaneously last a long time and harbor galaxies and stars. That universe is ours. We have seen that it has a density that is extremely close to the critical density (three hydrogen atoms per cubic meter), that is, the density of a universe that will only stop expanding after the passage of an infinitely long time. The latest observations suggest that the density of our universe is only about one-fifth of the critical density. It may be equal to the critical density if the universe contains innumerable elementary particles born in the first moments of its existence, or if numerous massive monsters, like the "Great Attractor," are roaming the depths of space. In any case, it is extraordinary that the current density of the universe should be so close to the critical density, when it could have been thousands, or even billions, of times smaller or larger. Calculations show that the difference between the actual density and the critical density increases with time. This difference has been multiplied by a factor of one trillion times since the era of primordial nucleosynthesis*. The fact that these two densities are still almost equal after 15 billion years of evolution indicates that, at the time when hydrogen and helium were being made, about 3 minutes after the original explosion, the density of the universe did not differ by more than one trillionth from the critical density. The density of the universe had been adjusted with exquisite precision to allow the emergence of galaxies, those havens of life, in the cosmic desert, and to give cosmic evolution enough time to rise to intelligence. The universal landscape corresponding to the critical density is flat and devoid of any curvature. If there had been enough matter to shape this landscape into hills and valleys, the universe would have collapsed after just a few seconds.

There are other adjustments that are just as precise. The universe is extremely homogeneous and isotropic. Its properties are almost the same everywhere and in every direction. The temperature of the background radiation does not differ by more than 0.001% between one point on the sky and another. This is lucky for us. A universe that was initially too chaotic and disorganized would not contain a single galaxy. The random motions would be dissipated as heat, warming the universe, and preventing the formation of galaxies. A homogeneous universe is required, but not too homogeneous. It should contain seeds of inhomogeneity here and there, which will later grow into galaxies harboring life. Once again, everything needs to be in balance: The universe should be neither too homogeneous to allow the existence of galactic seeds, nor too inhomogeneous to allow them to grow. Once more, the initial fine tuning is astoundingly precise: The difference between the rates at which the early universe initially expanded in different directions (at the Planck time of 10^{-43} second) could not be more than 10^{-40}. (This is an extremely small number: the 1 is preceded by 40 zeros.) The accuracy of this adjustment could be compared with that achieved by an archer whose arrow hits a target one centimeter square in area, located on the other side of the universe, 15 billion light-years away.

Of course, when we speak of life and intelligence, we are making the rather arguable assumption of life and intelligence like our own, based on the biochemical activity of giant molecules containing hydrogen, carbon, oxygen, etc. The biochemical way may not be the only path to consciousness. As we have seen, the Anglo-American physicist Freeman Dyson has suggested the idea of a brain made of a cloud of microscopic dust grains or collections of electrons and positrons. But these suggestions still belong to the realm of science fiction. For the time being, the biochemical path to life is the only one we know of. Perhaps we are showing excessive self-centeredness and exaggerated anthropocentrism, of which the phantom of Copernicus would have certainly disapproved. But what else can we do in the absence of additional information?

In any case, there is no doubt that the fundamental natural constants and the initial conditions had been fine tuned with exquisite precision for the universe to go through all the stages leading from elementary particles to biochemical life, with planets, stars, and galaxies on the way. Any slight modification, and the universe would have been lifeless and devoid of observers. What are we to think of this astonishing combination of circumstances? Some see it as simply the outcome of chance. In that case, the universe is merely an accident. The fact that the physical constants and the initial conditions were such as to allow life is just a happy coincidence, without any great significance. Mankind is lost "in the universe's unfeeling immensity out of which he emerged only by chance." This view, which would meet with approval from the phantom of Copernicus, provokes despair.

For others, this combination of circumstances is not accidental, but significant. If the universe exists as such, it is for the emergence of consciousness and intelligence. At its very beginning, the universe contained the seeds of the conditions required for the appearance of an observer. It tended towards self-consciousness through the creation of intelligence. "The universe in some sense must have known that we were coming."[17] The advocates of this point of view, headed by the Australian astrophysicist Brandon Carter, have even elevated this to the status of a principle: the anthropic principle. (The adjective "anthropic," which implies that only Mankind has the privilege of consciousness and intelligence, is, once again, the result of excessive anthropocentrism and ill-placed human vanity. Amoebas, chimpanzees, and whales would be quite right to protest. The Franco-Canadian astrophysicist Hubert Reeves has suggested that "anthropic principle" be replaced by the more general "complexity principle," which he states as follows: "Since the furthest reaches of time that we have been able to explore, the universe has possessed the requisite properties for matter to climb up the rungs of the complexity ladder."[18] This statement puts Mankind and chimpanzees on an equal footing.)

The anthropic principle totally defeats the phantom of Copernicus. In the light of modern cosmology, Mankind regains its preeminent place—not the central place in the Solar System and the universe it occupied before Copernicus, but in the destiny of the universe. Mankind no longer needs to fear the universe's immensity, which is there to accommodate him. The universe is vast, because it has to be sufficiently old for Mankind to appear on the scene. The age in question must be greater than several billion years, which is why the universe is some fifteen billion light-years across, rather than the size of the Solar System as it was thought in Copernicus' time.

Parallel Universes*

The anthropic principle has endowed the universe with the glow of finality: It tends towards consciousness. How are we to react to this? The scientist feels a highly justifiable suspicion about any finalist argument. Modern science has arisen from the systematic and categorical rejection of the explanation of natural phenomena in terms of "final causes," or of a "grand design," concepts that belong more to religious doctrines. Saying, like the writer Bernadin de Saint-Pierre, that "pumpkins are large because they are made for a whole family to eat" is to be like an ostrich and bury one's head in the sand to avoid seeing the existence of more profound causes. Similarly, we could have said, "The universe is so flat and homogenous because its final goal was Mankind, and such conditions are necessary to achieve that

17. F.J. Dyson, *Disturbing the Universe,* New York, Harper and Row, 1979.
18. H. Reeves, *The Hour of Delight,* New York, Freeman, 1992.

goal," and let it go at that, without seeking any further. We would never have discovered the inflationary era, which occurred in the first few fractions of a second of the universe's existence, and which naturally explains the properties of homogeneity* and flatness*, without invoking any final cause. Before its exponential growth (the "inflation"), the universe was so small that all its infinitesimal parts were in contact with one another, which allowed them to homogenize their properties. After the inflation, the regions were no longer in contact, but "remembered" that they once were. As for the geometry of space, it became flat during the inflation, just as the surface of a balloon flattens when it is inflated. By invoking too much the idea of a "grand plan," we run the risk of missing major discoveries.

What should a scientist do who refuses to attribute to chance the extraordinarily precise adjustment of the physical constants and initial conditions for the emergence of consciousness, if he equally denies any finality to the universe? He summons quantum mechanics to the rescue, and introduces the notion of parallel universes.

As we have seen, quantum mechanics has introduced fuzziness in the world of atoms and particles. An electron does not calmly follow a well-defined orbit around the atomic nucleus, like the Earth around the Sun, but instead twists and turns everywhere at once in the great ballroom of the atom. When I am not observing it, the electron discards its mask as a particle, and appears as a wave. It is impossible to state precisely where it is located. I can only give the probability of finding it here or there. The wave representing an electron, like waves in the sea, has high-amplitude regions (the crests of the waves) and low-amplitude regions (the troughs). The probability of encountering the electron is proportional to the square of the amplitude; so it is greatest at the crests of the wave, and least in the troughs. If I do not use any measuring apparatus to observe the electron, I can only describe its existence in terms of probabilities. Determinism is banished. Suppose I switch on the apparatus and try to make an observation of the electron at some specific location within the atom, let us say at the crest of a wave. The electron turns into a particle, and appears—or does not—at the location in question. If I find it there, I may wonder if it was there before the observation showed it to be so. If the answer is "Yes," I face a serious conceptual problem: If the electron was there and nowhere else, the probability of its being anywhere else within the vast ballroom of the atom should be zero. But, according to quantum mechanics, when it takes on the appearance of a wave, the electron has a nonzero probability of being at all the other locations. Can quantum mechanics be flawed?

Physicists like Albert Einstein or Erwin Schrödinger, who did not like the idea of God playing dice with nature, made every attempt to find examples of situations where quantum mechanics could be at fault. (The irony is that it was Schrödinger who showed us how the wave form of the electron and other particles could be calculated, and who taught us that the probability of finding a particle is proportional to the square of the amplitude of the wave.) Schrödinger, in particular, devised a situation in which the paradoxes of quantum mechanics found in the subatomic world would find their way into the everyday, macroscopic world. Imagine, he said, a cat shut in a box, which also contains a bottle of cyanide. Hanging over the bottle of poison, like the Sword of Damocles, is a hammer controlled by a radioactive substance, that is, by a substance whose nuclei spontaneously disintegrate after some time. As soon as the first nucleus disintegrates, the hammer falls, breaking the bottle and releasing the contents, which poison the cat. So far, there is nothing particularly unusual about this. But problems arise as soon as we try to predict the cat's fate. Its life depends on the first disintegration. However, that can only be described in terms of probabilities: There is a 50% chance that a nucleus disintegrates (or does not disintegrate) after an hour. Provided we do not open the box—in other words, as long as we do not intervene as observers—all we can say is that at the end of an hour the animal is the combination of 50% dead cat and 50% living cat (Fig. 59). The radioactive material has brought the indeterminancy of the subatomic world

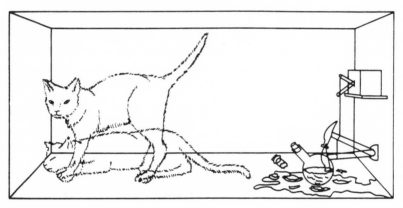

Fig. 59. *A cat hanging between life and death, or the paradoxes of quantum mechanics.* Quantum mechanics describes reality in terms of probabilities. Some noted physicists (including Einstein) could not accept this nondeterministic interpretation of the real world. The Austrian physicist Erwin Schrödinger conceived the following thought experiment to prove it wrong. A cat is shut in a box with a bottle of cyanide, which is underneath a hammer, whose position is controlled by a radioactive substance. As soon as the first atom disintegrates, the hammer is released, breaking the bottle, releasing the poison and killing the cat. According to the laws of quantum physics, we cannot say exactly when the first disintegration will occur. We can speak only in terms of probabilities: There may be, for example, a 50% chance of a disintegration occurring (or not occurring) at the end of the first hour. As far as the cat is concerned, all that we are able to say (without opening the box) is that after 1 hour the animal is a combination of 50% dead cat and 50% live cat. Schrödinger found this description of reality unacceptable. The concept of parallel (or multiple) universes was introduced to avoid this sort of paradox. According to this theory, there is one universe in which the cat is dead, and another, parallel to the former but completely disconnected from it, where the cat is alive.

into the macroscopic world. Schrödinger could not bring himself to accept this description of reality. To him, the cat is either alive or dead. To speak of a cat hanging between life and death, which only decides to live or die when the observer opens the box, appears totally absurd. The conceptual difficulties do not stop there: Even if we accept the observer's crucial role and his power to dispel the quantum uncertainty surrounding electrons or cats, what would happen if the observed object were the whole universe? By definition, the universe includes everything. There cannot be observers outside the universe. (Unless there is a God; we shall consider this possibility further in the next chapter.) Does this mean that the universe, because of the lack of an external observer, cannot be but fuzzy and blurred, and that it does not consist of a single reality, but is a combination of a multitude of realities, all as valid as one another? Such a concept cannot be correct: The things that surround us do undoubtedly possess a unique and concrete reality. Atomic fuzziness does not exist in the everyday world.

To resolve these conceptual difficulties and avoid cats hanging between life and death, in 1957, the American physicist Hugh Everett advanced the idea of parallel universes*. According to him, the universe splits into two almost identical copies every time there is an alternative action, choice, or decision. There would be one universe where the cat is alive and another where it is dead. Both universes would be as real as one another. They would both contain identical observers, who would have themselves split in two. These universes would be completely disconnected from one another: The observers in one would never be able to examine anything in the other. As for the electron in its guise as a wave, and spinning around inside the vast ballroom of the atom, the notion of parallel universes puts the observed position on an equal footing with all the other positions where the probability (given by the square of the amplitude of the wave) is nonzero. The electron occupies all the positions, observed or not, but each position is in a different, parallel universe. From this multitude of universes, the observer selects a specific universe, where the electron has a specific position. Because each universe is as real as the next, the presence of an external observer, needed to dispel the quantum uncertainty, is no longer required.

You will, quite rightly, protest that this idea of parallel universes is extremely odd. Your common sense rebels against the idea that the whims of a single electron can cause the universe to split into innumerable, nearly identical copies, in which your mind and personality are infinitely reproduced. But can common sense really serve as a guide in the strange world of quantum mechanics? At any rate, the notion of multiple universes is not contradicted by any of our laboratory experiments. Until proved wrong, we cannot reject it. In principle, therefore, the universe may indulge in an orgy of divisions, a frenzy of splitting, every time there is an alternative. A single atom in some star in a distant galaxy, itself lost in the immensity of space, changes its state, and the whole world around us divides into two similar copies. Some universes would not be very different from our own. They might differ, for example, only in the position of a single electron in one individual atom. It would be extremely hard to tell the difference. Other universes might differ more—in their population of live cats, for example. Yet other universes would exist where Judas did not betray Christ, the American revolution not happen, the Civil War was won by the South, Hitler never existed, or the Berlin Wall never fell, etc. Yet others would differ in more fundamental ways: They would have different physical constants and initial conditions, with different physical laws. Anything that could occur, does occur in this vast array of universes. But none of these parallel universes, whether they differ by just a single electron or have completely different physical laws, is accessible to us in any way. They are as real as our own, but are forever prohibited to us.

If we accept this idea of parallel universes, there is no more point in talking about our universe in terms of a grand design, nor need we attribute to it Mankind as its aim. The fabulous precision of the adjustments of the physical constants and the initial conditions is no longer surprising. Those physical constants and initial conditions were selected from an infinite number of choices by the very fact that we exist. The overwhelming majority of parallel universes would not begin with the conditions necessary for the emergence of life, and would therefore be devoid of consciousness. There would not be anyone in those inhospitable universes to wonder about things. It is because our universe has the appropriate "fertile" parameters that we are here to talk about it. In this new light, Mankind is not the ultimate goal of the universe (the "strong" form of the anthropic principle) but is a simple observer. The properties of the universe should be compatible with his existence (the "weak" form of the anthropic principle).

This takes us well into the realms of science fiction. The existence of these parallel universes can never be verified because they are not accessible to our instruments. Yet the idea of multiple universes continues to rear its head in other ways in physics. According to the American physicist John Wheeler, the universe goes through an endless series of cycles of expansions and contractions. After each contraction, when the temperature and density become infinite, the universe is reborn, like the phoenix, from its ashes, to undergo another cycle of expansion (see Fig. 56). With each new cycle, however, the universe begins with a different set of physical constants and initial conditions, and even different physical laws. Most of these cycles will not have the conditions necessary for the emergence of intelligence. By chance, our cycle happens to have the required conditions, which is why I am here writing these words and you are reading them. Wheeler has substituted an infinite succession of universes for Everett's frenzied duplication of universes, but the idea remains the same: an infinite number of universes, where the physical constants, the initial conditions, and even the physical laws may vary at random. Once again, these universes are completely unconnected with one another. The scientific foundation for Wheeler's cyclic universes is even weaker than for Everett's parallel universes. We do not even know if the universe contains enough matter for it to collapse eventually back on itself. As for what happens at extremely high temperatures and densities behind the Planck barrier—can the universe begin anew with new physical constants?—our current physics are powerless to say.

You will recall that the idea of multiple universes also came up when we discussed the universe's inflationary phase. In that scenario, our universe is just a tiny bubble, lost in a meta-universe, a much larger bubble. That meta-universe is itself lost in a multitude of other meta-universes. Although our knowledge of the initial instants of the universe's existence is still very sketchy, it is interesting to note that these multiple universes arise naturally from the Grand Unified Theories* that attempt to unify the fundamental forces of nature into a single force. Because these multiple universes have not been specially invoked to resolve paradoxes in quantum mechanics (as was the case with Everett's universes) or to explain the fine tuning of the properties of the universe (as with Wheeler's), they are perhaps more credible. But they still suffer from the greatest possible disadvantage there can be in science: They cannot be verified by observation, being all disconnected from one another, and in particular from our own universe.

Chance Revisited

What are we to think of the anthropic principle? It is difficult to deny that there must have been the most extraordinary adjustment of the universe's physical parameters to allow consciousness (based on carbon chemistry) to emerge. Voltaire's philosopher Pangloss did not know how right he was when he said, "Everything is for the best in the best of all possible worlds." Everything is for the best because the universe has been precisely adjusted for our appearance on the scene. The phantom of Copernicus can do no more than admit defeat. Our very existence is a remarkable fact, and one that conveys considerable information. Astronomers know that the universe they will be observing tonight has specific properties, which must be compatible with the very fact that they exist. But does this knowledge help them to comprehend the fundamental principles that govern the universe any better? Does the anthropic principle have its place in the scientific process? Science makes predictions. Initial conditions are stated (for example, the initial position and velocity of a ball thrown in the air), physical laws are applied (Newton's law of universal gravitation), and predictions are made (when and where the ball will hit the ground). These predictions may be verified by accurate measurements. The anthropic principle operates in the opposite sense. It is a statement made a posteriori. We know we exist. What can we deduce from that about the initial conditions and the laws that govern the universe? Is this principle sterile and devoid of any predictive properties (an accusation that has frequently been leveled against it)?

We must recognize that, since it was proposed by Brandon Carter in 1974, the anthropic principle has not brought about a harvest of scientific discoveries. It seems to me that, by itself, the principle cannot reveal any major truth. It can, however, serve as a guide for our intuition, and point us on the right path for unveiling Nature's secrets. An example comes to mind. In 1961, 13 years before the principle was formally stated by Carter, the American physicist Robert Dicke discovered a strange coincidence: The universe and stars have approximately the same age, roughly 10 billion years. Dicke realized immediately that this strange coincidence was not accidental, but necessary: The universe must be old enough to have had the time to produce carbon, which is required for the existence of Mankind in general, and of physicists in particular. Because the universe could not build anything more complex than helium in the first 3 minutes, it was necessary to wait for the arrival of stars, their creative alchemy, and their explosive agony, which would scatter the combustion products into interstellar space. All this took a few billion years. On the other hand, the universe cannot be very much older than a star. If the universe were too old, it would contain only stellar corpses: white dwarfs, neutron stars, and black holes. Physicists would not be able to live in it. The age of the universe is thus not an accident, but is determined by our presence. Dicke could have used this argument against the Steady-State theory*, which was the main rival to the Big Bang theory in the 1950s. The Steady-State theory predicted that the universe had neither beginning nor end, and that time extended indefinitely into the past and future. In this theory, there would be no obvious connection between the age of the universe and that of a star. Dicke could have argued (but he did not) that the similarity between the two ages tipped the balance in favor of a universe with a beginning—that is, in favor of the Big Bang. But he could not have left it at that. Simply saying that the Big Bang theory was the correct one because, in that theory, our presence naturally explains the coincidence between the age of the universe and that of stars would not have convinced anyone. Our intuition may be guided and our observations made more pertinent by taking into account the fact that our very existence is extraordinary. Dicke's argument, if it had been taken more seriously, might have brought forward the discovery of the fossil background radiation, which, in 1965, was to cause the Big Bang theory to emerge the winner. The anthropic principle may guide our ideas, but it is no substitute for research into the great physical laws that govern Nature, or for the observations that verify them.

Jacques Monod believed that the various stages of cosmic evolution leading to Mankind were no more than a series of lucky throws of the dice, a succession of happy coincidences that might well have never occurred. As a biochemist, he was thinking of the chance encounters between quarks that formed atomic nuclei; between atoms in the cores of stars that lighted their fire; between the atoms created by stellar combustion that formed interstellar molecules and planets; and between the organic molecules in the primitive oceans that gave birth to the intertwined double helix of DNA. Mankind emerged as a result of this long chain of lucky accidents, but it could just as well have never appeared. The notion of multiple parallel universes and the anthropic principle provide a completely different interpretation of chance than that envisaged by Monod. Chance does not lie in the encounters between particles, quarks, atoms, and molecules, but in the choice of the physical constants and initial conditions. Once these are fixed, matter contains within it all the seeds required for the emergence of consciousness, and the process of cosmic gestation will inexorably lead to ourselves.

Science and Finalism

The concept of multiple universes has enabled scientists to set aside their distrust of finalism—the doctrine that everything is determined by final causes—and to get around the idea that the universe has a "grand design." But every time they are soothed by one of Bach's fugues or a Mozart sonata, filled with admiration for Van Gogh's sunflowers or Cézanne's apples, awestruck by the fiery pastel shades of a sunset, entranced by the beauties of a landscape, doubt creeps in again. Might there not be an ultimate plan after all? Stating without any proof that there is not is no more scientific and is as dogmatic as maintaining that one does exist. Also postulating that there is an infinite number of unverifiable universes is not very satisfying to the mind. Why should nature indulge in an orgy of creation, producing innumerable infertile universes, simply to create one that is fertile? That seems to be purely and simply wasteful, and does not agree with the simplicity and economy of known natural laws.

To talk about a Grand Design is to talk about a supreme creator, about God. Daring to mix God and science is the worst of all heresies in the eyes of some scientists, such as Monod or Weinberg. To them, the scientific method has no direct answers to the questions of the existence of an ultimate plan or a supreme being, and they are quite right. God cannot be demonstrated like a mathematical theorem. He cannot be verified by observations at the telescope, measurements in the laboratory, or calculations with a computer. One either has faith or not. Recent discoveries in cosmology have, however, cast new light on this most fundamental and oldest question of all. And any serious consideration of the existence of God must take these factors into account. After all, the questions that cosmologists ask are astonishingly close to those that preoccupy theologians: How was the universe created? Was there a beginning to space and time? Will the universe come to an end? Where did it come from and where is it going? The realm of God is that of the mysterious and the invisible, that of the infinitely small and the infinitely large. This realm is no longer the sole prerogative of theologians; it also belongs to scientists. Science is there, accumulating discoveries and changing viewpoints. Theologians no longer have the right to be indifferent to it. Let us confront the religious and philosophical arguments concerning the existence of God with the new scientific view of the universe. Let us examine the encounter between God and modern cosmology.

| 8 |

God and the Big Bang

Is a First Cause Necessary?

One of the arguments most frequently used to demonstrate the existence of God, and the one that has been the subject of attention by the greatest philosophers and theologians through the ages, from Plato and Aristotle, through Saint Thomas Aquinas, to Immanuel Kant, is that of the chain of causes: Everything has a cause. Such a chain cannot be infinite. Sooner or later, we must encounter the first cause, responsible for all the content of the universe. That first cause is God.

This argument rests evidently on the Western concept of linear time. An event A occurs, which causes B, which in turn causes C, and so on. In certain Oriental philosophies and religions, such as Buddhism, time is not linear, but cyclic. Event A causes B, which causes C, which in turn causes A. The loop is closed, and there is no necessity for a first cause. Once again, the bizarre properties of quantum mechanics do not exclude this possibility. In fact, in the 1960s, one theory of physics proclaimed that there was *no* elementary particle, and that every particle consisted of every other, that every particle possessed some of the properties of the others. In this way, A would consist of B and C; B of A and C; and C of A and B. This theory has now run out of steam, because the concept of a hierarchy of particles that are more and more elementary seems to explain the properties of nature in a more satisfactory manner. Matter is made of molecules, which contain atoms, which consist of electrons and atomic nuclei, which are built from protons and neutrons, which in turn consist of quarks. For the time being, the chain stops there.

Quantum uncertainty has shattered the argument for a first cause. In the subatomic world of elementary particles, causal relations and determinism no longer hold. Virtual particles may come into the real world, as we have seen, in a sudden and unpredictable way, by drawing on Nature's energy reserves. We cannot tell for sure where and when they will appear; the most we can do is to assign a probability that they will spring up at such and such a place. They will be more likely to emerge in the curved space, filled with gravitational energy, near the radius of no return of a black hole somewhere in the Milky Way, than in the flat space of the room where you are reading this book. Nevertheless, thanks to the magic of quantum mechanics, the probability of a virtual particle appearing suddenly next to your hand may be infinitesimal, but it is not zero. Such virtual particles do not have a specific cause, and their behavior cannot be predicted in advance.

We have also seen that physicists believe that what is true for an elementary particle is also true for the entire universe at its beginning. Quantum uncertainty allows time and space, and then the whole universe, to arise spontaneously from a vacuum. At the Planck time (10^{-43} second), the universe was only 10^{-33} centimeter across, 10 million billion billion times smaller than an atom, and quantum mechanics, which governs the subatomic world, could do its work. The universe does not require a first cause. It appears thanks to a quantum fluctuation. Once created, inflation expands it exponentially during the first few fractions of a second of its existence, quarks and antiquarks enter the scene, and then begins the whole cosmic evolution, governed by the laws of physics and biology, that will eventually lead to us. This description of the creation of the universe strangely resembles the creation ex nihilo evoked by many religions. The major difference is that the appearance of the universe, thanks to the magic of quantum mechanics, no longer appears to require a first cause or the existence of God. Its emergence may be explained by purely physical processes.

God and Time

Quantum mechanics has rendered the notion of a first cause obsolete. But that is not all. Even the concept of "cause and effect" loses its usual meaning where the universe is concerned. Such a concept presupposes the existence of time: The cause comes before the effect. Mother and father come before a child, not the other way around. Time flows from cause to effect. Modern physics, on the other hand, believes that time and space were created simultaneously with the universe itself. This is not a new idea: Saint Augustine wrote, as early as the fourth century, that the universe was made not in time, but with time. He regarded as ridiculous the idea of God waiting an infinite time and then suddenly deciding to create the universe. What meaning, then, can we ascribe to the phrase "And God created the universe," if time did not yet exist, and it was created together with the universe? The act of creation makes sense only within time. It is as absurd to think that God existed "before" the universe as it is to ask what happened "before" the Big Bang. "Before" is meaningless, because time had not yet appeared.

To avoid these logical contradictions, some physicists have proposed theories that are ever more bizarre. The American physicist John Wheeler (the man who proposed the cyclic universe) suggests inverting the order of causality with respect to time: The cause does not come before, but after, the event; and the universe no longer requires a creator, because it is Mankind that is responsible for the appearance of the universe. By some mysterious causal connection between Man and the universe, which acts in the opposite sense to the normal direction of time, our existence will retroactively cause the emergence of the universe. This is the anthropic principle* taken to its extreme. This inversion of causality with respect to time can lead to far more serious logical contradictions—you could, for example, prevent your parents from meeting and your own birth—and the majority of physicists reject it. But certain physical theories do not exclude retroactive causality. Particles called "tachyons," which always travel faster than the speed of light, could influence the past. Luckily for our sanity, they still only exist in the unchecked imaginations of physicists. In some situations, quantum mechanics (again!) seems to imply that the act of observation may retroactively alter a past situation. This occurs on the subatomic level, however, and it is not clear if the universe as a whole is subject to this type of retroactive causality.

There are other conceptual difficulties associated with the idea of a God existing within time.[19] The passage of time is indicated by changes. But can we speak of a changing God, if He is the first cause of all changes in the universe? Who can change God? On the other hand, time, as Einstein demonstrated so clearly, is not universal. It may vary from place to place within the universe. Time near a black hole is not the same as it is on Earth. It is elastic, and human volition can alter it. Press down on the accelerator, and time slows down. Time may even change direction or come to a halt if the universe collapses back unto itself. A God existing within time would no longer be all powerful. He would be subject to the vagaries of time caused by black holes, neutron stars, other gravitational fields, or human actions. That would be the end of His omnipotence.

The solution to these dilemmas would be a God outside time, a God who transcends time. But that would also raise difficulties: Such a distant, impersonal God would no longer be able to help us. The God to whom we address our prayers is a God capable of feeling emotions, who is satisfied or not with human moral progress, who decides to grant our prayers or punish us, and who can plan and alter our future. In short, a God who acts within time. A God outside time could no longer come to our aid. Moreover, if God transcends time, He already knows the future. Why then, should He be concerned with the progress of the human struggle against evil? He knows the

19. P. C. W. Davies, *God and the New Physics*, New York, Simon and Schuster, 1983.

outcome in advance. A God outside time would no longer indulge in thought, because thinking is also an activity carried out within time. God's knowledge would no longer change with time. He would be aware in advance of all the changes with time, down to that of the very least atom.

Modern physics, then, presents us with a choice between a personal, but not omnipotent, God, or an omnipotent, but impersonal, God. Time, by becoming elastic, no longer allows a God who is personal and omnipotent at the same time.

God and Complexity

Who of us has not experienced the supreme aesthetic pleasure derived from listening to exquisite music, from contemplating a splendid work of art, from seeing the beauty of a woman, or from admiring a magnificent landscape? At such privileged moments, we find it hard to accept that the universe is devoid of meaning and sense of purpose. We feel that so much beauty and complexity cannot be the fruits of pure chance, that there must be a creative principle at work. In doing so, we instinctively make our own one of the theologians' favorite arguments for proving the existence of God: Only a Creator can be responsible for Nature being so complex and well organized. The English archbishop William Paley wrote in 1802: "In crossing a heath, suppose I pitched my foot against a stone, and were asked how the stone came to be there, I might possibly answer, that . . . it had lain there forever. . . . But suppose I had found a watch upon the ground. . . . [I am forced to conclude] that the watch must have had a maker."[20] The organization and complexity of the watch are proof of the existence of a clockmaker. Similarly, the organization and complexity of the universe demonstrate the existence of God.

This argument is certainly very convincing. Unfortunately, it does not quite agree with modern science. The latter tells us that extremely complex systems may be the result of perfectly natural evolutionary processes that follow well-understood physical or biological laws, and that there is no need to invoke God as watchmaker. Complexity does not necessarily presuppose a creator and a grand design. We have seen in our history book of the universe how the latter, by following well-established physical and biological laws, has been able, over a period of 15 billion years, and beginning from a very simple primordial state (a soup of elementary particles), to produce the wonderful cosmic tapestry of galaxies, and, in one of those galaxies, an intelligence capable of observing it. Once set on its course, the universe has no need of a watchmaker to develop its complexity fully. At first sight this appears to be a flagrant contradiction of the second law of thermodynamics (the physics of energy and heat), which states that the total disorder in the universe must always increase: Majestic cathedrals left to themselves fall

20. W. Paley, *Natural Theology—or Evidences of the Existence and Attributes of the Deity Collected from the Appearances of Nature,* Oxford, Clarendon Press, 1938.

into ruins, roses wilt, things and people age and wither with time, the greatest beauty fades away, and order tends towards chaos. The emergence of life, the ultimate in organization and complexity, from a state of zero complexity in a vacuum filled with gravitational energy, appears to defy physics.

As we have seen, however, what appears to be a major paradox is not one in reality. If you draw up a balance sheet of order and disorder in the universe (disorder is measured by a quantity physicists call entropy*), disorder always has the upper hand, and, without exception, increases with time. Thermodynamics does not forbid order from dominating in some special, privileged places, structures and complexity from appearing, nor consciousness from emerging, provided that in other locations greater compensatory disorder occurs, and that the balance sheet always shows an increase in disorder. The order represented by life on Earth is only possible thanks to the greater disorder created by the Sun in converting atoms of hydrogen into energy, light, and heat. All the structures in the universe, be they galaxies or planets, owe their existence to two factors: the expansion of the universe and the creation of disorder by stars. The expansion of the universe is essential to ensure the cooling of the background radiation, thereby creating an imbalance between the temperatures of the stars and the space surrounding them. This imbalance allows, in turn, the stars to become systems for creating disorder. These export their high-temperature disorganized radiation into the cooler, more highly organized regions around them. Disorder is transferred from the hot radiation to the cooler radiation, the total disorder increases, and pockets of order may appear without violating the second law of thermodynamics. Moreover, the expansion of the universe is essential for the very existence of stars. By diluting the universe, it restricted primordial nucleosynthesis* to the lightest elements, leaving the universe with three-quarters of its baryonic mass (protons and neutrons) in the form of hydrogen, and the rest in helium, at the end of the first 3 minutes. Without the expansion, the universe would have converted all matter into iron, which would mean saying goodbye to the reserves of hydrogen that fuel the fires of stars and their disorder. Without that disorder, farewell to order, life, and consciousness!

So, complexity and organization may arise spontaneously in a universe that is expanding and contains stars. The hand of a Creator no longer seems necessary.

God and Life

You remain skeptical. "OK," you say, "We will admit that the structures in the universe have natural causes and do not require the intervention of God. But what about life? Doesn't that require a supernatural cause, that is, God? After all, a human being is only a collection of 30 billion billion billion inanimate particles. The sum of a multiplicity of inanimate entities can only be inanimate itself. If human beings, animals, and plants are alive, it is because God added an essential ingredient, life, to the collection of atoms."

This argument does not take into consideration the fact that the whole may be greater than the sum of its parts, that it may acquire properties on the macroscopic scale that are absent at the microscopic level. Imagine you are examining a painting by the pointillist master Georges Seurat. The innumerable flecks of paint, all vibrant with color, mean nothing if you look at them individually. It is only by stepping back and contemplating the painting as a whole that the figures and landscapes become recognizable and take shape, and the painting acquires meaning. Similarly, individual musical notes leave us cold. It is only when the genius of a Beethoven or a Mozart brings them together into a symphony or sonata that the music moves us. Words in a dictionary are cold and impersonal, but in the poems of Rimbaud or Baudelaire, they affect us deeply. The whole has properties that are not found in the parts. Similarly, we can envisage perfectly inanimate atoms that combine according to completely natural laws of physics, and that, once a certain threshold of organization and complexity has been crossed, create life, without recourse to divine intervention. Life is the result of a collective (or holistic) phenomenon. It cannot be reduced to a collection of cells, DNA helices, or chains of atoms.

We are still utterly ignorant of the processes that begat life from inanimate atoms. What minimum degree of complexity and organization had to be reached for life to emerge? How did physics and chemistry lead to that complexity? Life probably began its ascent, as we have seen, in the Earth's primitive atmosphere. In 1953, the American chemists Stanley Miller and Harold Urey, in a famous experiment, reproduced the primitive terrestrial atmosphere in a test tube. This mixture of ammonia, methane, hydrogen, and water was subjected to electrical discharges to simulate the effect of the thunderstorms that raged on Earth 4.6 billion years ago. After several days, the basic building blocks of life, the amino acids, made their appearance. Miller and Urey were on the right path for unravelling the mystery of life. But it is a long way from amino acids to the double helix of DNA capable of replicating itself. The origin of life remains one of the greatest scientific enigmas. All we can say is that such an origin is not incompatible with known natural laws, and that it does not necessarily require divine intervention.

God and Consciousness

Once created, life embarked on an accelerated evolution to arrive at intelligence and consciousness, reason and thought. The first forms of life, the first living cells, appeared on Earth 3.5 billion years ago. For 3 billion years (i.e., about three-quarters of the time since the first primitive life until now), evolution was extremely slow, and did not go beyond unicellular forms. Then, in less than 1 billion years, evolution switched into overdrive: Multicellular organisms (such as mollusks, fish, reptiles, and mammals) invaded the Earth. Subsequently, in less than 100 million years (i.e., less than 3% of the age of living things) three species endowed with primary intelligence made their entrance: primates, dolphins, and rats. And then, about 2 million years ago, *Homo sapiens* appeared on the scene, blessed with consciousness and a "soul."

It is difficult to say with precision at what stage consciousness emerged. Orangutans, gorillas, and chimpanzees seem to be capable of feeling the very human emotions of love, sadness, and joy. They even seem to possess the rudiments of language. But are they capable of abstract thought? Whatever the reason, we do not see them feverishly composing symphonies, writing plays and novels, or creating paintings and sculptures. Once again, the question arises: Does consciousness require divine intervention? Does a soul have to be "introduced" into a body consisting of billions of atoms? Does the mind have to be "grafted" onto a brain consisting of billions of neurons?

Once again, the answer to these questions is that they do not make sense. Formulating questions in these terms is to confuse descriptive concepts on completely different planes.[21] Speaking of atoms in a body, or neurons in a brain, is like talking about notes of music, or words in a dictionary. When we speak of life and consciousness, we are moving to a different plane, we are abandoning the reductionist description for a collective, holistic description; we are referring to the melody in a symphony, or to the plot in a novel. Body and soul are not concepts that can be described in the same terms. Putting them on equal footing (as Descartes did when he discussed the dual nature of body and soul) is to lay oneself open to absurd questions such as: Where is the soul in space and time? (Descartes thought that the pineal gland inside the brain was the seat of the soul; Teilhard de Chardin thought that consciousness was spread throughout every atom in the body. Neither of these ideas has any experimental basis.) Where was the soul before the body arrived in the world? Where does it go after the body is destroyed? Does God have a reserve stock of souls, from which He draws to graft them onto collections of atoms? These questions are nonsensical, because the spirit and the body are not two distinct material substances, one within the other. They cannot be discussed on the same plane.

When the soul and the body are described on two different planes, nothing prevents the perfectly natural, spontaneous appearance of consciousness, if evolution goes beyond a certain threshold of organization and complexity. The divine spark is not required. This conclusion has rather disagreeable implications for our self-esteem. The brain is no more than a thinking machine, nothing but a collection of components that constitute a sort of society. And it is the relationships within this "society" that form what we call "the mind."[22] This means that machines, if they become sufficiently complex, may be able to think and feel. Machines will have a heart. Of course, although the mental capacity of modern, intelligent machines surpasses our own in many respects—they can calculate far more quickly and

---

21. P. C. W. Davies, *God and the New Physics*, New York, Simon and Schuster, 1983.
22. M. Minsky, *The Society of the Mind*, Simon and Schuster, New York, 1987.

without errors, and are able to beat us at chess, for example—their sensorial abilities are still extremely limited. They cannot see very well, have difficulty in recognizing to whom they are talking, understand only about 10,000 words (provided one speaks very slowly and distinctly), and speak in a rather husky voice. Still, these intelligent machines have only been in existence for the past 15 years or so—and they have already attained a degree of complexity comparable with that of insects—whereas human beings are the product of millions of years of evolution! This picture of future intelligent machines is perhaps not very appealing, but it is not ruled out by current research into artificial intelligence.

God and the Extraterrestrials

If we accept the hypothesis that life and consciousness emerged naturally on Earth, with no divine assistance, we must envisage the possibility of the existence of other forms of intelligence elsewhere in the universe. After all, the observable universe contains 100 billion galaxies, each of which contains some 100 billion stars. If each star has, like the Sun, about 10 planets, the total population of planets in the universe amounts to some 100,000 billion billion (10^{23}) planets. Why should our planet be the only one to harbor life? (It does, however, appear to be alone in this respect as far as the nine planets of the Solar System are concerned. The exploration of Mars—which is the planet, after Earth, most suitable for life as we know it—by the American Viking probes did not reveal any Martians or indeed any other living organisms.) Why should the various rungs of the complexity ladder be climbed on Earth alone? That appears highly improbable, and the phantom of Copernicus would be incensed that the idea could even be considered. There is a heated debate on this subject. Some think we are alone in the universe because we have never picked up any messages from the cosmos. One might counter the argument by saying that perhaps we do not yet possess the technology or the knowledge to capture and decipher such interstellar messages, or that extraterrestrials have no desire to communicate with us, that they are observing us from afar like onlookers watching animals in cages at the zoo. The absence of proof is not proof of absence.

The existence of extraterrestrial civilizations would, in any case, raise some extremely interesting theological questions, in particular with respect to the Christian religion.[23] The latter maintains that we have inherited "original sin" from Adam and Eve. An extraterrestrial race that has arisen on another planet would not have that inheritance: Would the members of this race be therefore free from sin? On the other hand, God sent his Son, Jesus Christ, onto this Earth to redeem the human race. Would there be hordes of extraterrestrial Jesus Christs visiting each fertile planet to bring salvation to all the beings that had arisen out there? These questions may appear absurd

23. P. C. W. Davies, *God and the New Physics*, New York, Simon and Schuster, 1983.

at first sight, but are ones that theologians will have to grapple with if tomorrow we establish contact with an extraterrestrial civilization.

Giordano Bruno raised these questions as long ago as 1600 A.D., when he advanced the idea of an infinite universe, containing an infinite number of worlds inhabited by an infinite variety of life forms, all proclaiming the glory of God. However, the Church, instead of attempting to resolve the problem, preferred to silence Bruno by condemning him to be burned at the stake.

The Final Wager

Modern cosmology has profoundly altered our ideas concerning the nature of time and space, the origin or matter, the development of life and consciousness, order and disorder, and causality and determinism. It has tackled subjects that for many years were the exclusive property of religion, and has thrown a completely new light on them. By attacking the wall surrounding physical reality with the powerful tools of physical and mathematical laws, cosmologists and astronomers have found themselves face to face with theologians.

At first sight, modern physics seems to have abolished the necessity for God. The two favorite arguments advanced to demonstrate the existence of God appear not to hold anymore. The argument of the first cause has lost its strength, because quantum uncertainty has abolished determinism in the subatomic world. The argument of God as a watchmaker, the creator of beauty, organization, and complexity in the universe, is no longer infallible. Order may appear at oases in the cosmic immensity without divine intervention if greater disorder is created elsewhere. The stars, by exporting their high-temperature radiation into the universe, cooled by its expansion, are the sources of that disorder. Life and consciousness may emerge from matter without any divine spark, because they are collective and holistic phenomena.

Yet a doubt persists. As we have seen, cosmic evolution had to be very precisely adjusted to lead to us. If the fundamental constants, the initial conditions, or the physical laws were to differ by even the slightest fraction, we would no longer be here to talk about it. The whole cosmic gestation may be explained by natural laws—disparate laws, which physicists hope eventually to combine in a single super-law, a grand unifying principle—but there still remain the important questions of fixing the values of the physical constants and setting down the laws (or grand principle). Faced with this situation, there can be two possible attitudes. We can invoke a supreme being, who is the author of these laws, and who has planned and fine tuned everything, or else we can come down firmly on the side of chance, by introducing the concept of multiple, parallel universes. According to this hypothesis, there would be an infinite number of universes, exhibiting every possible combination of physical constants and laws. The majority of these universes would possess losing combinations and would thus be infertile. Ours, by chance, had a winning combination, and we were the grand prize. What appears to be precise adjustment is thus no more than pure coincidence.

Which attitude should we adopt? Faced with this dilemma, scientists, despite all their knowledge, find themselves just as unprepared and ill equipped as their neighbors. Science is no great help when it is a question of faith. Scientists have to weigh the risks and take the plunge. They have to make a wager, just like Pascal. For myself, I am prepared to bet on the existence of a supreme being. The hypothesis of a multiplicity of imaginary, unverifiable universes violates my sense of simplicity and economy. Without going as far as the Austrian philosopher Karl Popper, for whom "Anything that cannot be observed or refuted is in the realm of magic or mysticism, but does not belong to the sphere of science,"[24] I find the hypothesis of innumerable universes, all inaccessible to observation, to be rather unwarranted. And, besides, I like to have free will. In a world of parallel universes, everything that can occur does happen. Suppose that one Friday evening, I have the choice between going to the cinema or staying at home to watch TV. The theory of multiple universes says that I do both. The universe and I would both split. In one universe, I would go to the cinema, and in the other I would stay at home. Because all alternatives do happen, there is no more choice. In such a world of parallel universes, a criminal would have a field day for pleading clemency from a jury.[25] The laws of quantum mechanics forced him to commit the crime. Had he not committed the offence in this universe, one of his doubles would certainly have committed it in a parallel universe. Finally, betting on chance implies nonsense and despair, as witness the cries of distress by Monod and Weinberg. Why not, then, bet rather on sense and hope?

24. K. Popper, *The Logic of Scientific Discovery,* New York, Harper and Row, 1965.
25. P. C. W. Davies, *God and the New Physics,* New York, Simon and Schuster, 1983.

| 9 |

The Secret Melody

The Steady-State Universe and Continuous Creation

Nature is not silent. She delights in continuously sending us her notes of music. She does not, however, tell us the organization of these notes, nor does she reveal the secret of their melody. It is up to us to uncover the secret, discover the melody, and write out the score. The Big Bang's melody has gained recognition and acceptance from the majority of cosmologists. To a small number of cosmologists, however,the Big Bang does not ring true. They hear the same musical notes, but the melody that they construct from them is completely different. There are many rival cosmological theories that have tried, so far without success, to dethrone the Big Bang theory. Let us listen to some of these alternative melodies.

Besides the Big Bang, the cosmological theory that has been developed the most is undoubtedly the Steady-State theory. This theory was devised in 1948 by the British astronomers Hermann Bondi, Thomas Gold, and Fred Hoyle. It exerted considerable influence on cosmological thought in the 1950s and 1960s and stimulated numerous studies aiming at distinguishing between it and the Big Bang.

The Steady-State theory originated from aesthetic, philosophical, and religious considerations, and also from an observation that subsequently proved to be false. The aesthetic factor stems from the cosmological principle*, one of the fundamental postulates of general relativity, which was to be confirmed in spectacular fashion later by the study of the fossil background radiation. The principle states that the universe should be homogeneous and isotropic, that is the same everywhere and in every direction of space. Why then not go even further? Why not extend the concept of homogeneity and isotropy to time itself? The universe would then follow the perfect cosmological principle: It would be unchanging in both space and time. The universe would always look the same: It would exist in a steady state. This theory therefore rejects the idea of evolution and change inherent in the Big Bang theory. The Aristotelian view of the immutability of the heavens is resurfacing in another guise.

In the Steady-State theory, the universe has neither a beginning nor an end. And this is where the philosophical and religious considerations play a role. For scientists in the 1950s and 1960s, the idea of creation and beginning (and, consequently, that of a supreme being) smacked of heresy. Many cosmologists were not prepared to consider it and were happy to adopt a theory that offered them, on a silver platter, a universe that had existed forever. This allowed them to sweep the problem of creation under the carpet with good conscience. Moreover, a universe with an infinite age would solve the problem posed by the ages of the universe and the Earth as determined at the time. In the Big Bang theory, the universe was only 2 billion years old, according to Hubble, whereas measurements of the radioactive elements in the Earth's crust gave a value of 4.6 billion years. How could the Earth be older than the universe? Bondi, Gold, and Hoyle concluded that the Big Bang model must be wrong. We now know that it was not the Big Bang model that was at fault, but Hubble. He had misjudged the intrinsic brightness of his cosmic candles, the Cepheid stars. Nowadays, the age of the universe is estimated to be about 15 billion years, so there is no more danger of the contents being older than the container, of the Earth being older than the universe, in the context of the Big Bang theory.

But how to reconcile the idea of a universe unchanging in time with its observed expansion? If the galaxies were continually receding from one another, and more and more empty space were created between them, the universe could not remain the same as time passes. To rescue the perfect cosmological principle*, Bondi, Gold, and Hoyle had to postulate a continuous creation of matter (and hence of galaxies) that compensated exactly for the empty space resulting from the expansion of the universe. The level of water in a reservoir that is emptying can remain the same if fresh water is added. Your common sense rebels against the idea: You do not see matter pop up spontaneously on every street corner. The rate at which matter has to be created is so low, however (one hydrogen atom per liter of space per billion years), that it is perfectly imperceptible and unmeasurable. The mechanism behind this continuous creation of matter was not explained, but after all, neither was the creation of matter in the Big Bang in the 1950s. (We have seen earlier how there is now a possible explanation in the context of the Big Bang, thanks to particle physics.) In trying to avoid the Great Creation, Bondi, Gold, and Hoyle had to have recourse to an infinite series of smaller creations (Fig. 60).

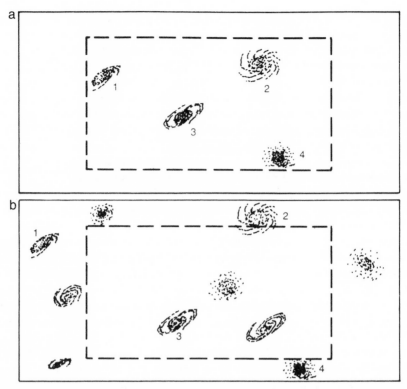

Fig. 60. *The Steady-State and Big Bang theories.* Until the 1960s, the cosmological theory, which was the most serious rival to the Big Bang, was the Steady-State theory. This diagram shows how the Steady-State theory must postulate the continuous creation of matter to ensure that the universe always looked the same, despite its expansion. (a) There are four galaxies, numbered 1 to 4, which participate in the expansion of the universe. At a later stage, the area within the dashed lines has expanded to fill the whole of (b). The four numbered galaxies have moved away from one another, leaving new empty space between them. In the Big Bang theory, the void between galaxies continually increases (if the universe is open, or until the time of maximum expansion if the universe is closed). In the Steady-State theory, new galaxies (un-numbered) are created, so that the number of galaxies within the region enclosed by the new dashed lines remains constant. The rate of matter creation (one hydrogen atom per liter of space per billion years) is so low that it cannot be measured. It was the discovery of the 3 K fossil background radiation in 1965 that gave the coup de grâce to the Steady-State theory.

Observations soon cast doubt on the Steady-State theory. Quasars, which were discovered at the beginning of the 1960s, appeared to die out and dwindle in numbers as the universe grew older. There was a flourishing population of quasars some 10 billion years ago, whereas it is greatly reduced today. (Astronomers are able to study the demography of quasars with time by observing them at greater and greater distances, and by making use of the principle "looking far is looking back in time.") Quasars undergo evolution, an observation that the theory could not accommodate. The coup de grâce, however, was the discovery in 1965 of the fossil background radiation. Since the Steady-State theory rejected the idea of a hot, dense initial state, it could not offer any natural explanation for the radiation that bathes the entire universe. The melody of the Steady-State universe had to be relegated to the graveyard of dead music.

What If the Universe Were Not Expanding?

Other melodies were suggested. Some tried to shake one of the two cornerstones on which the Big Bang was built: the expansion of the universe. (The background radiation constitutes the other cornerstone.) You will recall that the expansion of the universe is revealed by the motion of the galaxies: by the fact that they are receding from one another. A human lifetime, or even the few thousand years of civilized man, is far too short for observing changes in the positions of the galaxies directly, as we might observe the movements of people or vehicles. Hubble demonstrated the motion of the galaxies by means of the Doppler effect, where light changes color when it is emitted by an object in motion. Because the light of the galaxies is red shifted, they are moving away from us. In addition, the recession velocity of a galaxy (obtained by measuring the change in color) is proportional to its distance, which implies a universal expansion.

Some theories question the interpretation of the red shift of light from the galaxies as being caused by a receding motion. They suggest that the loss of energy that produces this red shift—a weakening of the photons that carry the radiation—is caused not by the expansion of the universe, but is a result of the photons becoming increasingly "tired" in the course of their long intergalactic and interstellar travels through a perfectly static and motionless universe before they reach our telescopes. A static universe model (without expansion) with "tired light" may be devised (Hubble and the American physicist Richard Tolman suggested one in 1935) that reproduces the observed relationship between the red shift and the distance of a galaxy. However, the "tired-light" theory has many weaknesses. First, there is no known physical mechanism that would cause light to become tired. The French astrophysicist Jean-Claude Pecker has indeed suggested that the photons lose energy during their travels because they interact with a new particle, whose mass should be less than that of the electron. This mechanism has not received general acceptance by astrophysicists, because, so far, the proposed new particle has not been observed either in the laboratory or in the universe.

Quite independently of the exact cause of light's loss of energy, the "tired-light" theory suffers from many other problems. It predicts certain relationships between the properties of galaxies (such as their size and apparent brightness) and their red shift, which are not in agreement with ob-

servation. The principal stumbling block for the "tired-light" theory, however, as for all other rival theories to that of the Big Bang, is the fossil
radiation: It has no way of explaining its existence.

Noncosmological Red Shifts

There have been many other attacks on the Big Bang theory, and on the so-
called "cosmological" interpretation, which attributes the red shift of light
from the galaxies to the expansion of the universe. It was not purely by
chance that these attacks by a minority of astronomers became even more
vociferous just after the discovery of quasars in 1963. These objects, which
have a starlike appearance, had far larger red shifts than those observed in
stars and galaxies until that time. These large red shifts, if interpreted in
terms of the Doppler shift, implied enormous recession velocities and thus
distances, if the quasars participated in the general expansion of the universe
described by Hubble's law. They would be at the very edge of the universe.
Yet quasars have an extremely high apparent brightness, despite being so
distant. The amount of energy released within them must be immense,
equivalent to the energy released by 100,000 billion Suns, or 1000 galaxies.
In addition, this huge energy comes from a region that is just a few light-
months across, less than one hundred-thousandth of the size of a galaxy, and
scarcely 100 times as large as our Solar System. (Astronomers can estimate
the size of quasars because their light varies. It increases or decreases over
a period of a few months. The size of a quasar is the distance traveled by
light during that interval of time, in other words, a few light-months.)

The hearts of quasars are therefore extremely dense and energetic. We
have seen that a monstrous black hole, with the mass of 1 billion solar
masses, that rips apart and devours the ill-fated stars from the host galaxy
that are within its reach, can produce such a tremendous amount of energy
in such a small a volume. However, invoking the presence of such massive
black holes, the existence of which is predicted by relativity theory, but has
not been confirmed observationally, was not to the taste of some astronomers. They prefer to abandon the cosmological hypothesis for quasars. If
the large red shifts of quasars are not caused by the expansion of the universe, then the relationship between red shift and distance, which has been
established for galaxies, cannot be used to derive the distances of quasars.
In what is known as the "local" hypothesis, quasars might be very close, in
which case their energies would be perfectly ordinary, and massive black
holes would no longer be required to explain them.

The attack on the cosmological interpretation of quasars has primarily
been led by the American astronomer Halton Arp. He maintains that there
are numerous examples where galaxies and quasars occur together in pairs
or groups, and are thus at the same distance from us, yet have completely
different red shifts (the latter being of course far greater for the quasars than
for the galaxies). The red shifts therefore cannot originate in the expansion

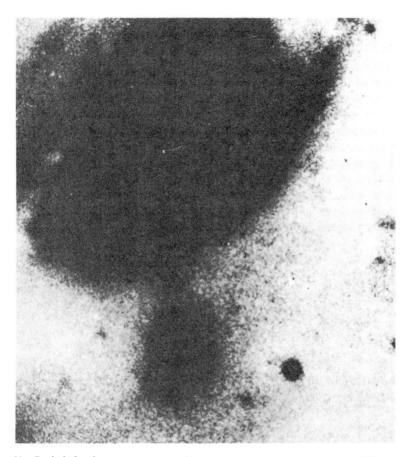

Fig. 61. *Red shifts that are not caused by the expansion of the universe?* The American astronomer Halton Arp has questioned what is known as the "cosmological" interpretation of the red shift of light from galaxies and quasars. According to him, Hubble's law, one of the cornerstones of the Big Bang theory, which attributes the red shifts to the expansion of the universe, cannot be applied to certain galaxies and quasars. Hubble's law implies that objects lying at the same distance from us should have the same red shift. Arp maintains, however, that he has found pairs of objects, which, according to him, are at the same distance , but which have completely different red shifts. An example is shown here. This photograph shows a galaxy (*top*) and a quasar (*bottom*). The galaxy (NGC 4319) has a small red shift corresponding to a recession velocity of 1700 kilometers per second. The quasar (Markarian 205) has a red shift 12 times as large, corresponding to a recession velocity of 20,250 kilometers per second. According to Arp, the galaxy and the quasar, despite having such different red shifts, are at the same distance, because they are joined by a bridge of light (visible between the two bodies in the photograph). The reality of this bridge of light has, however, been challenged by other astronomers, who attribute them to misleading photographic effects, or to reflection nebulae seen in projection (Fig. 62). The only way of resolving this controversy is to measure velocities across the bridge of light. If measurements were to show that the velocity increased along the bridge from that of the galaxy to that of the quasar, there would no longer be any shadow of a doubt about the existence of noncosmological red shifts (i.e., not caused by the expansion of the universe). Unfortunately, the brightness of such bridges of light is so low that velocity measurements are impossible with current instruments (photograph: H. Arp).

of the universe, because if that were so, objects at the same distance from us would have the same red shifts. But how can one show that these pairs or groups of objects are really at the same distance, and that their apparent proximity to one another on the sky is not just a projection effect? Two bodies that are at greatly differing distances may appear very close to one another in the sky if they happen to lie along the same line of sight (Fig. 32). In normal circumstances, it is precisely the red shift that is used to indicate cosmic distances and to map the universe in depth. In this case, however, this method is of no use, since it is the very relationship between red shift and distance that is called into question. Arp has to fall back on other techniques for demonstrating the spatial association of his galaxies and quasars. He points to bridges of light that connect them (Fig. 61) or draws attention to various galaxy–quasar alignments, which indicate, according to him, a common origin for the two types of objects. The reality of the bridges of light has been challenged by the supporters of the cosmological interpretation. They attribute them either to misleading photographic effects or to clouds of gas and dust in the Milky Way (known as "reflection nebulae"), which reflect light from nearby stars, and are seen in projection against Arp's objects (Fig. 62). Moreover, the alignments are, according to Arp's critics, accidental and do not reflect any real connection between the objects.

What causes the red shift of quasars if the expansion of the universe is not responsible? (The cosmological interpretation of the red shift of galaxies is not generally questioned, although Arp favors a noncosmological component for certain galaxies.) Because of the alignments he believes he has discovered, Arp envisages immense explosions taking place in the centers of parent galaxies, which eject galaxies and quasars into intergalactic space. The red shift is still interpreted as a Doppler effect (i.e., caused by motion). This movement, however, is not caused by the expansion of the universe, but by an explosion followed by an ejection. To me, the weak point of this theory is the general absence of any blue shift. In any scheme based on ejection, there should be as much chance of a galaxy or quasar being ejected toward us, which would give a blue shift, as away from us, which would produce a red shift.

The debate, or rather the controversy (can one call a dialogue of the deaf a debate?) on the nature of the red shift in the light from quasars continues, although the voices in the noncosmological camp appear to be getting weaker and weaker. I believe that the existence of noncosmological red shifts will not be finally established or refuted until our instruments are sensitive enough to measure the red shift of the very faint radiation from the bridges of light. If such measurements were to show red shifts increasing along the bridge of light, from the red shift of the galaxy to the far greater red shift of the quasar, the spatial association would be confirmed, and the existence of noncosmological red shifts would be established. In the meantime, the demonstration of their existence is far from being convincing enough to shake the magnificent edifice of the Big Bang theory.

Other voices have been raised to question the nature of the red shift. The

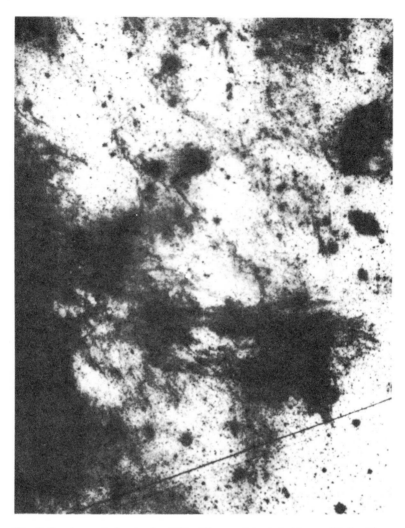

Fig. 62. *Reflection nebulae in the Milky Way and Arp's bridges of light.* The reality of Arp's bridges of light connecting galaxies and quasars with very different red shifts (*see* Fig. 61) has been disputed by many astronomers. One explanation that has been advanced is that these bridges of light are reflection nebulae (clouds of gas and dust that reflect starlight) at high latitudes in the Milky Way, seen in projection against the galaxy and quasar, thus creating the illusion of a bridge of light connecting them. This photograph shows some examples of such reflection nebulae with their characteristic filamentary structure. They may be seen only on extremely sensitive photographic plates (photograph: A. Sandage).

American astronomer William Tifft asserts that the differences in velocity between the galaxies in a pair or cluster (as measured from their red shifts) are invariably multiples of a very specific value, 73 kilometers per second! These results have been received by the astronomical community with great skepticism, and have not provoked enough enthusiasm for other workers to

verify them. As long as Tifft's results have not been confirmed by independent verification, they cannot be taken seriously.

Faster than Light, or a Variable Gravity?

Other attacks have taken place, not directed specifically against the Big
Bang theory, but against general relativity, which constitutes its theoretical
basis. There was great concern when some radio sources (objects that emit
a large fraction of their energy in the form of radio waves) were discovered
to exhibit motions that were "superluminal" (i.e., that were faster than
light). Einstein's general relativity would have been discredited, because one
of its fundamental postulates is that nothing can move faster than light.
Luckily, it was a false alarm. The phenomenon of apparently superluminal
velocities may be explained perfectly well within the context of general relativity as an illusion, caused when radio sources are ejected from the (visible) galaxies with which they are associated at speeds close to, but not exceeding, the velocity of light, and in a direction that is almost exactly aligned
with that of the Earth.

General relativity is also based on the implicit hypothesis that gravity is
constant, and does not vary with time or space. Some cosmological theories
challenged this assumption. One came about because of a strange coincidence. You will recall that gravity has no part in the atomic world, where
electromagnetism plays the leading role. The ratio of the electromagnetic
and gravitational forces between an electron and a proton is an extremely
large number, approximately 10^{40} (1 followed by 40 zeros). The British physicist Paul Dirac, one of the founding fathers of quantum mechanics, noted
in 1937 that this number also occurs in a totally different context. It turns
out also to be the ratio of the age of the universe to the time taken for light
to travel a distance equal to the diameter of a proton. According to Dirac,
this could not be coincidence, but the sign of some hidden law of nature, of
some still mysterious connection between the worlds of the infinitely large
and the infinitely small.

But there is a snag. The age of the universe is not constant—it increases
with time. The coincidence therefore only exists at the present time. To preserve the coincidence for all times, Dirac suggested that gravity or, more
exactly, the physical constant G, which controls the intensity of the gravitational force, decreases with time. The ratio of the electromagnetic and
gravitational forces between an electron and a proton would then increase
with time, thus remaining equal to the ratio between the age of the universe
and the time taken for light to cross the proton. The required change in G,
however, considerably exceeds the observational limits. The theory predicts
changes in the orbits of the planets in the Solar System that disagree with
observations. If G decreases, the gravitational force that holds the planets
in their orbits around the Sun should decrease. The planets ought to be receding from the Sun. But they are not.

The Modification of Newton's Laws

There have also been attacks against Newton's law of universal gravitation. These attacks have been prompted by the mystery of the universe's dark matter. You will recall that astronomers, studying with Newton's laws the motion of stars and gas within a spiral galaxy, or of galaxies in a cluster, have been forced to conclude that we live in an iceberg universe, where 90 to 98% of the mass is invisible. In addition, the nature of this invisible mass is totally unknown. Faced with this uncomfortable situation, a few astronomers have suggested modifying Newton's laws. After all, they say, although the latter have been verified numerous times in laboratories and on the scale of the Solar System by the study of the orbits of planets, they have not been directly confirmed on the scale of galaxies, precisely where the dark mass is encountered. If the laws of gravity were different, and if the gravitational force between two objects did not vary inversely as the square of the distance between them, the motions of the galaxies would no longer necessarily imply the presence of enormous quantities of invisible matter, and the extremely disagreeable sensation of living in a universe that, for the most part, eludes us would no longer have to be.

This suggestion has not met with a wave of enthusiasm from the astronomical community. First, the dark mass is not inferred solely from the motions of stars and gas in galaxies, and of galaxies in clusters. We have seen that the abundances of elements created in the Big Bang (hydrogen, helium, and deuterium) also imply a comparable mass of invisible matter. Second, the predictions of the theories with a modified law of gravity have been, until now, vague and difficult to verify experimentally. But the greatest objection is the fact that calling Newton's laws into question also casts doubt on general relativity, the theoretical basis for the Big Bang, because the latter theory reduces to Newton's theory when velocities are much smaller than that of light. However, no modified theory of general relativity has ever been constructed. In science, it is not enough to destroy a theory: Another must be offered in its place.

A Matter–Antimatter Cosmology

Let us turn now to matter–antimatter cosmologies. The names that are associated with this concept are those of the Swedish physicists Hannes Alfvén and Oskar Klein, and that of the French physicist Roland Omnès. According to Alfvén and Klein, there should be a perfect symmetry between matter and antimatter, such that there ought to be as many examples of you and me as of anti-you and anti-me in the universe. The latter would have begun as a giant metagalaxy with equal quantities of matter and antimatter. It would have collapsed under the effect of gravity, and as soon as the density at its center became sufficiently high, annihilation of matter and antimatter would have occurred, resulting in enormous quantities of radiation.

This radiation would have succeeded in reversing the contracting motion of the remaining matter and antimatter, causing it to become an expansion.

Matter–antimatter cosmologies are inconsistent with a number of observations. First, cosmic rays—showers of charged particles released in the explosive death throes of massive stars that reach us from the furthermost regions of the Galaxy—consist almost exclusively of matter (protons and electrons). Second, the enormous quantities of x rays that should have resulted from the matter–antimatter annihilation have not been observed. Third, particle physics tells us that nature has an infinitesimally (one billionth) greater preference for matter than antimatter, and that all ideas of perfect symmetry between matter and antimatter is false. Finally, there is the supreme hurdle: There is no natural explanation for the background fossil radiation in a matter–antimatter cosmology.

This brings us to the end of our wanderings among possible alternative melodies. None has proved capable of dethroning the Big Bang. None has shown the persuasive beauty and seductive simplicity of the theory of an initial explosion. None has known to accord as well with Nature's sinuous contours. Because these rival theories have not received general acceptance by the scientific community, they are, for the most part, the work of a single person, or a single team of researchers. These theories are therefore not as well developed as one might like. They do not have the powers of conviction the Big Bang theory possesses, either because they remain hidden behind a veil of abstract mathematics and do not allow direct confrontation between theory and observation, or because their predictions are in complete contradiction to observations—the fossil background radiation being a stumbling block for all of them—or yet again because they introduce new physical laws in a perfectly arbitrary manner.

The Scientist's Wager

Although they have not dethroned the Big Bang theory, this does not mean that the rival theories—the other melodies—are devoid of usefulness, and that they play no part in the scientific process. In science, as in every other field, it is essential to beware of fashion. A theory that meets with majority approval is not necessarily the right one. Most of those who accept it have not done so after a long, critical investigation of its merits, but perhaps either from a tendency to conform, or from intellectual inertia, simply because the theory was vigorously defended by a few particularly eloquent leaders. Heretical, unorthodox theories therefore play a particularly important role: They prevent the defenders of the orthodox theory from resting on their laurels. They force them to be constantly on the qui vive, on the watch for any cracks or possible faults in the structure that has been erected. If a crack becomes too wide, and cannot be patched, the edifice crumbles, and a new one takes its place. This is how scientific revolutions proceed. Quantum mechanics arose because classical mechanics was incapable of explaining the properties of atoms. However, we must be careful not to go to the other extreme and destroy everything at the first signs of trouble. Building on top of ruins is extremely difficult. We should not rush to propose that red shifts are noncosmological, just because we do not like the explanation that quasars are powered by massive black holes, nor to change the law of gravity because the nature of the invisible mass remains a mystery. Are these difficulties not faults in our imagination rather than defects in the Big Bang structure?

Faced with all these rival theories, and confronted with a multitude of different melodies, cosmologists weigh the pros and cons and make their choice. I am betting—as readers will have gathered—like most of my colleagues, on the Big Bang theory (Fig. 63). Apart from its simplicity and elegance, it has a quality essential in any good theory: It has great powers of prediction. The most important of its predictions (the background fossil radiation and the abundance of hydrogen and helium) have been spectacularly confirmed by observations. Thanks to the contribution of ideas coming from elementary particle physics, thanks to the union of the infinitely large with the infinitesimally small, it has become even richer, and may even be able to answer the deepest and most fundamental questions that can be posed: How was the universe created? What is the origin of matter?

The heretical Swedish cosmologist Hannes Alfvén (of the matter–antimatter universe) has accused the Big Bang of being "A myth, and perhaps a marvellous myth that merits a place of honor in a zoo containing already the Hindu myth of a cyclic universe, the Chinese cosmic egg, the Biblical myth of the Creation in six days, the myth of Ptolemy's cosmology, and many others."[26] I believe that Alfvén is wrong. In the light of what has been said, there can be no doubt that the Big Bang theory is now far more than a myth. It has acquired the nobility of a science. It is a theory with robust health, which has so far resisted all attacks, and which, for the time being, gives the best explanation of the universe. If one day the Big Bang is to be supplanted by a more sophisticated cosmological theory, the latter will have to incorporate all the achievements of the Big Bang theory, just as Einsteinian physics had to incorporate all the achievements of Newtonian physics.

The Big Bang universe is the latest in a long line of successive universes that began with the magical universe, and passed through the mythical, mathematical, and geocentric universes. It will, undoubtedly, not be the last universe. It would be surprising if we were to have the final word, and were the chosen ones finally to reveal the secret melody. There will be, in the future, a whole series of universes, which will come progressively closer to the true Universe (with a capital "U" to distinguish it from the universes created by the human mind). But will we ever attain the final goal; will we ever reach the ultimate Truth, where the Universe will be revealed in all its splendor, and the melody will have yielded all its secrets? To reply to this

---

26. *La Recherche* **69,** 1976, p. 610.

a

b

c

Fig. 63. *The three cornerstones of the Big Bang theory and their discoverers.* The Big Bang theory rests primarily on three foundations: two observational and one theoretical. The first observation concerns the red shift of light from the galaxies, interpreted as caused by their velocity of recession. The American astronomer Edwin Hubble discovered in 1929 that this recession velocity increased with increasing distance. This law, known as Hubble's Law, is generally considered to be the most convincing proof of the expansion of the universe. (a) Hubble in front of the 100-inch telescope at Mount Wilson in California, which he used to make his great discovery (photograph: Mount Wilson and Palomar Observatories). The second observation is that of the 3 K fossil radiation that bathes the whole universe. This radiation was emitted when the universe was just 300,000 years old, when it was still very hot and dense. (b) The American radio astronomers Arno Penzias and Robert Wilson in front of the radio telescope that helped them discover the background radiation in 1965. This telescope, located at what was then called the Bell Telephone Research Laboratories in New Jersey, is relatively small when compared with the largest radio telescopes used today (like that shown in Fig. 64). The observation of diffuse radiation that is present everywhere does not require a large telescope, just an extremely sensitive and accurate detector. Just such a detector had been developed by Penzias and Wilson for picking up the signals from Telstar, the first communications satellite. Telephones thus provided the Big Bang with its second cornerstone. Most theories have difficulty in explaining these two observational results in a natural manner. Only general relativity, the theory proposed in 1915 by Albert Einstein [caught in photograph (c) in an impish mood], passes both tests with flying colors. Unifying space and time, it is the third pillar supporting the Big Bang.

question, we must examine in detail the various stages that lie along the road to knowledge, from the moment when we capture nature's signals, its notes of music, to the instant when understanding and knowledge burst out.

Digitized Images

Over breakfast you open your morning paper. On the science page, two headlines catch your attention: "Astronomer observes star being born" and "Physicists at CERN—the European center for nuclear research—discover a new elementary particle." Immediately, your mind evokes the image of an astronomer seated in the dark, following the birth of a star through his telescope, like a father watching the delivery of his child in a hospital room. Then comes the image of a group of physicists, all dressed in white, huddling over their instruments and marvelling at the beauty of the new elementary particle, just as proud parents huddle over the cradle of their newborn child and coo over his beauty.

Nothing could be further from the truth. The "discoveries" of the star in the process of being born, or of the new elementary particle, were made months, or even years, after the initial observation or experiment. Numerous stages are required before the information acquired at the telescope or in the particle accelerator becomes knowledge. The first stage consists of converting the information into numbers, of "digitizing" it. In the case of the star, this is done via an electronic detector attached to the telescope, which the astronomer controls from a well-lighted, separate room. The detector captures the image of the star and converts it into numbers: The brightest points in the image are assigned the highest numbers, and lower numbers are ascribed to fainter points. (The same technique is at work in a compact disc, which gives such pure and crystalline sounds, with the sole difference being that here it is sound, rather than light, that is being digitized.)

As for the elementary particle, it is so small [one ten thousand-billionth (10^{-13}) of a centimeter] and has generally such a short lifetime [of the order of one hundred-thousand-billion-billionth (10^{-23}) of a second] that no measuring instrument can detect it directly. The most powerful modern electronic microscopes are not able to see anything smaller than about one hundred-millionth (10^{-8}) of a centimeter. Even the fastest electronics fail when it comes to events that are shorter than one billionth (10^{-9}) of a second. How, then, can we "visualize" the elementary particle? It may be sent into a chamber full of liquid. The particle interacts with the atoms of the liquid and leaves behind it a string of tiny bubbles (whence the name "bubble chamber" used for these large tanks of liquid). These bubbles expand rapidly until they are large enough to be photographed. The image of the path left by the particle is then "digitized" just like the image of the star.

These "digitized" images are recorded on magnetic tape just as you might record one of Mozart's sonatas on the magnetic tape in your cassette recorder. Then the astronomers or physicists leave the telescope or accelerator with a harvest of magnetic tapes. In their universities or laboratories, they spend days, months, or even years "visualizing" the image of the star or the track of the particle on their television screens, just as you might view a film with your video recorder. They are able to vary the contrast and play at their leisure with colors to bring out one or other specific detail. Using powerful computers, they can "process" the images and eliminate "parasitic" features, to improve the quality of the images. Finally, the images have to be "interpreted." This means bringing the whole arsenal of available models and theories into play. Astronomers will use theories of star formation, while physicists will call on their knowledge of interactions between elementary particles and their properties. Finally, after much thinking, and numerous discussions with colleagues, a conclusion is reached and understanding dawns. The findings are described and summarized in a paper that is submitted to a scientific journal for publication. It is only then that the discovery is released to the media and reaches you through your morning paper.

The reader may have gathered by now that scientists do not compose their melody with the pristine notes sent out by nature. Such notes are inevitably altered by the instruments used for observation and analysis (telescope, bubble chamber, computer, etc.) and by the brains of the observers who interpret them.

The Interpretive Eye

But does this mean that we can capture Nature's virginal notes and thus access raw reality, devoid of all artificiality, if we abandon all telescopes and intermediary instruments, and rely just on our eyes? Even quite recently it was believed that this was possible, and that the eye functioned just like a television camera, simply transmitting images without added interpretation: The eye scans a scene and registers a multitude of individual luminous points; signals are sent to the brain, which faithfully reconstructs the scene, producing the sensation of seeing. In this scenario, the sharpness of the details observed in a landscape would depend on the size of the cells in the retina that registers the light: The larger the cells, the more fuzzy the image. Monkeys do not perceive reality with as much detail as humans. Does this mean that their retinas are less finely divided? Recent studies of the mechanism of vision show that this is not the case. The eye consists of a large number of conical cells, filled with a purple pigment. A cone loses its purple color, becoming colorless when struck by a particle of light. It takes a certain time to regain its purple color and, during that time, loses its ability to "see." This momentary inactivity of the colorless cones implies in principle that the human eye cannot form a detailed and continuous image. The latter ought to be very coarse, with regions of light (corresponding to the active cones) interspersed with dark regions (corresponding to the inactive cones). And yet the images of the external world that reach us are not broken by dark areas, and the exquisite details of a painting by Monet or a sculpture by Rodin enchant us. Biologists explain this apparent paradox by the way in which the cones are organized. They are interconnected in such a manner that the signals sent to the brain conspire to give us the impression of a continuous luminous image, despite the areas of shadow produced by the cones. Thus, even at the most elementary visual level, reality has already been interpreted, and the musical notes have been altered.

Visualizing the Invisible

Naked-eye astronomy is dead. Ever since Galileo turned the first astronomical telescope onto Jupiter and its satellites in 1610, the situation has never been the same. Helped by gigantic telescopes, controlled by powerful computers, astronomers' eyes have continued to improve, detecting fainter and fainter objects, farther and farther away, and with finer and finer detail. Astrophysicists have even conquered the invisible. They have invented telescopes that can capture radiation undetectable by the human eye. Radio and infrared telescopes have revealed a universe filled with unexpected beauty and extraordinary richness. Astronomers have then freed themselves from the shackles of Earth's gravity and lifted "satellite" eyes into orbit. Telescopes sensitive to ultraviolet light, x rays, or gamma rays, carried on satellites that cut through space, hundreds or thousands of kilometers above the Earth, have enabled them to contemplate a universe full of extreme violence above the opaque veil of the Earth's atmosphere. Astronomers have now tamed all forms of radiation; they have mastered the whole electromagnetic spectrum (see Appendix A).

Just like visible images, the invisible images are digitized (with the strongest radio or x-ray signals being assigned the highest numbers, and weaker signals lower numbers) and recorded on magnetic tape. They are displayed on a television screen, and images of radio or x-ray galaxies appear in all their splendor, displaying vibrant (false) colors and hues: The invisible has been visualized (Fig. 64).

Machines May Fool Us

You will have gathered that, as their instruments become more complex and sophisticated, as the invisible is conquered, astronomers are getting further and further away from the raw reality and the pristine musical notes. Reality is filtered through a nightmarish web of electronic circuits; it is manipulated, digitized, and reconstituted by powerful computers and complex mathematical treatments. Galileo had, at the beginning, a very hard time convincing his colleagues of the reality of the wonders visible through his telescope. They thought that the satellites of Jupiter and the craters on the Moon were no more than optical illusions caused by the lens of the telescope. After all, lenses bend light and amplify images. Why should they not create optical illusions?

Fig. 64. *Visualizing the invisible.* Astronomers have conquered the invisible by constructing telescopes that can capture radiation inaccessible to the human eye. (a) The Very Large Array (VLA) of radio telescopes at Socorro, New Mexico. The largest telescope of its kind in the world, the VLA consists of 27 telescopes, each 25 meters in diameter, placed on railway tracks (allowing them to be moved), which form an immense Y, with arms 21 kilometers long each (photograph: NRAO). The 27 telescopes are controlled by powerful computers, which combine the radio signals collected by each individual telescope into a single signal, "digitize" it, and manipulate it with complex mathematical operations, before displaying an image on a television screen. (b) The radio image of a galaxy (called Centaurus A) as it appears on the television screen. The invisible has been visualized (photograph: NRAO). In the majority of cases, the radio emission comes from a galaxy that is also emitting visible light. This applies to the galaxy just mentioned. (c) The radio image, superimposed on the visual image. It is obvious that the visual image differs considerably from the radio image. The visible galaxy is an elliptical galaxy, crossed by an enormous band of dust (which appears dark because the dust absorbs light), whereas the radio emission is concentrated into two long jets, originating at the center of the visible galaxy, and perpendicular to the band of dust. Astronomers believe that these jets are created by a massive black hole (with a mass of hundreds of millions of solar masses) at the center of the visible galaxy. "Radio eyes" therefore complement "visual eyes" in revealing the richness and complexity of the universe (photograph: NRAO and Cerro-Tololo Observatory).

The problem of the veracity of images is a thousand times worse in modern astronomy. I always feel an eery sense of unreality when I use the giant radio telescope in New Mexico. It is not just that the instrument is gigantic (27 telescopes, each 25 meters in diameter, spread over an area 27 kilometers across—as large as the city of Paris and its suburbs), but I am observing invisible objects that are emitting at radio wavelengths, and with an instrument that I do not control directly. I am entirely at the mercy of the computers. They point the 27 telescopes, combine the radio signals received by each telescope, digitize them, and manipulate them using complex mathematical operations, before displaying a color image on a television screen in front of me. There have been so many steps between the raw signals and the final image that it is quite legitimate to wonder what "objective truth" remains in the image. Galileo's colleagues were right to be skeptical. In science, a result—or an observation—is not accepted until it has been verified independently by other workers, using other techniques or other measuring instruments. It is highly unlikely that the same error would be repeated each time, or that the machines should fool us on every occasion.

Sifting and Transforming Reality

In principle, the technical difficulties are surmountable. All that is required is extreme attention at every stage, and great care in building the measuring instruments and programming the computers. In short, avoid having human error creep in. If we could rely upon machines alone, reality could, in theory, be rendered as objective as possible. But what cannot be avoided is the human brain. Man cannot observe nature in an objective manner. There is a constant interaction between his inner world and the outer world. Changes in the inner world influence the perception of the outer world, and, conversely, contact with the outer world alters the inner world. The inner world of the scientist is full of concepts, models, and theories acquired during his professional training. This inner world, when projected onto the outer world, prevents the scientist from seeing the "bare" objective facts, free from any interpretation. Even the most objective researcher will have "prejudices." Such prejudices (which have been given the collective name of "paradigm" by the American historian of science Thomas Kuhn) are, in fact, the very engine that drives the scientific process. Without any preconceived opinion, and with no paradigm, how is the scientist to choose among the multiplicity of information nature sends him, among the avalanche of facts he is bombarded with, which are the most important, the most likely to reveal laws and principles, and convey the most information? Sifting reality is an essential part of the scientific process, and the greatest scientists are those who are able to go straight to the essentials while neglecting what is insignificant.

Reality is therefore inevitably transformed by the inner world, and we only see what we want to see. Who has not felt an ineffable lightness of being when love strikes us, when everything that looked so sad and ugly the day before suddenly seemed endowed with an indescribable beauty? Contemplating a madeleine, Proust sees the whole of his childhood pass before him. An ordinary street, a simple tune suddenly stirs you to the depths of your soul because they are associated with important events in your life. Scientists are no exception to the rule. The British naturalist Charles Darwin, father of the theory of evolution, told a charming story: He spent a whole day on a river bank and saw nothing but stones and water. Eleven years later, he returned to the same spot, searching for traces of earlier glaciation. This time, the evidence stuck out like a sore thumb. Not even an extinct volcano could have left more visible traces of its past activity than this old glacier. Darwin discovered what he was looking for as soon as he knew how to see.

"Reality" therefore depends intimately on the scientist's conceptual baggage. It is sorted out and transformed by his "prejudices." In astrophysics, the most striking example of this close dependence of "reality" and the conceptual arsenal is, to my mind, the problem of the invisible mass. This mass does not emit any form of radiation and is therefore completely inaccessible to direct observation, even if astrophysicists have mastered all the subtleties of electromagnetic radiation. This invisible matter cannot be further removed from tangible reality. Yet the majority of astrophysicists believe that between 90 and 98% of the mass in the universe consists of this invisible material, because, if we accept Newton's law of universal gravitation, the motions of stars and gas in galaxies, and of galaxies in clusters, would be completely different if this mass did not exist. Similarly, the edifice of the Big Bang rests on general relativity. Without that theoretical support, it would crumble. Man creates the universe by projecting his inner world onto the outer world.

This interaction of the inner and outer worlds also explains why science arose in Europe and not elsewhere. The scientist does not work in isolation. The surrounding cultural milieu modifies his inner world and the direction of his research. According to the Belgian chemist Ilya Prigogine, "The emergence of science is a function of the idea that Man has of the universe. People who are persuaded that a supreme Creator is responsible for the existence of the world and its future also believe that there must be laws and a discernible future. It is then up to Man to decipher these divine laws."[27] Newton, a man steeped in the Christian religion, is the embodiment of this Western science, this urge to seek the reflection of God in the laws of nature. Science did not arise in China, even though gunpowder and the compass were invented there, because the notion of a divine Creator ruling over the universe with his laws was absent. (According to Confucius, the world arose from the opposition of two forces, the Yin and the Yang.)

27. Interview by G. Sorman in *Le Figaro* Magazine, March 5, 1988.

The Inner Fermentation

But if the scientist only sees what he is looking for, if he interprets every new fact within a conceptual framework that already exists, how is he to make progress, how can he acquire any new knowledge? Researchers are conservative by nature, and novelty emerges with difficulty. Any attempt to sow trouble and to shake the existing edifice always meets with considerable resistance. And, frequently, facts that will later be shown to be "abnormal" (i.e., not consistent with the existing conceptual framework) have to bend to ideas currently in vogue, when they are not simply ignored. The most extreme example of this conformity is that of Ptolemy, who added epicycle upon epicycle to explain the motion of the planets around an Earth fixed at the center of the universe. This conservatism is not as harmful as it might first appear: It acts as a safety valve against constant challenges, against perpetual revolutions. It ensures that science progresses smoothly in normal times and protects it against a state of permanent chaos that would otherwise paralyze it.

It is essential for science to progress, however. Revolutions occur when new facts accumulate that do not fit within the old scheme of things, but also when men of genius appear who glimpse new connections between facts that previously appeared totally unrelated. The whole universe is interconnected, and every part reflects the whole, as we shall see later. The process of science consists of sifting and breaking down reality. At any one stage, science cannot but decipher a small fraction of the whole. Every time a new interconnection is found, science makes a great leap forward. Newton discovered universal gravitation by establishing a relationship between the fall of an apple and the motion of the Moon around the Earth. General relativity came to Einstein as soon as he saw the interconnection of space and time.

These interconnections are as much acts of creation and imagination as those at play when Picasso painted his *Demoiselles d' Avignon,* when Beethoven composed his symphonies, or Proust recaptured his childhood in his *Remembrance of Things Past.* Great scientific discoveries do not arise by chance. They are the product of an intense, inner fermentation, fed by apparently disparate factors from the external world, which regurgitates and returns these factors, transformed and transfigured, unified and connected.

Foucault's Pendulum

Science is constantly in pursuit of new interconnections. Indeed, the universe that it purports to describe appears to be totally interconnected. The whole universe appears to be mysteriously present everywhere and at any time. Each part reflects the whole. Two very different experiments tell us so.

The first of these involves Foucault's pendulum. The French physicist Léon Foucault wanted to demonstrate that the Earth rotated. In 1851, in a famous experiment, he hung a pendulum (which simply consisted of a heavy weight at the end of a long cord) from the vault of the Pantheon in Paris. Once set in motion, the pendulum's behavior was remarkable: The plane in which it swung (known as the plane of oscillation) pivoted around the vertical axis as the hours passed. If it began swinging in the north–south direction, for example, it oscillated in the east–west direction after a few hours. (At the Earth's poles, the plane of oscillation would make one complete rotation every day. At Paris, because of a latitude-dependent effect, it only rotates by a fraction of 360° every 24 hours.)

Why did the plane of oscillation of Foucault's pendulum rotate? Foucault replied that the motion was only apparent, and that in reality the plane was fixed, and it was the Earth that was turning. He left it there. This reply was, however, incomplete, because a motion may only be described by reference to an absence of motion. The Earth rotates with reference to something that does not rotate. Absolute motion does not exist. If there were only one unique object in the universe, we would not be able to speak of the object's motion, because its movement could not be compared with anything else. In describing the Earth's rotation, Foucault should have specified reference objects that were not in motion. But how could such objects be found, and how could we ascertain that they were not moving? Foucault's pendulum itself comes to our rescue. Because, once set in motion, its plane of oscillation is fixed, it would suffice to orient the pendulum toward the object whose stationary nature we wish to establish. If the body is fixed, it will remain in the plane of oscillation. If, on the other hand, it is in motion, there will be a slow drift away from the plane of oscillation.

Let us try all the astronomical objects we know of in succession, beginning with the closest and gradually moving farther and farther away. Let us align the plane of oscillation with the Sun. Not much use. After a month, the Sun has already left the plane. The closest stars, within a radius of several light-years, remain longer in the plane of oscillation, but they also drift away after several years. What about the Andromeda Galaxy, 2.3 million light-years away? It does rather better, but does eventually also move out of the plane. Then comes the Local Supercluster, some 40 million light-years away. The time spent by the test objects in the plane of oscillation increases considerably as they are farther and farther away, but the slow drift persists. Finally, in desperation, we align the pendulum with a distant cluster of galaxies, billions of light-years away, and visible only in the largest telescopes. Our patience is finally rewarded: There is no more drift.

These results are perfectly comprehensible. We have seen that there is a fantastic cosmic ballet superimposed on the expansion of the universe. The Earth orbits the Sun, which orbits the center of the Milky Way. The latter is moving toward the center of the Local Group, which in turn is moving toward the center of the Local Supercluster. The latter appears to be moving in turn toward a large massive grouping of galaxies called "the Great Attractor" (Fig. 27). It is only on scales much larger than that of the Local Supercluster that the balletlike motions cease and that the expansion of the universe can be contemplated in its purest form. The ballet motions cause the test objects to drift away from the plane of oscillation of Foucault's pendulum. The drift disappears only when the objects are so far away that they no longer take part in the cosmic ballet.

The conclusion to be drawn from these experiments is an extraordinary one. Foucault's pendulum swings in blissful ignorance of its local surroundings: It does not pay attention either to the Earth, the Sun, the Local Group, or even the Local Supercluster. Its behavior is influenced by the most distant galaxies or, since most of the visible mass in the universe is in galaxies, by the universe as a whole. In other words, what happens here is determined in the vast expanse of space, and what takes place on our minuscule planet is dictated by the whole hierarchy of structures in the universe. Every part contains the whole, and depends on everything else. The universe is interconnected. The Austrian philosopher and physicist Ernst Mach (after whom supersonic velocities are named) arrived at this conclusion at the end of the nineteenth century, when he was considering the problem of rotating bodies. He believed that the mass of an object, the quantity that measures its inertia or resistance to motion, is the result of the effect of the whole universe on that object. According to him, when you are struggling to push your car after it has broken down, or to move a piece of furniture in a room, the resistance that you encounter is produced by the whole cosmic immensity. Mach never described in detail this mysterious universal influence that affects your car or your furniture, and no one has done so since.

In any case, the behavior of Foucault's pendulum forces us to conclude that there exists a sort of interaction totally different from those described by recognized physics, a mysterious interaction that does not involve any force, nor exchange of energy, but that connects the whole universe together.

The Universe's Indivisibility

Foucault's pendulum has shown how the universe is interconnected on the macroscopic scale. We shall now see how the subatomic world is also indivisible. This evidence rests on a famous experiment proposed in 1930 by Albert Einstein, Boris Podolsky, and Nathan Rosen, in an attempt to disprove the probabilistic interpretation of reality offered by quantum mechanics. Imagine, they said, that a particle disintegrates spontaneously into two photons A and B. Nothing allows us to say priori in which directions these two photons will propagate. There is one certainty, however: Because of symmetry, they will leave in opposite directions. If A goes toward the west, B will go toward the east. Let us set up our instruments and check this: Yes, A goes west and B goes east. It is just as we predicted.

But this does not take into account the indeterminacy of the subatomic world. Quantum mechanics tells us that A had no precise direction before it is captured by the measuring instrument. It was wearing its guise as a wave and could take any direction. It is only after it has interacted with the detector that A turns into a particle and "learns" that it is going west. If A did not "know" what direction to take before being captured by the measuring instrument, how could B "guess" in advance the direction of A, and arrange its trajectory so that it would be captured at the same time in the opposite direction? This does not make sense. Einstein and his colleagues concluded that quantum mechanics had therefore gone wrong, that reality could not be described in terms of probabilities and that God did not play dice. A "knew" in advance which direction it was going to take, and communicated that direction to B before they separated, so that it could go in the opposite direction. A and B have a well-determined objective reality, independent of the act of observation. But Einstein was wrong. Laboratory experiments have always confirmed quantum mechanics, and the theory does truly account for the behavior of atoms. A does not "know" in which direction it will be going, God does play dice, and particles do not have an objective reality, independent of the act of observation. How, then, are we to resolve the Einstein–Podolsky–Rosen (or EPR) paradox?

This paradox does not really exist unless we assume that reality is "localized" on each of the two particles of radiation, that the latter are distinct and separate, and that they cannot affect one another. The paradox is no more if we accept the idea that the two photons, even if they are separated by billions of light-years, are part of a single whole before they are recorded by the measuring instruments, and that they are in permanent contact with each other by some sort of mysterious interaction. B therefore "knows" instantly everything that is happening to A; they have no need to exchange signals. Reality is no longer local, but global. There is no longer a distinct "here" or "there." Everything is interconnected and "here" is identical to "there."

Foucault's pendulum and the EPR experiment have compelled us to go beyond our usual concepts of space and time. We are forced to conclude that the universe does possess an overall, indivisible order, both at the subatomic and macroscopic scales. An omnipresent and mysterious influence acts to make every part contain the whole, and the whole reflect every part. Every living being in the universe, all the matter, the book that you are holding, the furniture around you, the clothes that you wear, and all the objects that we identify as fragments of reality, all contain the totality of the universe within them. We each "hold infinity in the palm of our hand."[28]

Western science, which by necessity is reductionist—to make progress, fragments of reality need to be isolated and studied in detail—appears therefore to be converging more and more towards a global (or "holistic") view of the universe. Current attempts to unify the laws of physics are only one manifestation of this. Science has taught us that we share a common history with all the matter in the universe, that we are children of the stars, brothers

---

28. William Blake, *Auguries of Innocence,* in *Complete Writings of William Blake,* ed. Geoffrey Keynes, New York, Oxford University Press, 1989.

of the wild beasts, and cousins of the red poppies in the fields. It also tells us that we carry the universe within us, and that we are indivisibly part of it. Will this feeling of cosmic belonging prevent us from committing nuclear or ecological suicide? In any case, the scientist does not have a choice. To paraphrase a famous saying by André Malraux: "Science in the twenty-first century will be spiritual, or it will not be."

The Secret of the Melody

Let us return to our initial question: Will the Universe one day be revealed in all its glorious reality? Will we ever succeed in unlocking the secret of the true melody? In the light of what has been said, this would seem to be extremely difficult, if not impossible. Quantum mechanics teaches us that the very act of observation alters reality. This "reality" is subsequently modified and interpreted by our eyes, our measuring instruments, and our "prejudices." Even more seriously, we can never escape from our bounds in space and time, and so we can never study more than an infinitesimal part of the vast universe, which is all interconnected. Through fantastic efforts of imagination and creativity, some men of genius will discover more and more of the interconnections, and science will progress. But *all* the interconnections will never be revealed. (In this respect, we may mention the work of the Austrian mathematician Kurt Gödel, who showed in 1931 that, within a mathematical system, there will always be certain propositions that cannot be proved. Just as it is impossible to prove everything in mathematics, the human mind will never be able to comprehend the whole of the Universe.)

The Universe will remain forever inaccessible. The melody will remain secret forever. But is this any reason for being discouraged, for abandoning the search? I think not. Man will never be able to escape from the urgent need to organize the external world into a coherent and unified whole. After the universe of the Big Bang, he will continue to create others, which will come closer and closer to the true Universe, without ever attaining it, and which will illuminate and magnify his existence.

These notes are intended for readers who would like to have more quantitative information on certain subjects mentioned in the text. They are not necessary for its understanding. Generally, instead of giving complete equations, I shall simply indicate the relationship between some quantities and others. Thus I shall frequently use the symbol α, which means "varies as." Thus $a \propto b^2$ means that if b is doubled, a will be multiplied by the square of 2 (i.e., 4).

APPENDIX A

Light and the Doppler Effect

Imagine you throw a stone into a pond. Circular waves propagate from the point where the stone entered the water out to the edge of the pond. The distance between successive crests is known as the wavelength. The number of crests that arrive at the edge of the pond per second (obtained by dividing the speed of propagation of the wave by the wavelength) is called the wave's frequency. Similarly, a luminous source emits concentric waves of light, which propagate at a velocity of 300,000 kilometers per second; however, the waves are not circular, but spherical. Light is also characterized by a wavelength l and a frequency f, which varies inversely in proportion to l ($f = c/l$, where c is the velocity of light). The wavelength and frequency of the light determine its energy E: The higher the frequency or the smaller the wavelength, the more energetic the light ($E \propto f$ or $E \propto 1/l$).

The range of electromagnetic waves is obtained by varying the wavelength or the frequency of the light. Gamma rays have the shortest wavelength (about one ten-billionth of a centimeter), the highest frequency [the number of waves arriving in one second is about 300 billion billion (3×10^{20})] and the highest energy. In decreasing order of energy or frequency—or, equivalently in increasing order of wavelength—we then have x rays, ultraviolet light, visible light, infrared radiation, and radio waves. Visible light occupies a tiny region between wavelengths of about 3 to 7 hundred-thousandths of a centimeter, and the frequency with which it enters our eyes is about one million billion (10^{15}) per second. As for radio waves, their wavelength may attain 1 kilometer. Such radio waves arrive at the relatively low frequency of just 300,000 per second (Fig. A1). The Earth's atmosphere is not transparent to every form of radiation. Only visible light, a small range of infrared radiation, and radio waves are able to pass through it and be captured by telescopes on Earth. The other types of radiation cannot be detected except above the Earth's atmosphere, with telescopes carried on balloons or satellites.

The wavelength of the radiation (or its frequency or its energy) alters when the luminous source is in motion with respect to the observer (or when the observer moves in relation to the luminous source—only the relative motion counts). If the luminous source is moving away, the radiation takes longer to reach the observer: The observed wavelength of the radiation, $l_{(observed)}$—which is the distance between two successive crests as they arrive at the observer—is increased relative to the wavelength of the emitted radiation, $l_{(emitted)}$, and is given by:

$$\frac{l_{(observed)}}{l_{(emitted)}} = 1 + \frac{v}{c}$$

where v is the relative velocity of the luminous source and the observer and c is the velocity of light. The frequency and energy of the observed radiation are less than those of the emitted radiation. If, instead of receding, the source is approaching, the observed wavelength decreases relative to the wavelength of the emitted radiation. In the equation given, the $+$ sign would be replaced by a $-$ sign. The frequency and energy of the observed radiation would increase instead of decreasing.

This phenomenon is known as the Doppler effect, after the name of the physicist who first studied it. It is by using a radar set and the equation that a policeman is able to determine the speed of your car on the freeway. He knows the frequency (or wavelength, $l_{(emitted)}$) of the radio waves emitted by the radar. He can measure the frequency (or wavelength, $l_{(observed)}$) of the waves reflected by the back of your vehicle. He only needs to use the formula to obtain your speed! The Doppler effect also applies to sound. Doubt-

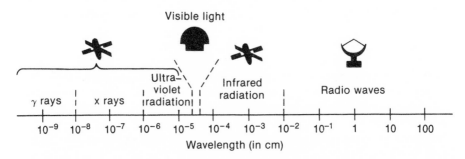

Fig. A1. *The electromagnetic spectrum and the instruments used to detect the different types of radiation.*

less everyone has noticed that the sound of a vehicle's horn has a higher pitch (its frequency is increased) when it is approaching, and a lower pitch (its frequency is decreased) when it is going away.

In the case of visible light, red light has a longer wavelength (as well as lower frequency and energy) than blue light. This is why the term "red shift" is used to describe the phenomenon of an increase in the wavelength of the emitted radiation caused by the relative recession of the luminous source from the observer—such as the recession of the galaxies caused by the expansion of the universe—even when the radiation is not in the visible region. This red shift z is defined as the fractional change in the wavelength relative to that when it was emitted:

$$z = \frac{l_{(observed)} - l_{(emitted)}}{l_{(emitted)}} = \frac{v}{c}$$

Conversely, the decrease in wavelength of the emitted radiation caused by the relative approach of the luminous source is known as a "blue shift." The equation is the same, except that the numerator becomes $l_{(emitted)} - l_{(observed)}$.

The equations just given are correct only when the velocity v is small relative to the velocity of light c. When v approaches c, the equations becomes those given by special relativity:

$$\frac{l_{(observed)}}{l_{(emitted)}} = \frac{1 + v/c}{(1 - v^2/c^2)^{1/2}}$$

which gives:

$$z = \frac{l_{(observed)} - l_{(emitted)}}{l_{(emitted)}} = \frac{1 + v/c}{(1 - v^2/c^2)^{1/2}} - 1$$

From this equation it is obvious that when v approaches c, the denominator approaches zero, and the red shift z may become very large. The objects with the greatest red shifts in the Universe are the quasars. The greatest red shift currently known is that of a quasar with $z = 4.8$: Its light is shifted toward the red by 480%. For comparison, the value of z for a nearby galaxy in the Virgo Cluster, at 44 million light-years, is only 0.003. Its light is red shifted by only 0.3%. Hubble's law, which links the red shift to distance, implies that the quasars, with their extreme red shifts, must be very distant, at the very edge of the universe. Since looking far away is also looking back in time, observing a quasar with a z of 4.8 is looking back to the epoch when the universe was just 2 billion years old, less than one-seventh of its current age. Objects with high z are thus like archaeological relics that help us decipher the universe's past.

Until quite recently, the known red shifts for galaxies were far less than those of quasars: They did not exceed 1. However, very distant galaxies have been discovered with a z of around 4, comparable with those of quasars. These observations imply that galaxies were born very early in the history of the universe, at 2 to 3 billion years after the Big Bang. Any model of the universe that tries to describe the formation of galaxies has to take this into account. The model of a universe filled with cold dark matter (Chapter 5), which is the one favored by most cosmologists at present, does not appear to meet this constraint, because it predicts that galaxies will form relatively late, about 5 billion years after the Big Bang.

APPENDIX B

The Elasticity of the
Space–Time Couple

With Einstein's publication of special relativity in 1905, time and space lost their universal and absolute character, which they had possessed since the time of Newton. Subsequently they became specific to the observer and dependent on his motion. An observer at rest would see the time t_0 of an observer in motion dilate by a factor γ, defined by:

$$\gamma = \frac{1}{(1 - v^2/c^2)^{1/2}}$$

where v is the velocity of the observer in motion and c is the velocity of light. In other words:

$$t = \gamma t_0$$

Thus if Jules is inside a spaceship traveling at 87% of the speed of light, Jim, his twin brother, who remains on Earth, will see Jules's time pass at half the rate of his own: The hands of a clock on board the spaceship will move at half the speed, and Jules will take twice as long to brush his teeth. Biological processes will also slow down: Jules will age only by 6 months for every year that passes on Earth. The time dilation is the greater, the closer to the speed of light the spaceship moves. If Jules were moving at 99.99% of the speed of light, 10 years on board would be equivalent to 707 years on Earth. This time dilation has been verified numerous times, in particular in particle accelerators: The elementary particles live far longer when they are accelerated to velocities close to the speed of light.

But you would object quite rightly that all motion is relative. If Jim sees Jules moving at a velocity v in his spaceship, Jules could as well consider himself at rest, and that it is Jim who is receding at a velocity v. Jules will then see Jim's time on Earth dilate in exactly the same way as Jim sees Jules's time dilate. If the spaceship is moving at 87% of the speed of light, 1 year for Jules will be equivalent to 6 months on Earth. Both points of view are equally valid, and there are no problems, provided Jules and Jim do not meet.

Suppose, however, that Jules, tired of interstellar wandering, decides to return home to see his twin Jim. Now comes the $64,000 question: Who will have aged more, Jules or Jim? The answer is perfectly clear: Jim, who stayed on Earth, has aged more. If Jules traveled at 87% of the speed of light, and the length of his voyage, as measured by clocks on board the spaceship, was 10 years, Jules will have aged by 10 years, whereas Jim will be 20 years older. Why? Because the situation between Jules and Jim is no longer symmetrical, as it was when Jules was moving at constant velocity in his spaceship. To return, Jules had to decelerate, bringing his spaceship to rest and then accelerate in the opposite direction. It is Jules pinned to his seat, and not Jim, who feels the effects of acceleration and deceleration. Jules knows perfectly well that he is the one in motion: It would be absurd to suppose that at the precise moment when he maneuvers his spaceship to turn around, the Earth and the whole universe came to a halt and then accelerated in the opposite direction. Motion at a constant velocity is symmetrical for Jules and Jim, but not accelerated or decelerated motion. When Jules returns to Earth, he is younger than Jim. He is not simultaneously older and younger. Calculations that take into account the effects of acceleration and deceleration reach the same conclusion, whether made from the point of view of Jules or that of Jim.

Space and time are intimately linked in relativity theory. The latter states that a dilation of time must be accompanied by a contraction of space. Jim would not only see Jules's time dilate, he would also observe the latter's space contract. For example, Jules's spaceship would be considerably shortened. If l_0 is the length of the spaceship, Jim would measure a length l, which is given by:

$$l = l_0/\gamma = l_0 (1 - v^2/c^2)^{1/2}$$

It is the same factor γ that applies to time dilation and space contraction. If Jules were to move at 87% of the speed of light, Jim would see the spacecraft shortened by one-half. These simultaneous transformations may be considered as space being transformed into time. The contraction of space is converted into the dilation of time. The conversion rate is 300,000 kilometers of space for every second of time. Velocity is not the only factor that can deform time and space. Gravity is able to do so as well. The gravitational field of a black hole can fold space back onto itself and stop time. This, however, is the subject of general relativity, which we do not intend to discuss here.

At the same time as its time dilates and its length decreases, the mass m_0 of an object in motion increases according to:

$$m = m_0\gamma = \frac{m_0}{(1 - v^2/c^2)^{1/2}}$$

Once again, the same factor γ controls the change in mass. Jim would see Jules's spaceship double in mass if it were moving at 87% of the speed of light. (Jules would not detect any change in the mass.) The mass increases more and more as the speed increases, becoming infinite when the velocity reaches that of light. The energy required to accelerate the greater mass also increases, until it eventually becomes infinite. This illustrates the impossibility of accelerating any object to the velocity of light: Not even all the energy in the universe would suffice.

This increase in mass with velocity means that any collision with an object that is moving very fast is extremely dangerous, however small the rest mass of the latter might be. A grain of sand accelerated to a speed close to that of light would have the same mass as a planet. It would break our planet into thousands of pieces, were it to collide with the Earth.

Black Holes

How does one make a black hole? All we need to do is to compress a mass—any mass: a coin, chair, person, mountain, star, or galaxy—into a sufficiently small volume for its gravity to become so great that not even light, which moves at the highest possible velocity (300,000 kilometers per second) can escape. The radius below which the body must be compressed for it to become a black hole is known to scientists as the Schwarzschild radius (after the physicist who discovered it). Whoever crosses this radius of no return will not be able to get back out. This radius R varies in proportion to the mass of the compact object and is given by:

$$R \propto M$$

(The exact equation is $R = 2GM/c^2$, where G is the gravitational constant and c is the velocity of light.)

The Sun, which has a mass of 2×10^{33} grams, has a Schwarzschild radius of 3 kilometers; that of the Eiffel Tower, which has a mass of 6900 metric tons, is one billion-billionth (10^{-18}) centimeter. A 70-kilogram person would become a black hole if he (or she) were compressed to less than 10^{-23} centimeter. Given the infinitesimal black hole radii of everyday objects, it is easy to understand why they do not all collapse into black holes. Their masses, and thus their gravities, are not large enough to overcome the resistance from the electromagnetic forces—which link the atoms together, and are responsible for the solidity of the things in life—and compress such objects to such extremely minute sizes. Only stars and galaxies have enough mass and gravity to collapse into black holes.

The density of a black hole is:

$$d = \frac{\text{Mass}}{\text{Volume}} = \frac{3M}{4\pi R^3} \; \alpha \; \frac{M}{M^3} \; \alpha \; \frac{1}{M^2} \; \alpha \; \frac{1}{R^2}$$

Because the density decreases inversely as the square of the mass, a very massive black hole is not necessarily very dense. The density of a black hole of one solar mass is quite high (1.8×10^{16} grams per cubic centimeter), but a black hole with a mass of 4.2 billion solar masses would have a density equal to that of the air we breathe (0.001 gram per cubic centimeter).

The British astrophysicist Stephen Hawking has shown that black holes are not, after all, perfectly black. Quantum mechanics allows them to evaporate slowly, turning into radiation. The evaporation rate is controlled by the temperature T of the black hole. This temperature depends inversely on the mass M of the black hole:

$$T \; \alpha \; \frac{1}{M}$$

The time that a black hole takes to evaporate completely, $t_{(\text{evaporation})}$, varies as the cube of the mass:

$$t_{(\text{evaporation})} \; \alpha \; M^3$$

The temperature of a black hole with the mass of the Sun is one ten-millionth (10^{-7}) of a Kelvin, and its evaporation time is 10^{65} years. The temperature of a galactic black hole of one billion (10^9) solar masses is 10^{-16} K, and its evaporation time is 10^{92} years, while the temperature of a hypergalactic black hole of one thousand billion (10^{12}) solar masses is 10^{-19} K, and its evaporation time is 10^{100} years.

Naturally, just as a cup of coffee can evaporate only if the surrounding air is cooler (heat travels only from hot to cold), black holes are able to evaporate only if their temperature is higher than that of the background fossil radiation bathing the whole universe. This condition is not met in the present-day universe, which has a temperature of 3 K after some 15 billion years of evolution. Knowing that the temperature $T_{(\text{universe})}$ of the universe varies as:

$$T_{(\text{universe})} \; \alpha \; \frac{1}{R_{(\text{universe})}} \; \alpha \; \frac{1}{t^{2/3}}$$

(see Appendix E), where t is the age of the universe, we can deduce that a black hole with the mass of the Sun will not begin to evaporate until $t = 10^{20}$ years, when the temperature of the universe will have fallen to 10^{-7} K. A galactic black hole will have to wait until $t = 10^{34}$ years, and a hypergalactic black hole until $t = 10^{39}$ years, before they can begin to evaporate.

In principle, there is no upper limit to the mass of black hole. This mass cannot, however, be less than 20 micrograms (2×10^{-5} g). A black hole with this minimum mass (known as the Planck mass) would have a temperature of 10^{32} K (the Planck temperature) and would evaporate in 10^{-43} second (the Planck time). These numbers describe the universe at its very beginning, at the Planck barrier, at the instant in time when known physics breaks down. Hawking has suggested that at that time, the universe was so dense that it collapsed into innumerable mini black holes (or primordial black holes) having the Planck mass, which would have evaporated and been reborn every 10^{-43} second in an infernal cycle of life and death. Such primordial black holes have never been detected.

The Uncertainty Principle in Quantum Mechanics

Chance and indeterminacy prevail in the submicroscopic world of atoms. I cannot describe the motion of an electron within an atom as I would describe the path of a ball thrown into the air, or the track of a ship moving across the sea. The motion of an electron escapes me, because I cannot, at any particular instant, accurately measure its position and its velocity simultaneously, as I can with the ball or the ship. This inaccuracy, or uncertainty, cannot be eliminated, no matter how sophisticated my measuring instruments may be. It is inherent to the very act of measurement. To measure the position of an electron, for example, I must illuminate it somehow, and in doing so, I bombard it with photons that alter its velocity. The German physicist Werner Heisenberg expressed this uncertainty in the atomic world by an inequality. If Δx is the uncertainty in the measurement of the position x, and Δv is the uncertainty in the measurement of the velocity v, their product must always be greater than a very small number equal to $h/2\pi$ (where h is Planck's constant, equal to 6.63×10^{-27} erg seconds):

$$\Delta x \, \Delta v \geqslant \frac{h}{2\pi}$$

If, therefore, I reduce the positional uncertainty (Δx becoming close to zero), the velocity uncertainty becomes enormous by way of compensation, ensuring that the product $\Delta x \, \Delta v$ always remains larger than $h/2\pi$. I cannot reduce Δx and Δv simultaneously.

A similar inequality links the uncertainty in energy ΔE of an elementary particle to the uncertainty Δt in its lifetime:

$$\Delta E \, \Delta t \geqslant \frac{h}{2\pi}$$

The shorter a particle's lifetime, the greater the uncertainty in its energy. It is this rule that allows quantum mechanics to violate the principle of the conservation of energy that applies in the macroscopic world, and create "phantom" or virtual particles. Because a virtual particle has an extremely short lifetime, it is able to borrow the energy ΔE from Nature's reserves and pop into existence for a fleeting instant. Most of the time, Nature reclaims its energy very swiftly, and the virtual particle disappears after a time Δt. Occasionally, the virtual particle is able to find a generous benefactor that helps it repay its energy debt, and thus gains access to the real world. Gravity has energy to spare, and this is why virtual particles may materialize in an intense gravitational field, such as that around a black hole, just outside the radius of no return. Similarly it is also the energy indeterminacy that allowed the whole universe to emerge from a subatomic void.

The Cosmological Parameters and the Evolution of the Universe

The evolution of the universe (its past, present, and future) is governed by the struggle between two opposing forces: the force resulting from the initial explosion that is responsible for the expansion of the universe, and the gravitational force exerted by all the matter (visible and invisible) contained within it. In 1922, the Russian mathematician Alexander Friedmann used the theory of general relativity, published by Albert Einstein in 1915, to obtain equations describing this epic struggle. These equations, in their simplest version, are based on three parameters, known as "cosmological parameters." The first characterizes the expansion of the universe, while the other two, which are intimately linked, describe the way it slows down.

The first cosmological parameter, H, is also called the "Hubble constant" or "Hubble parameter," because it enters in the law discovered by Edwin Hubble in 1929 that describes the expansion of the universe:

$$v = Hr$$

where v is the recession velocity (in kilometers per second) of any point in the universe relative to any other, and r is the distance (in million light-years) between them. Observations show that the Hubble constant currently has a value of between 15 and 30 kilometers per second per million light years (see Chapter 4). The inverse of the Hubble constant gives the age of the universe. This age is approximate; it would be exact if the universe did not decelerate:

$$\text{Age} \propto \frac{1}{H} \begin{cases} = 10 \text{ billion years for } H = 30 \text{ km s}^{-1} (10^6 \text{ light-years})^{-1} \\ = 20 \text{ billion years for } H = 15 \text{ km s}^{-1} (10^6 \text{ light-years})^{-1} \end{cases}$$

The age of the universe remains uncertain by a factor of 2. It is obvious that the Hubble parameter, being inversely proportional to the age of the universe, varies with time. It was very large in the early universe and has decreased as the universe aged.

The second cosmological parameter q describes the *deceleration* of the universe as a result of the braking exerted by the gravitational force of the matter contained within it. The deceleration is, by definition, the slowing down per unit time (for example, in 1 second):

$$q \propto \frac{\Delta v}{\Delta t}$$

whereas the velocity is the change in distance per unit time:

$$v = \frac{\Delta r}{\Delta t} = \dot{r}$$

where the symbol · indicates the derivative with respect to time. We thus have:

$$q \propto \frac{\Delta \dot{r}}{\Delta t} = \ddot{r}$$

The exact equation given by Friedmann is:

$$q = -\frac{\ddot{r}\, r}{\dot{r}^2}$$

The minus sign indicates that the universe is decelerating (a plus sign would correspond to an accelerating universe). Depending on whether q is less than, equal to, or greater than ½, the universe is open (the expansion is eternal), flat (the expansion will cease only after an infinite time), or closed (the universe will collapse back onto itself).

The third cosmological parameter d is the *mean density of matter* (visible and invisible) in the universe. It is linked to the deceleration parameter q, because it is the force of gravity exerted by the matter that slows down the expansion of the universe. Friedmann's equations show that gravity will be strong enough to stop the expansion if d is greater than a certain critical density, given by:

$$d_{\text{(critical)}} = \frac{3H^2}{8\pi G}$$

where G is the number that controls the strength of the force of gravity. For $H = 15$ kilometers per second per million light-years, $d_{\text{(critical)}} = 3$ atoms of hydrogen per cubic meter, or 4.5×10^{-30} gram per cubic centimeter. If H is 30 kilometers per second per million light-years, $d_{\text{(critical)}}$ becomes 12 atoms of hydrogen per cubic meter or 1.8×10^{-29} gram per cubic centimeter. When $d > d_{\text{(critical)}}$, the universe is closed. If $d = d_{\text{(critical)}}$, the expansion of the universe will come to an end only after an infinite length of time: We have a flat universe. If $d < d_{(critical)}$, the universe will continue to expand forever: It is an open universe. As yet we have been able to account for a density of matter (both visible and invisible) equal only to one-fifth of the critical density. Until further notice we are living in an open universe, with an eternal expansion (see Chapter 6).

The deceleration parameter q is linked to the density parameter d by the relationship:

$$q = \frac{d}{2d_{(critical)}} = \frac{4\pi Gd}{3H^2}$$

If we can measure all three cosmological parameters H, q, and d accurately—some methods of measurement are described in Chapters 4 and 6—we should be able to verify if this relationship does hold, and thus test the validity of general relativity on the scale of the whole universe.

Let us review now the major stages in the evolution of the universe:

1. There was first the *inflationary era* ($10^{-35} < t < 10^{-32}$ second), when the universe dilated exponentially as a function of time:

$$r \propto e^{Ht}$$

where r is the distance between any two points in the universe and H is the Hubble parameter during the inflationary phase. This phase, which arose because of crystallization phenomena in the universe, is not described by Friedmann's equations. These do, however, describe the subsequent phases accurately.

2. The *radiation era* (1 second $< t < 300,000$ years), when the density of the radiation $d_{(radiation)}$ is higher than the density of matter, $d_{(matter)}$, and controls the evolution of the universe. During this phase the universe expands in proportion to the square root of time:

$$r \propto t^{1/2}$$

The temperature T decreases inversely with distance:

$$T \propto \frac{1}{r} \propto \frac{1}{t^{1/2}}$$

The density of matter decreases inversely with the volume:

$$d_{(matter)} \propto \frac{1}{r^3} \propto \frac{1}{t^{3/2}}$$

However, the density of the radiation decreases even more swiftly:

$$d_{(radiation)} \propto \frac{1}{r^4} \propto \frac{1}{t^2}$$

The difference comes about because the mass energy of the matter within the universe is conserved while the latter expands, whereas the radiation energy decreases as $1/r$. As the universe ages, the difference between the radiation and matter densities decreases. The two densities are equal around the year 300,000.

3. The *matter era* ($t > 300,000$ years) began at this epoch and still prevails today. During this phase, matter dominates the universe and controls its expansion. The latter is now described by:

$$r \propto t^{2/3}$$

The temperature and densities of matter and radiation continue to decrease with time:

$$T \propto \frac{1}{r} \propto \frac{1}{t^{2/3}}$$

$$d_{(matter)} \propto \frac{1}{r^3} \propto \frac{1}{t^2}$$

$$d_{(radiation)} \propto \frac{1}{r^4} \propto \frac{1}{t^{8/3}}$$

The temperature of the universe today is 3 K. The density of matter is about 10^{-30} gram per cubic centimeter, while the density of radiation is about 7×10^{-34} gram per cubic centimeter, that is, about 1430 times smaller.

Glossary

absolute zero: zero on the Kelvin temperature scale, equal to $-273°C$, and corresponding to the total absence of any particle motion; it is the lowest attainable temperature.

active galaxy: a galaxy that emits most of its light and energy from a very compact central region, known as the "nucleus," a few light-hours to a few light-months across, or billions of times smaller than the entire galaxy. The energy is thought to be powered by the activity of a black hole tens of millions times more massive than the Sun, which lies at the center of the galaxy and devours stars passing nearby.

adiabatic fluctuation: density fluctuation where matter and radiation vary together, such that there are always 1 billion photons for every baryon (see Fig. 36).

Andromeda Galaxy: twin galaxy to the Milky Way, also known as Messier 31, 2.3 million light-years away. Together, the two galaxies comprise the bulk of the mass of the Local Group.

anthropic principle: the idea that the universe has been very precisely adjusted to permit the emergence of life and consciousness.

antiparticle: an elementary particle that is a constituent of antimatter, with the same properties as an ordinary particle, but the opposite electrical charge. The antiparticle of the electron is the positron; that of the proton is the antiproton. Some particles, such as photons, are their own antiparticles. When particles and antiparticles come into contact, they become radiation by annihilating each other. We live in a matter-dominated universe. Antimatter is extremely rare, and is observed only in cosmic-ray showers and in high-energy particle accelerators.

asteroid: a rocky irregularly shaped body, whose size may reach 1000 kilometers. Its gravity is not sufficiently high to sculpt it into a spherical shape, as with planets.

baryon: an elementary particle that is subject to the strong nuclear interaction. The proton and neutron are baryons.

Big Bang: the cosmological theory that holds that all the matter and energy in the universe was concentrated in an immensely hot and dense point, which exploded 10 to 20 billion years ago. The universe has been expanding and cooling ever since.

binary: a pair of stars or galaxies that are gravitationally bound together and that orbit one another.

black dwarf: invisible remnant of a white dwarf that has radiated all the energy of motion of its electrons into space.

black hole: region of space produced when, for example, a star of more than 5 solar masses collapses, creating a gravitational field so powerful that matter and radiation cannot escape. *See* radius of no return.

brown dwarf: a failed star, with a mass less than one one-hundredth of a solar mass. Like a planet, a brown dwarf is not massive, hot, or dense enough for energy-producing nuclear reactions to occur in its core.

Cepheids: variable stars whose light varies periodically in a very specific fashion; the interval of time between two maxima (or minima), known as the period, depends on the star's intrinsic brightness. The brighter the Cepheid, the longer its period (which ranges from several days to 1 month). Astronomers exploit this property to use Cepheids as distance indicators. All that is required is to identify one of these stars in a nearby galaxy and observe its variations in brightness to obtain its period and thus its intrinsic brightness. The latter, when combined with the observed apparent brightness, gives the distance. Edwin Hubble used Cepheids to demonstrate the existence of galaxies lying well beyond the Milky Way. Unfortunately, Cepheids are not bright enough to be visible from the ground at distances greater than 13 million light-years (see Fig. 18).

closed universe: a universe in which the density of matter is greater than the critical density and that will thus collapse onto itself in the future.

cluster of galaxies: a dense grouping of several thousand galaxies bound by gravity, with an average diameter of some 60 million light-years, and an average mass of a few million billion solar masses (see Fig. 30).

"cold" matter: matter consisting of elementary particles of relatively high mass that are moving relatively slowly (the term "cold" indicates a low temperature and thus a small energy of motion). The axion and photino are examples of "cold" matter.

comet: a body of ice and dust, with a nucleus about 10 kilometers in diameter. It is visible only when it travels close enough to the Sun to reflect light. As solar heating evaporates the ice on the nucleus surface, a comet develops a tail hundreds of millions of kilometers long that always points away from the Sun.

complementarity, principle of: enunciated by the Danish physicist Niels Bohr, the principle says that matter and radiation may exhibit the properties of both waves and particles, these two descriptions of nature being complementary to one another. It forms one of the cornerstones of quantum mechanics.

convergent point: the point on the sky towards which the stars in a galactic cluster appear to converge. The point of convergence is used to determine the distance of the cluster (see Fig. 16).

cosmic background radiation: the microwave radiation that bathes the entire universe, and that dates from the epoch when the universe was just 300,000 years old.

Its temperature of 3 K does not vary by more than 0.001% at any point over the sky. It is homogeneous and isotropic, in accordance with the cosmological principle. With the expansion of the universe, the background fossil radiation constitutes one of the two pillars of the Big Bang theory.

cosmic rays: particles (mostly protons and electrons) that have been accelerated to very high energies by supernovae and the magnetic fields in the interstellar medium.

cosmic string: a dislocation in the fabric of space, which is very thin (10^{-28} cm) and exceptionally long. According to grand unification theories, cosmic strings may be produced during the first fractions of a second of the universe's existence. Cosmic strings have never been directly observed, although some astronomers believe that they can be the cause of the observed filamentary structures traced by galaxies (see Fig. 39). Cosmic strings have fallen somewhat into disrepute since the observations of the background fossil radiation by the satellite COBE. They would have caused larger temperature fluctuations than those observed by COBE.

cosminos: elementary particles having mass, but which are not governed by the strong nuclear force. Their existence is predicted by grand unification theories, but none has yet been observed, except for the neutrino. It is not yet known whether the latter has a mass.

cosmological horizon: delimits the observable region of the universe. The distance to the horizon is equal to the distance traveled by light since the Big Bang. The observable universe increases as it ages, and light has more time to reach us. On average, ten new galaxies enter the observable universe every year.

cosmological parameter: a number in a cosmological theory that determines the fate of the universe, its past, present, and future.

cosmological principle: a hypothesis that the universe is similar to itself everywhere (it is homogeneous) and in every direction (it is isotropic). This principle has been confirmed in spectacular fashion by the observation of the cosmic background radiation.

cosmology: the study of the origin, evolution, and structure of the universe.

critical density: the density of matter corresponding to a flat universe, devoid of curvature, and equal to three hydrogen atoms per cubic meter. A universe with the critical density would cease to expand only after an infinite time. A universe with a density greater than the critical density would have positive curvature and would collapse into itself in the future (it is said to be closed). A universe with a density less than the critical density would have negative curvature and would expand forever (it is said to be open). Observations appear to favor an open universe (see Appendix E).

cyclic universe: a universe that undergoes a series of expansions and contractions.

dark matter: matter of unknown nature that does not emit any radiation. The existence of this invisible mass is deduced from the study of the motions of stars and gas in galaxies and of galaxies in clusters of galaxies and from the relative abundance of the chemical elements created in the Big Bang. The invisible mass may make up between 90 and 98% of the mass of the universe.

deceleration parameter: a number that measures the slowing down in the expansion rate of the universe (see Appendix E).

density fluctuations: spatial variations in the density of the universe, detected as minute temperature differences in the cosmic background radiation (the satellite

COBE has measured them to be about 30 millionths of a degree K), and thought to be the seeds of protogalaxies (see Fig. 36).

deuterium: chemical element whose nucleus consists of a proton and a neutron, created mainly in the first 3 minutes of the universe's history. *See* nucleosynthesis.

Doppler effect: variation in the energy and color of light (or sound) caused by the motion of a source of light (or sound) relative to an observer. If the source is receding, the energy decreases and the light is shifted toward the red (the sound has a lower pitch). If the source is approaching, the energy increases, and the light is shifted toward the blue (the sound has a higher pitch). *See* red shift.

dwarf galaxy: a galaxy with a small size and mass. The average diameter is about 15,000 light-years, that is, about one-sixth of that of a normal galaxy. Masses range from 100 million to 1 billion solar masses, about 1000 to 10,000 times less than the mass of an ordinary galaxy. Dwarf galaxies may be elliptical or irregular, but dwarf spiral galaxies have not been observed (see Fig. 41). *See* Magellanic Clouds.

electromagnetic force: a force that acts only on charged particles, causing oppositely charged particles to attract and particles of like charges to repel. The electromagnetic force binds electrons to nuclei to form atoms and binds atoms together to form solid matter.

electron: the lightest of the subatomic particles with electrical charge. The electron has a mass of 9×10^{-28} gram and is negatively charged.

electronuclear force: the force resulting from the unification of the electromagnetic force with the two (strong and weak) nuclear forces.

electroweak force: the force resulting from the unification of the electromagnetic force and the weak nuclear force.

elementary particle: basic building block of matter and radiation. What is considered as "elementary" changes with time as knowledge increases. For example, the proton and neutron were once considered elementary, but are now thought to consist of three quarks. Electrons, neutrinos, photons, and quarks are considered elementary particles.

elliptical galaxy: galaxy observed as an oval-shaped system generally composed of old stars and containing little or no gas and dust (see Fig. 40).

entropy: a quantity measuring the degree of disorder in a system. The total entropy, and hence disorder of an isolated system, must always increase (second law of thermodynamics).

epicycle: circular planetary orbit whose center moves itself on a circle centered on the Earth. Epicycles were invented to account for planetary motions in a geocentric universe (see Fig. 5).

exclusion principle: discovered by the German physicist Wolfgang Pauli, the principle states that no two particles of the same type (such as electrons or neutrons) can be in the same state; that is, they cannot have the same position and velocity. This principle explains why white dwarfs and neutron stars do not collapse under the influence of their own gravity. The electrons in the white dwarf and neutrons in the neutron star resist gravitational compression because they cannot be packed too tightly, otherwise they would have the same position and velocity.

flatness: the geometry of a universe with no overall curvature, with a density of matter equal to the critical density (three hydrogen atoms per cubic meter). According to observations, the universe has a density very close to the critical density (about one-fifth).

galactic cannibalism: process by which the motion of a galaxy is braked by the gravitational attraction of a more massive galaxy, sending it spiraling in toward the larger galaxy. The less massive galaxy is consumed and loses its identity as its stars mingle with those of the "cannibal" galaxy.

galactic cluster: irregularly shaped grouping of a few tens of young stars. The stars are not bound together by gravity, but are in the same region of the sky because they were born from the collapse and fragmentation of a single interstellar cloud. A galactic cluster disperses after a few hundred million years (see Fig. 15).

galactic disk: flattened aggregation of stars, gas, and dust in a spiral galaxy. The average disk is some 90,000 light-years in diameter and 300 light-years thick. In the Milky Way, the stars complete one turn around the galactic center every 250 million years, at a velocity of 230 kilometers per second (see Fig. 20).

galactic halo: spherical region around a spiral galaxy populated by old stars and globular clusters. Observations suggest that it is surrounded by an invisible halo some 10 times larger and more massive (see Fig. 53).

galaxy: system of stars (10 million in a dwarf galaxy, 100 billion in an average galaxy like the Milky Way, 10 trillion in a giant galaxy) held together by gravity. Galaxies are the fundamental building blocks of the cosmic macrostructure.

gamma rays: most energetic form of electromagnetic radiation.

general relativity: gravitational theory proposed by Albert Einstein in 1915, which is more accurate than that of Newton. The two theories differ mainly in situations where gravitational fields are very intense, such as around a pulsar or black hole. General relativity constitutes the theoretical support of the Big Bang theory.

geocentric universe: a model of the universe with the Earth at the center, and with the Sun, planets, and stars orbiting it.

globular cluster: spherical aggregation of about 100,000 old stars held together by gravity (see Fig. 19).

Grand Unified Theory (GUT): theory that attempts to unify the electromagnetic force and the strong and weak nuclear forces into a single force. Such a unification is possible only under conditions of extremely high energies and temperatures, such as those prevailing in the early universe. Attempts have also been made to include the gravitational force, which, for the present, remain unsuccessful. *See* Theory of Everything.

gravitational force: force responsible for attraction between all matter. The weakest of the four forces, it also possesses the longest range.

gravitational lens: a body (a star, galaxy, quasar, or cluster of galaxies) located between the Earth and a more distant body, and aligned with them, and whose gravitational field deflects the light from the more distant body, to create a cosmic mirage. The image of the distant object is deformed (magnified or reduced), changed, or multiplied. Additional images appear alongside the true image. These gravitational "mirages" result from the interaction of the light from the distant body, not only with the gravitational field of the lens, but also with any gravitational field that the light has to cross on its way to Earth. The study of these cosmic images therefore provides information on the total amount of matter (visible and invisible) responsible for the gravity, and on its spatial distribution both in the lens and in intergalactic space.

Great Attractor: large grouping of galaxies with a total mass of 100 million billion solar masses, gravitationally attracting the Local Supercluster, which is moving toward it (see Fig. 27).

group of galaxies: collection of about twenty galaxies held together by gravity, some 6 million light-years across, and averaging between 1 and 10 trillion solar masses (see Fig. 29).

hadron: particle subject to the strong nuclear force. Hadrons result from the combination of several quarks. Baryons are examples of hadrons.

hadron era: period in the history of the universe between one one-millionth and one ten-thousandth of a second after the Big Bang, during which the contents of the universe were dominated by hadrons (protons, neutrons, and their antiparticles) in equilibrium with photons. The hadron era ended when the photons, weakened by the expansion of the universe, were no longer able to change into hadron–antihadron pairs.

heavy elements: all chemical elements with nuclei heavier than helium. Also known as "metals," these heavy elements are built up by nuclear fusion in the interiors of stars and supernovae.

heliocentric universe: a model of the universe in which the Sun occupies the central place.

helium: chemical element with a nucleus of two protons and two neutrons (helium-4). A second, far less abundant isotope has two protons and one neutron (helium-3). Produced during the first 3 minutes of the universe, helium comprises about 25% of the mass of the universe. *See* nucleosynthesis.

homogeneity: property of the universe to be similar everywhere. Thus the number of galaxies per unit volume is, on average, the same wherever we look. *See* cosmological principle and cosmic background radiation.

"hot" matter: matter consisting of elementary particles of low mass that are moving very fast (the temperature is a measure of the energy of motion; the term "hot" indicates a high temperature and thus a high energy of motion). The neutrino is an example of "hot" matter.

Hubble constant (or Hubble parameter): the constant of proportionality between the recession velocity of a galaxy and its distance in Hubble's law. The inverse of the Hubble constant (the Hubble time) would be equal to the age of the universe, if there was no slowing down of the expansion. Because of the deceleration, the Hubble time is an upper limit to the age of the universe. Measurements of the Hubble constant give an age between 10 and 20 billion years for the universe (see Appendix E).

Hubble's law: the law discovered in 1929 by the American astronomer Edwin Hubble, which states that the distance of galaxies varies in proportion to their red shift, and thus, because of the Doppler effect, to their velocity of recession. The law gave birth to the idea of an expanding universe. Hubble's law and the cosmic background radiation are the two cornerstones of the Big Bang theory (see Fig. 21). *See* Hubble constant.

Hyades: a galactic cluster in the Milky Way, which plays a fundamental role in the determination of the distance scale of the universe.

hydrogen: lightest of all chemical elements, consisting of one proton and one electron. Hydrogen makes up 75% of the mass of the universe. *See* nucleosynthesis.

inferior planet: a Solar System planet lying between the Sun and the Earth. Mercury and Venus are the inferior planets.

inflation: the period between 10^{-35} and 10^{-32} second after the Big Bang, during which the universe expanded in exponential fashion, tripling in size every 10^{-34} second. This inflation, which is predicted by grand unification theories, is caused by the injection of energy resulting from the breakup of the electronuclear force into the electroweak and strong nuclear forces.

interstellar grains: solid microscopic particles of dust, about one one-millionth of a centimeter in size, which condense in the outer atmospheres of red giant stars. They absorb the blue light of stars, causing them to dim and redden.

interstellar medium: the space between the stars in a galaxy. The interstellar medium is made of atomic gas (primarily hydrogen), molecular gas (carbon monoxide, molecular hydrogen, water, etc.—about 100 molecules are known), and dust (interstellar grains).

irregular galaxy: galaxy that is neither spiral nor elliptical. Often it is a dwarf galaxy and contains many young stars and large amounts of gas and dust (see Fig. 41). *See* Magellanic Clouds.

isothermal fluctuation: a density fluctuation where only the matter content varies, and where the radiation remains the same (see Fig. 36).

isotropy: the property of the universe to be similar in every direction. *See* cosmological principle and cosmic background radiation.

Kelvin: the Kelvin temperature scale measures the motion of atoms. The coldest possible temperature, called "absolute zero," corresponds to no motion. To convert to the Celsius scale (°C), one needs to subtract 273: $T(°C) = T(K) - 273$. Thus absolute zero (0 K) corresponds to $-273°C$. Water freezes at 0°C or 273 K. It boils at 100°C or 373 K.

lepton: an elementary particle that is not subject to the strong nuclear force. The electron and neutrino are leptons.

lepton era: the period in the history of the universe that followed the hadron era, when the contents of the universe were dominated by leptons (electrons and neutrinos) in equilibrium with photons. It lasted from one ten-thousandth of a second to 1 second after the Big Bang, when the photons, weakened by the expansion of the universe, could no longer be converted into electron–positron pairs.

light-year: the distance traveled by light (which moves at a velocity of 300,000 kilometers per second) in 1 year, and equal to 9460 billion kilometers. Similarly, 1 light-day = 26 billion kilometers; 1 light-hour = 1.1 billion kilometers; 1 light-minute = 18 million kilometers; and 1 light-second = 300,000 kilometers.

Local Group: grouping of galaxies—extending over a region of space of about 10 million light-years—of which the Milky Way and Andromeda are the principal and most massive members (1 trillion solar masses each). It also includes dwarf galaxies ranging from 10 million to 10 billion solar masses.

Local Supercluster: huge flattened supercluster that contains the Local Group (whence the term "local"). The Local Group, containing the Galaxy, lies at the edge of the supercluster, while the Virgo Cluster of galaxies is at its center. The Local Supercluster is also known as the "Virgo Supercluster" (see Fig. 31).

Mach's principle: the idea advanced by the Austrian physicist Ernst Mach that the mass of an object—which measures its "inertia," or resistance to motion—is determined by all the mass distribution in the universe.

Magellanic Clouds: two dwarf irregular galaxies satellites of the Milky Way Galaxy, lying at a distance of about 150,000 light-years. The Small Magellanic Cloud contains about 100 million solar masses, while the Large Magellanic Cloud contains about 1 billion. First described by Magellan, the navigator, the two galaxies may only be seen from the Southern Hemisphere (see Fig. 17).

matter era: the period in the history of the universe that followed the radiation era, lasting from 300,000 years to the far distant future. In this era, the density of matter is greater than that of radiation and governs the evolution of the universe.

Milky Way: figurative name for our Galaxy, referring to the pale white band of stars, unresolved to the naked eye, that may be seen arching across the sky.

nebula: celestial object with a diffuse nonstellar appearance. The term is used to designate clouds of gas and dust in the Milky Way and was once used for what are now known as galaxies.

neutrino: neutral particle governed only by the weak nuclear force (and, if it has mass, by the force of gravity). Produced in huge quantities in the first moments after the Big Bang, and to a lesser extent, in stellar cores and supernovae, neutrinos could account for the bulk of the mass of the universe if their mass were equal only to one one-millionth of that of the electron. With a mass of only one ten-thousandth that of the electron, their gravity would be able to halt the expansion of the universe and cause it to collapse back on itself. It is not yet known whether neutrinos have mass.

neutron: neutral particle composed of three quarks. Like the proton, it is a constituent of atomic nuclei. The neutron is 1838 times more massive than the electron and slightly more massive than the proton.

neutron star: extremely compact (its radius is about 10 kilometers) and super-dense (10^{14} grams per cubic centimeter) star composed almost entirely of neutrons, created when a dying star (with mass between 1.4 and 5 solar masses) has exhausted its nuclear fuel and collapses under its own gravity. Neutron stars spin rapidly and emit beams of radiation, one of which sweeps past the Earth at each rotation. Since the beam is detected as an energy pulse separated by regular intervals, neutron stars are commonly known as pulsars (see Fig. 47).

nonbaryonic matter: matter composed of elementary particles that are not subject to the strong nuclear force. The neutrino is an example of nonbaryonic matter.

nucleon: component of an atomic nucleus, which may be either a proton or a neutron.

nucleosynthesis: fusion of two atomic nuclei to form a heavier nucleus through nuclear reactions either in the Big Bang (primordial nucleosynthesis, responsible for the light elements such as hydrogen and helium), within the cores of stars (elements heavier than helium and lighter than iron), or in supernovae (where elements heavier than iron are formed).

nucleus, atomic: core of an atom, consisting of protons and neutrons bound together by the strong nuclear force. It has a positive electrical charge equal to the total number of proton charges. 100,000 times smaller than the atom (its size is about 10^{-13} centimeter), the nucleus occupies only one thousand-trillionth of the atomic volume.

observable universe: that portion of the universe from which light has had time to reach us, and which is bounded by the cosmological horizon.

Olbers paradox: the question posed by the German astronomer Heinrich Olbers: "Why is the sky dark at night?" We now know that the sky is dark at night because the universe had a beginning and does not contain an infinite number of stars, which would otherwise fill it with light.

open universe: an universe in which the density of matter is less than the critical density, and which will thus expand forever.

parallax: the angle corresponding to the apparent motion of a celestial object relative to distant stars, as observed from two different positions (for example, from two diametrically opposite positions of the Earth in its orbit around the Sun) (see Fig. 14).

parallel (or multiple) universes, theory of: theory in which the universe splits into two every time there is a choice or alternative. There is no possibility of communication between these parallel universes.

perfect cosmological principle: extension of the cosmological principle from space to time: the universe is not only the same at all points and in every direction in space, it is also similar for all times. The universe does not evolve either in space or time. This forms the basis of the Steady-State theory.

photons: massless elementary particle of radiation, which moves at 300,000 kilometers per second, the highest possible velocity. Depending on the energy it carries, the particle may be, in order of decreasing energy, a gamma-ray, x-ray, ultraviolet, visible, infrared, or radio photon (see Appendix A).

Planck time: equal to 10^{-43} second, this time indicates the moment when known physics breaks down, when the limits of current knowledge are reached. To go beyond the Planck time, a quantum theory of gravity needs to be developed, where the force of gravity is unified with the other forces.

planet: a spherical celestial object more than 1000 kilometers in diameter, without its own source of nuclear energy, and in orbit around a star the light of which it reflects.

planetary nebula: shell of gas ejected when a dying star of less than 1.4 solar masses collapses into a white dwarf. The nebula glows as it is illuminated by radiation from the white dwarf (see Fig. 45).

positron: the antiparticle of the electron.

primordial black hole (or mini black hole): a black hole with a very small mass compared to the Sun. According to the British astrophysicist Stephen Hawking, mini black holes may have formed in the very early universe at the Planck time (10^{-43} second), when the density was extremely high. These primordial black holes would have had masses of 20 micrograms and a radius of no return of 10^{-33} centimeters. They have never been observed.

primordial elements: chemical elements created in the Big Bang during the first 3 minutes of the universe. They are primarily hydrogen (which makes up three-quarters of the mass of the universe), helium (which forms about one-quarter of the mass), and a small amount of deuterium and lithium. *See* nucleosynthesis.

proton: positively charged particle composed of three quarks, which, together with the neutron, forms atomic nuclei. The proton is 1836 times more massive than the electron.

pulsar: *see* neutron star.

quantum mechanics: physical theory developed at the beginning of the twentieth century, which describes the properties of matter and radiation on the subatomic scale. According to this theory, matter and radiation may exhibit both wavelike and particlelike characteristics and may be described only in terms of probabilities. The particle of light (the photon) is also known as the "quantum of energy," hence the name of the theory. *See* principle of complementarity, exclusion principle, and uncertainty principle.

quantum vacuum: space filled with virtual particles and antiparticles that appear and disappear in life and death cycles of extremely short duration, and that owe their existence to quantum uncertainty.

quark: elementary particle and a building block of protons and neutrons. Quarks have fractional electrical charges that are either one-third or two-thirds of the electron charge and are subject to the strong nuclear force. Quarks have yet to be observed.

quasar: celestial object with a starlike appearance (the name comes from the contraction of "quasistellar object"). Its light shows a very large red shift, implying a very great distance according to Hubble's law. Quasars are the most distant and most luminous objects in the universe. They are at the centers of galaxies, and their enormous energy is probably generated by black holes of one billion solar masses that "feed" on stars from the host galaxy. A small minority of astronomers think that the large red shifts of quasars bear no relation to their distances and that they are relatively nearby (see Fig. 48).

radiation era: the period in the history of the universe that followed the lepton era, lasting from about 1 second after the Big Bang until the 300,000th year. During this period radiation had greater density than matter and governed the evolution of the universe. Atoms, planets, stars, and galaxies had not yet appeared.

radius of no return: also known as the "Schwarzschild radius" (after the German astronomer Karl Schwarzschild, who studied it), it defines the boundary of a black hole. Once this radius is crossed, no particle of matter or radiation can ever escape from the black hole.

red giant: star that burns helium because it has exhausted its central hydrogen fuel. The outward pressure produced by helium burning swells the star's outer layers to several tens of times its initial size (hence the name of giant). At the same time, the star's surface layers cool, reddening its light.

red shift: change of color in a luminous source that is moving away (through the Doppler effect). This change is proportional to the velocity of recession. The red shift of a galaxy varies in proportion to its distance (Hubble's law).

retrograde motion: apparent reverse motion of a planet relative to background stars.

special relativity: theory of relative motion proposed by Albert Einstein in 1905, which showed that space and time are intimately linked, and that they are not universal, but depend on the motion of the observer.

spectroscope: an instrument that separates light into its different components.

spiral galaxy: flattened, disklike system of stars and interstellar gas and dust, with a spherical collection of stars, known as the bulge, at its center. Bright, young stars outline exquisitely shaped spiral arms in the plane of the disk (see Fig. 20).

star: a sphere of gas consisting of 98% hydrogen and helium and 2% heavy elements, in equilibrium under the action of two opposing forces: the compressive force of gravity and the outward radiation pressure from the nuclear fusion reactions in its core. The Sun has a mass of 2×10^{33} grams, and masses of stars range between 0.1 and 100 solar masses.

Steady-State theory: cosmological theory based on the perfect cosmological principle that maintains that the universe is unchanging not only in space, but also in time. To compensate for the empty space created by the expansion of the universe, the theory postulates a continuous creation of matter.

strong nuclear force: strongest of the four fundamental forces, it binds quarks into protons and neutrons, and protons and neutrons together to form atomic nuclei. Its sphere of influence is that of an atomic nucleus (10^{-13} centimeter). It does not act on photons or electrons.

supercluster: aggregation of tens of thousands of galaxies held together by gravity and gathered into groups and clusters. Superclusters have the shape of flattened pancakes with an average diameter of 90 million light-years and masses of 10,000 trillion (10^{16}) solar masses.

superior planet: a Solar System planet that lies farther from the Sun than the Earth. Mars, Jupiter, Saturn, Uranus, Neptune, and Pluto are the superior planets.

supernova: the explosive death of a star that has exhausted its nuclear fuel. The explosion may reach the brightness of 100 million Suns. The outer envelope of the star is ejected while the core collapses into a neutron star (in the case of stars with masses between 1.4 and 5 solar masses) or a black hole (for stars with masses greater than 5 solar masses). Numerous, highly energetic particles (protons and electrons), known as cosmic rays, are also expelled into space.

Theory of Everything (TOE): a unification theory that would bring together all four fundamental forces into a single force, and that would require a quantum theory of gravity. Physicists are still a long way from being able to devise such a theory.

tired light, theory of: a cosmological theory in which the red shift of the light of galaxies is not caused by the expansion of the universe, but by a loss of energy of unknown cause, suffered by light as it travels in intergalactic and interstellar space.

uncertainty principle: principle developed by the German physicist Werner Heisenberg, which states that quantum uncertainty prevents the simultaneous accurate measurements of both the velocity and position of a particle, however sophisticated the measuring instrument might be. The uncertainty principle also applies to the energy of an elementary particle with a very short lifetime. Virtual particles and antiparticles can appear in the quantum vacuum thanks to quantum uncertainty (see Appendix D).

virtual particle: phantom elementary particle that populates the quantum vacuum and that owes its existence to the uncertainty principle. Its lifetime is so short that it cannot be observed directly. It may become a "real" particle when there is injection of energy, as, for example, in the very early universe. A virtual particle always appears in the company of its virtual antiparticle. (Charge must be conserved: It was zero before; it must remain zero after.)

void: region of the universe devoid of galaxies that extends for several tens of millions of light-years.

wavelength: the distance between two successive crests (or troughs) in a wave of radiation (or of matter, as with waves in the sea) (see Appendix A).

weak nuclear force: fundamental force responsible for the disintegration of atoms via radioactive decay. It acts only on the subatomic scale (10^{-15} centimeter).

white dwarf: small (hence the name of dwarf), dense (between 10^5 and 10^8 grams per cubic centimeter) celestial object with a diameter of about 10,000 kilometers (about the size of Earth) created when a star of less than 1.4 solar masses exhausts its nuclear fuel and collapses under its own gravity. The electrons of the white dwarf, in accordance with the exclusion principle, cannot be too closely compressed, and thus exert a pressure that opposes gravity, preventing the star from collapsing further. The motion of the electrons heats the stellar remnant, which radiates its white light into space.

x rays: the most energetic form of electromagnetic radiation, after gamma rays.

Bibliography

Barrow, J. D., and Tipler, F. J., *The Anthropic Cosmological Principle,* Oxford University Press, New York, 1986.

Davies, P. C. W., *God and the New Physics,* Simon and Schuster, New York, 1983.

Davies, P. C. W., *Superforce,* Simon and Schuster, New York, 1984.

Dyson, F. J., *Disturbing the Universe,* Harper and Row, New York, 1979.

Ferris, T., *Coming of Age in the Milky Way,* Morrow, New York, 1988.

Gribbin, J. R., and Rees, M. J., *Cosmic Coincidences,* Bantam Books, New York, 1989.

Harrison, E. R., *Masks of the Universe,* Macmillan, New York, 1985.

Hawking, S. W., *A Brief History of Time,* Bantam Books, New York, 1988.

Islam, J. N., *The Ultimate Fate of the Universe,* Cambridge University Press, Cambridge, 1983.

Kuhn, T. S., *The Structure of Scientific Revolutions,* University of Chicago Press, Chicago, 1962.

Lederman, L. M., and Schramm, D. N., *From Quarks to the Cosmos,* Scientific American Library, Freeman, New York, 1989.

Luminet, J. P., *Black Holes,* Cambridge University Press, Cambridge, 1992.

Minsky, M., *The Society of the Mind,* Simon and Schuster, New York, 1987.

Monod, J., *Chance and Necessity,* Knopf, New York, 1971.

Popper, K., *The Logic of Scientific Discovery,* Harper and Row, New York, 1965.

Reeves, H., *Atoms of Silence,* MIT Press, Cambridge, Massachusetts, 1984.

Weinberg, S., *The First Three Minutes,* Basic Books, New York, 1977.

Index